Advances in the Physics of Particles and Nuclei
Volume 28

Advances in the Physics of Particles and Nuclei

Series Editors

Douglas H. Beck
Department of Physics
University of Illinois at Urbana-Champaign
1110 West Green Street
Urbana, IL 61801-3080
USA

Dieter Haidt
DESY
Notkestraße 85
22603 Hamburg
Germany

John W. Negele
William A. Coolidge Professor of Physics
Massachusetts Institute of Technology
Center for Theoretical Physics
77 Massachusetts Ave. NE25-4079
Cambridge MA 02139
USA

Advances in the Physics of Particles and Nuclei

Edited by

Douglas H. Beck
Dieter Haidt
John W. Negele

Volume 28

Contributions to this Volume:

A. Quadt
Top Quark Physics at Hadron Colliders

 Springer

Arnulf Quadt
Georg-August-Universität Göttingen, Friedrich-Hund-Platz 1, 37077 Göttingen, Germany
Physikalisches Institut, Universität Bonn, Nußallee 12, 53115 Bonn, Germany
University of Rochester, New York, c/o Fermilab, P.O. Box 500, Batavia, IL, 60510, USA
e-mail: arnulf.quadt@cern.ch, aquadt@uni-goettingen.de

Originally published in Eur. Phys. J. C 48, 835-1000 (2006)
© Springer-Verlag / Società Italiana di Fisica 2006

Library of Congress Control Number: 2007924601

ISBN 978-3-540-71059-2 Springer Berlin Heidelberg New York

Springer is a part of Springer Science+Business Media

springer.com

© Springer-Verlag Berlin Heidelberg 2007

Typesetting and Production: LE-TEX Jelonek, Schmidt & Vöckler GbR, Leipzig, Germany
Cover Design: eStudio Calamar S.L., F. Steinen-Broo, Girona, Spain

SPIN 11775119   54/3100YL - 5 4 3 2 1 0   Printed on acid-free paper

Contents

Top quark physics at hadron colliders

A. Quadt

Abstract. The top quark, discovered at the FERMILAB TEVATRON collider in 1995, is the heaviest known elementary particle. Today, ten years later, still relatively little is known about its properties. The strong and weak interactions of the top quark are not nearly as well studied as those of the other quarks and leptons. The strong interaction is most directly measured in top quark pair production. The weak interaction is measured in top quark decay and single top quark production, which remains thus far unobserved. The large top-quark mass of about $175 \, \mathrm{GeV}/c^2$ suggests that it may play a special role in nature. It behaves differently from all other quarks due to its large mass and its correspondingly short lifetime. The top quark decays before it hadronises, passing its spin information on to its decay products. Therefore, it is possible to measure observables that depend on the top quark spin, providing a unique environment for tests of the Standard Model and for searches for physics beyond the Standard Model.

This report summarises the latest measurements and studies of top quark properties and rare decays from the TEVATRON in Run II. With more than $1 \, \mathrm{fb}^{-1}$ of luminosity delivered to each experiment, CDF and DØ, top quark physics at the TEVATRON is at a turning point from first studies to precision measurements with sensitivity to new physics. An outlook onto top quark physics at the Large Hadron Collider (LHC) at CERN, planned to begin operation in the year 2007, is also given.

1 Introduction

There are six known quarks in nature, the up, down, strange, charm, bottom, and the top quark. The quarks are arranged in three pairs or "generations". Each member of a pair may be transformed into its partner via the charged-current weak interaction. Together with the six known leptons (the electron, muon, tau, and their associated neutrinos), the six quarks constitute all of the known luminous matter in the universe. The understanding of the properties of the quarks and leptons and their interactions is therefore of paramount importance.

The top quark is the charge, $Q = 2/3$, and $T_3 = +1/2$ member of the weak-isospin doublet containing the bottom quark. It is the most recently discovered quark, which was directly observed almost exactly ten years ago, in 1995 by the CDF and DØ experiments at the FERMILAB

TEVATRON, a proton-antiproton collider at a centre-of-mass energy of $\sqrt{s} = 1.8 \, \mathrm{TeV}$, located in the suburbs of Chicago. This discovery was a great success of the Standard Model of Elementary Particle Physics, which suggested the existence of the top quark as the weak-isospin partner of the b-quark already in 1977 at its discovery. Indirect evidence for the existence of the top quark became compelling over the years and constraints on the top quark mass, inferred from electroweak precision data, pointed exactly at the range where the top quark was discovered. Due to its relatively recent discovery, far less is known about the top quark than about the other quarks and leptons.

The strong and weak interactions of the top quark are not nearly as well studied as those of the other quarks and leptons. The strong interaction is most directly measured in top quark pair production. The weak interaction is measured in top quark decay and single top quark production. There are only a few fundamental parameters associated with the top quark in the Standard Model: the top quark mass and the three CKM matrix elements involving top.

Thus far, the properties of the quarks and leptons are successfully described by the Standard Model. However, this theory does not account for the masses of these particles, it merely accommodates them. Due to the mass of the top quark being by far the heaviest of all quarks, it is often speculated that it might be special amongst all quarks and leptons and might play a role in the mechanism of electroweak symmetry breaking. Even if the top quark turned out to be a Standard Model quark, the experimental consequences of this very large mass are interesting in their own. Many of the measurements described in this review have no analogue for the lighter quarks. This is not just a consequence of the large mass of the top quark, but also of its very short lifetime. In contrast to the lighter quarks, which are permanently confined in bound states (hadrons) with other quarks and antiquarks, the top quark decays so quickly that it does not have time to form such bound states. There is also insufficient time to depolarise the spin of the top quark, in contrast to the lighter quarks, whose spin is depolarised by chromomagnetic interactions within the bound states. Thus the top quark is free of many of the complications associated with the strong interaction. Also, top quarks are and will remain a major source of background for almost all searches for physics beyond the Standard Model. Precise understanding of the top signal is crucial to claim new physics.

This review summarises the present knowledge of the properties of the top quark such as its mass and electric charge, its production mechanisms and rate and its decay branching ratios, *etc.*, and provides a discussion of the experimental and theoretical issues involved in their determination. Earlier reviews on top quark physics at Run I or the earlier Run II can be found in [1–6]. An early, general review on hadron collider physics, including results from the CERN $Sp\bar{p}S$, is given in [7].

Since the TEVATRON at FERMILAB is today still the only place where top quarks can be produced and studied directly, most of the discussion in this article describes top quark physics at the TEVATRON. In particular, the focus is placed on the already available wealth of results from the Run II, which started in 2001 after a five year upgrade of the TEVATRON collider and the experiments CDF and DØ. However, the Large Hadron Collider, LHC, a proton-proton collider at a centre-of-mass energy of $\sqrt{s} = 14$ TeV, planned to start operation at CERN in 2007, will be a prolific source of top quarks and produce about 8 million $t\bar{t}$ events per year (at "low" luminosity, 10^{33} cm^{-2} s^{-1}), a real "top factory". Measurements such as the top quark mass are entirely an issue of systematics, as the statistical uncertainty is negligible. Prospects for top quark physics at the LHC in the near future are summarised at the end of this article.

This article is organised as follows:

- In the remainder of Sect. 1, a brief summary of the Standard Model is given, followed by the main arguments why the top quark as weak-isospin partner of the b-quark had to exist, indirect constraints on the top quark mass from electroweak precision data and a historic overview over searches for the top quark leading to its discovery.

- In Sect. 2, the top quark production mechanisms at hadron colliders and its decay in the Standard Model are discussed.

- Section 3 describes the experimental conditions, focusing in detail on the $p\bar{p}$ collider TEVATRON at FERMILAB with its two experiments, CDF and DØ, since the TEVATRON is presently the only source of top quarks and all available direct measurements of top quark properties have been made there. In the following sections, the TEVATRON measurements of top quark properties are discussed in detail.

- Sections 4 and 5 describe measurements of top quark production rates, in particular in strong $t\bar{t}$ production (Sect. 4) and in electroweak single-top production (Sect. 5).

- Section 6 summarises studies of the top quark interactions with gauge bosons, in particular studies on $t\bar{t}$ spin correlations, the top quark decay ratio $B(t \to Wb)/B(t \to Wq)$, the top quark decay $t \to \tau\nu X$, measurements of the helicity of the W-boson in top decay and searches for flavour-changing neutral current top quark couplings.

- In Sect. 7, measurements of the fundamental properties of the top quark such as its mass and its electric charge are described.

- Section 8 presents studies on anomalous top quark production via the measurements of the cross section ratio $\sigma_{\ell\ell}/\sigma_{\ell+jets}$, studies of the $t\bar{t}$ event kinematics, in particular the transverse momentum spectrum of the top quark, and the search for top quark production via intermediate, narrow resonances.

- In Sect. 9, the search for anomalous top quark decay, in particular the search for top quark decay to charged Higgs bosons is discussed.

- Section 10 places its focus on the search for new physics in events with $t\bar{t}$ topology.

- In the last Sect. 11, a brief introduction to the Large Hadron Collider (LHC) and the omni-purpose experiments ATLAS and CMS is given, followed by an outlook onto the expected precision and sensitivity for measurements and searches in and beyond the Standard Model in the sector of top quark physics.

1.1 Brief summary of the Standard Model

Quantum field theory combines two great achievements of physics in the 20th-century, quantum mechanics and relativity. The Standard Model [8–19] is a particular quantum field theory, based on the set of fields shown in Table 1, and the gauge symmetries $SU(3)_C \times SU(2)_L \times U(1)_Y$. There are three generations of quarks and leptons, labelled by the index $i = 1, 2, 3$, and one Higgs field, ϕ.

Once the gauge symmetries and the fields with their (gauge) quantum numbers are specified, the Lagrangian of the Standard Model is fixed by requiring it to be gauge-invariant, local, and renormalisable. The Standard Model

Table 1. The fields of the Standard Model and their gauge quantum numbers. T and T_3 are the total weak-isospin and its third component, and Q is the electric charge

				$SU(3)_C$	$SU(2)_L$	$U(1)_Y$	T	T_3	Q
$Q_L^i =$	$\begin{pmatrix} u_L \\ d_L \end{pmatrix}$	$\begin{pmatrix} c_L \\ s_L \end{pmatrix}$	$\begin{pmatrix} t_L \\ b_L \end{pmatrix}$	3	2	1/6	1/2	$+1/2$ $-1/2$	$+2/3$ $-1/3$
$u_R^i =$	u_R	c_R	t_R	3	1	2/3	0	0	$+2/3$
$d_R^i =$	d_R	s_R	b_R	3	1	$-1/3$	0	0	$-1/3$
$L_L^i =$	$\begin{pmatrix} \nu_{eL} \\ e_L \end{pmatrix}$	$\begin{pmatrix} \nu_{\mu L} \\ \mu_L \end{pmatrix}$	$\begin{pmatrix} \nu_{\tau L} \\ \tau_L \end{pmatrix}$	1	2	$-1/2$	1/2	$+1/2$ $-1/2$	0 -1
$e_R^i =$	e_R	μ_R	τ_R	1	1	-1	0	0	-1
$\nu_R^i =$	ν_R^e	ν_R^μ	ν_R^τ	0	0	0	0	0	0
$\phi =$	$\begin{pmatrix} \phi^+ \\ \phi^0 \end{pmatrix}$			1	2	1/2	1/2	$+1/2$ $-1/2$	$+1$ 0

Lagrangian can be divided into several pieces:

$$\mathcal{L}_{\mathrm{SM}} = \mathcal{L}_{\mathrm{Gauge}} + \mathcal{L}_{\mathrm{Matter}} + \mathcal{L}_{\mathrm{Yukawa}} + \mathcal{L}_{\mathrm{Higgs}}. \qquad (1)$$

The first piece is the pure gauge Lagrangian, given by

$$\mathcal{L}_{\mathrm{Gauge}} = \frac{1}{2g_S^2}\,\mathrm{Tr}\,G^{\mu\nu}G_{\mu\nu} + \frac{1}{2g^2}\,\mathrm{Tr}\,W^{\mu\nu}W_{\mu\nu}$$
$$- \frac{1}{4g'^2}B^{\mu\nu}B_{\mu\nu}\,, \qquad (2)$$

where $G^{\mu\nu}$, $W^{\mu\nu}$, and $B^{\mu\nu}$ are the gluon, weak, and hypercharge field-strength tensors. These terms contain the kinetic energy of the gauge fields and their self interactions. The next piece is the matter Lagrangian, given by

$$\mathcal{L}_{\mathrm{Matter}} = i\overline{Q}_L^i \slashed{D} Q_L^i + i\bar{u}_R^i \slashed{D} u_R^i + i\bar{d}_R^i \slashed{D} d_R^i$$
$$+ i\overline{L}_L^i \slashed{D} L_L^i + i\bar{e}_R^i \slashed{D} e_R^i. \qquad (3)$$

This piece contains the kinetic energy of the fermions and their interactions with the gauge fields, which are contained in the covariant derivatives. For example,

$$\slashed{D}Q_L = \gamma^\mu \left(\partial_\mu + ig_S G_\mu + igW_\mu + i\frac{1}{6}g'B_\mu \right) Q_L\,, \quad (4)$$

since the field Q_L participates in all three gauge interactions. A sum on the index i, which represents the generations, is implied in the Lagrangian.

These two pieces of the Lagrangian depend only on the gauge couplings g_S, g, g'. Their approximate values, evaluated at M_Z, are

$$g_S \approx 1, \qquad (5)$$
$$g \approx 2/3, \qquad (6)$$
$$g' \approx 2/(3\sqrt{3})\,. \qquad (7)$$

Mass terms for the gauge bosons and the fermions are forbidden by the gauge symmetries.

The next piece of the Lagrangian is the Yukawa interaction of the Higgs field with the fermions, given by

$$\mathcal{L}_{\mathrm{Yukawa}} = -\Gamma_u^{ij}\overline{Q}_L^i \epsilon\phi^* u_R^j - \Gamma_d^{ij}\overline{Q}_L^i \phi d_R^j$$
$$- \Gamma_e^{ij}\overline{L}_L^i \phi e_R^j + h.c.\,, \qquad (8)$$

where $\epsilon = i\sigma_2$ is the total antisymmetric tensor in 2 dimensions, related to the second Pauli matrix σ_2 and required to ensure each term separately to be electrically neutral, and the coefficients Γ_u, Γ_d, Γ_e are 3×3 complex matrices in generation space. They need not be diagonal, so in general there is mixing between different generations. These matrices contain most of the parameters of the Standard Model.

The final piece is the Higgs Lagrangian [20–22], given by

$$\mathcal{L}_{\mathrm{Higgs}} = (D^\mu\phi)^\dagger D_\mu\phi + \mu^2\phi^\dagger\phi - \lambda(\phi^\dagger\phi)^2\,, \qquad (9)$$

with the Higgs doublet ϕ as given in Table 1. This piece contains the kinetic energy of the Higgs field, its gauge interactions, and the Higgs potential, shown in Fig. 1. The coefficient of the quadratic term, μ^2, is the *only* dimensionful parameter in the Standard Model. The sign of this term is chosen such that the Higgs field has a non-zero vacuum-expectation value on the circle of minima in Higgs-field space given by $\langle\phi^0\rangle = \mu/\sqrt{2\lambda} \equiv v/\sqrt{2}$. The dimensionful

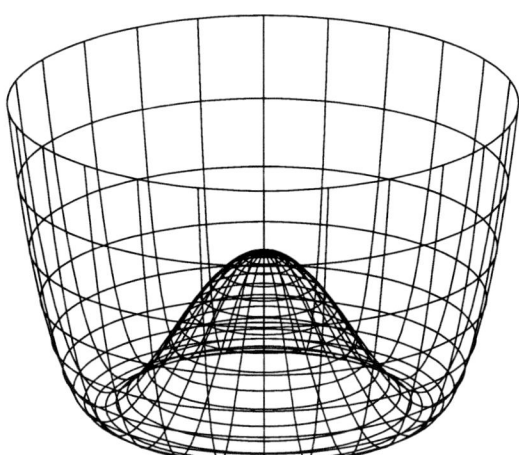

Fig. 1. The Higgs potential. The neutral component of the Higgs field acquires a vacuum-expectation value $\langle\phi^0\rangle = v/\sqrt{2}$ on the circle of minima in Higgs-field space

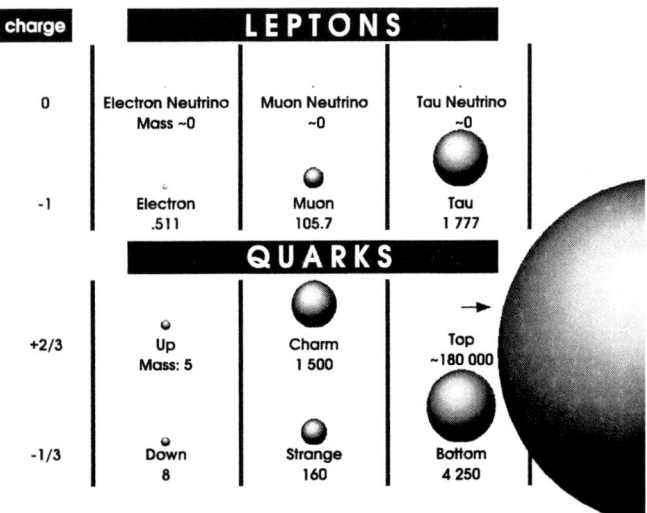

Fig. 2. Table of lepton and quark properties such as electric charge and mass (in MeV/c^2). The top quark is unique amongst all fermions due to its very large mass. (The size of the drawn spheres does not scale linearly with the fermion mass)

parameter μ is replaced by the dimensionful parameter $v \approx 246$ GeV.

The acquisition of a non-zero vacuum-expectation value by the Higgs field breaks the electroweak symmetry and generates masses for the gauge bosons,

$$M_W = \frac{1}{2} g v \,, \tag{10}$$

$$M_Z = \frac{1}{2} \sqrt{g^2 + g'^2} v \,, \quad \text{and for the fermions} \,, \tag{11}$$

$$M_f = \Gamma_{\rm t} \frac{v}{\sqrt{2}} \,, \tag{12}$$

with the Yukawa coupling $\Gamma_{\rm t}$. Diagonalising the fermion mass matrices generates the Cabibbo–Kobayashi–Maskawa (CKM) matrix [23,24], including the CP-violating phase. The CKM matrix elements related to the top quark are discussed in more detail in Sect. 2.3.

Figure 2 shows three lepton and quark families with their electric charge and approximate mass. While the neutrinos have non-zero, but very small masses of at least[1] $m_\nu > 45$ meV/c^2, the quark masses are much larger. The top quark, with a mass of ≈ 175 GeV/c^2, is by far the heaviest fermion. Theoretical and experimental consequences of the large value of the top quark mass are discussed in Sect. 2.

---

[1] In the Standard Model the neutrino masses are assumed to be zero. However, in recent years, experimental evidence for neutrinos to be massive has been accumulated. The mass of the heaviest neutrino cannot be less than $\sqrt{\Delta m_{\rm atm}^2}$, where $\Delta m_{\rm atm}^2 = 1.9$–$3.0 \times 10^{-3}$ eV^2 is the square of the neutrino mass difference as measured by SuperKamiokande. The nature of neutrinos (Dirac versus Majorana) and the origin of their mass are at present unknown. Several experiments looking for neutrinoless double-beta decay or using other techniques are trying to answer those questions.

1.2 Indirect evidence for the existence of the top quark

Why should one expect quarks to come in doublets? There are two main reasons for this. First it provides a natural way to suppress the experimentally not observed flavour-changing neutral current. The argument on which the GIM mechanism [11] is based applies just as well for three as for two quark doublets.

The second reason is concerned with the desire to obtain a renormalisable gauge theory of weak interactions[2]. The Standard Model of electroweak interactions can be proven to be renormalisable under the condition that the sum of the weak hypercharges, Y_i, of all left-handed fermions is zero, i.e.

$$\sum_{\substack{\text{left-handed} \\ \text{quarks and leptons}}} Y_i = 0 \,. \tag{13}$$

Since every lepton multiplet contributes a value of $y = -2$ and every quark multiplet a value of $+2/3$, the sum only vanishes if

1. there are three colours, i.e. every quark exists in three colour versions, and
2. the number of quark flavours equals the number of lepton species.

The general proof that gauge theories can be renormalised, however, can only be applied if the particular gauge theory is *anomaly free*[3]. This requires a delicate cancellation between different diagrams, relations which can easily be upset by "anomalies" due to fermion loops such as the one shown in Fig. 3. The major aspect is an odd number of axial-vector couplings. In general, anomaly freedom is guaranteed if the coefficient[4]

$$d_{abc} = \sum_{\text{fermions}} \text{Tr} \left[\hat{\lambda}^a \left\{ \hat{\lambda}^b, \hat{\lambda}^c \right\}_+ \right] = 0 \,, \tag{14}$$

where the $\hat{\lambda}^i$ are in general the generators of the gauge group under consideration. In the Standard Model of electroweak interactions, the gauge group $SU(2) \times U(1)$ is generated by the three Pauli matrices, σ_i, and the hypercharge $Y: \hat{\lambda}^i = \sigma_i$, for $i = 1, 2, 3$, and $\hat{\lambda}^4 = \hat{Y} = 2(\hat{Q} - \hat{T}_3)$.

---

[2] The gauge theory has to be consistent, i.e. anomaly-free, in order to be at least unitary. The requirement of the gauge theory to be renormalisable is stronger than to be consistent, but the former argument is more familiar to most readers. The important consequence of both requirements is that the gauge theory is anomaly-free.

[3] A gauge theory might be renormalisable, whether or not it is anomaly free. The general proof of renormalisability, however, cannot be applied if it is not.

[4] d_{abc} is the coefficient in the definition of the anomaly: $\left[\partial_\mu J_\alpha^\mu(x) \right]_{\text{anom}} = -\frac{1}{32\pi^2} d_{\alpha\beta\gamma} \epsilon^{\kappa\nu\lambda\rho} F_{\kappa\nu}^\beta(x) F_{\lambda\rho}^\gamma(x)$, with the current $J_\alpha^\mu(x)$, the field strength tensor $F_{\kappa\nu}^\beta$ and the total antisymmetric tensor $\epsilon^{\kappa\nu\lambda\rho}$.

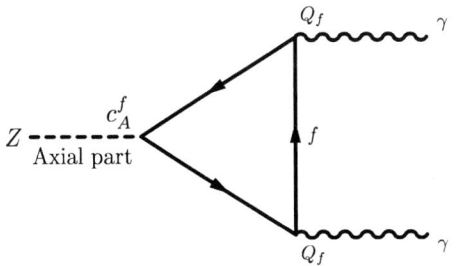

Fig. 3. A fermion (quark or lepton) triangle diagram which potentially could cause an anomaly

In the specific example shown in Fig. 3, one consequence of (14) is a relation where each triangle is proportional to $c_A^f Q_f^2$, where Q_f is the charge and c_A^f is the axial coupling of the weak neutral current. Thus, for an equal number N of lepton and quark doublets, the total anomaly is proportional to:

$$d \propto \sum_{i=1}^{N} \left(\frac{1}{2}(0)^2 - \frac{1}{2}(-1)^2 \right.$$
$$\left. + \frac{1}{2} N_c \left(+\frac{2}{3} \right)^2 - \frac{1}{2} N_c \left(-\frac{1}{3} \right)^2 \right) . \quad (15)$$

Consequently, taking into account the three colours of each quark ($N_c = 3$), the anomalies are cancelled. Since three lepton doublets were observed many years ago (the tau neutrino was experimentally only observed directly in the year 2000, but the number of light neutrino generations was known to be 3 from the LEP data on the Z-pole), the lack of anomalies such as the one shown in Fig. 3 therefore requires the existence of the three quark doublets.

There is a lot of indirect experimental evidence for the existence of the top quark. The experimental limits on flavour changing neutral current (FCNC) decays of the b-quark [25, 26] such as $b \rightarrow s\ell^+\ell^-$ and the absence of large tree level (lowest order) $B_d^0 \bar{B}_d^0$ mixing at the $\Upsilon(4S)$ resonance [27–30] rule out the hypothesis of an isosinglet b-quark. In other words, the b-quark must be a member of a left-handed weak isospin doublet.

The most compelling argument for the existence of the top quark comes from the wealth of data accumulated at the e^+e^- colliders LEP and SLC in recent years, particularly the detailed studies of the $Zb\bar{b}$ vertex near the Z resonance [31]. These studies have yielded a measurement of the isospin of the b-quark. The Z-boson is coupled to the b-quarks (as well as the other quarks) through vector and axial vector charges (v_b and a_b) with strength (Feynman diagram vertex factor)

$$Z \sim \left\{ \begin{array}{c} \bar{b} \\ b \end{array} \right. = \frac{-ig}{\cos\theta_W} \gamma^\mu \frac{1}{2} \left(v_b - a_b \gamma^5 \right) \quad (16)$$

$$= -i\sqrt{\sqrt{2}G_F M_Z^2} \gamma^\mu \left(v_b - a_b \gamma^5 \right), \quad (17)$$

where v_b and a_b are given by

$$v_b = \left[T_3^L(b) + T_3^R(b) \right] - 2e_b \sin^2\theta_W , \quad \text{and}$$
$$a_b = \left[T_3^L(b) + T_3^R(b) \right] . \quad (18)$$

Here, $T_3^L(b)$ and $T_3^R(b)$ are the third components of the weak isospin for the left-handed and right-handed b-quark fields, respectively. The electric charge of the b-quark, $e_b = -1/3$, has been well established from the Υ leptonic width as measured by the DORIS e^+e^- experiment [32–34]. Therefore, measurements of the weak vector and axial-vector coupling of the b-quark, v_b and a_b, can be interpreted as measurements of its weak isospin.

The (improved) Born approximation for the partial Z-boson decay rate gives in the limit of a zero mass b-quark:

$$\Gamma_{b\bar{b}} \equiv \Gamma(Z \rightarrow b\bar{b}) = \frac{G_F M_Z^3}{2\sqrt{2}\pi}(v_b^2 + a_b^2) . \quad (19)$$

The partial width $\Gamma_{b\bar{b}}$ is expected to be thirteen times smaller if $T_3^L(b) = 0$. The LEP measurement of the ratio of this partial width to the full hadronic decay width, $R_b = \Gamma_b/\Gamma_{\text{had}} = 0.21629 \pm 0.00066$ (Fig. 4), is in excellent agreement with the Standard Model expectation (including the effects of the top quark) of 0.2158, ruling out $T_3^L(b) = 0$. Figure 5 shows the sensitivity of R_b to the mass of the top quark. A top quark with a mass around $m_t \approx 175 \text{ GeV}/c^2$ is strongly favoured.

In addition, the forward-backward asymmetry in $e^+e^- \rightarrow b\bar{b}$ below [35] and at the Z pole [31],

$$A_{\text{FB}}^0(M_Z) = \frac{3}{4} \frac{2v_e a_e}{(v_e^2 + a_e^2)} \frac{2v_b a_b}{(v_b^2 + a_b^2)} , \quad (20)$$

measured to be $A_{\text{FB}}^{0,b} = 0.0992 \pm 0.0016$ (Fig. 6) is sensitive [31, 35] to the relative size of the vector and axial vector couplings of the $Zb\bar{b}$ vertex. The sign ambiguity for the two contributions can be resolved by the A_{FB} measurements

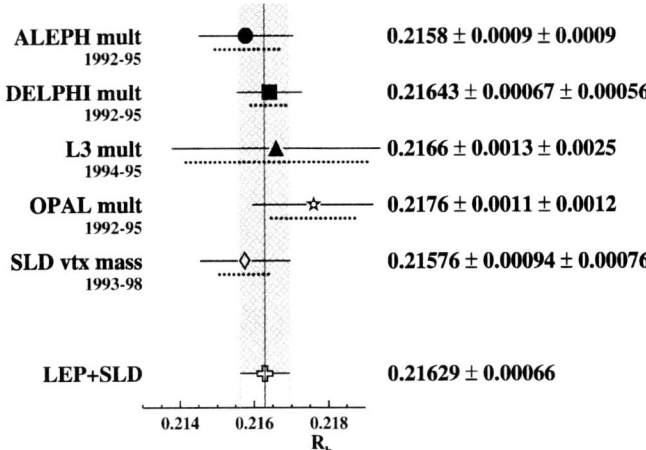

Fig. 4. R_b measurements used in the heavy flavour combination in the electroweak multi-parameter fit. The *dotted lines* indicate the size of the systematic error

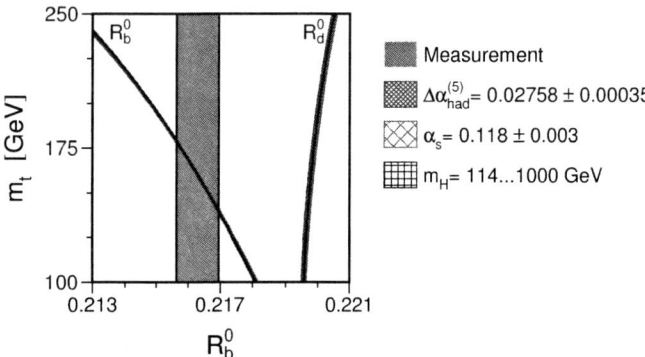

Fig. 5. Comparison of the LEP combined measurement of R_b^0 with the Standard Model prediction as a function of the mass of the top quark. From [31]

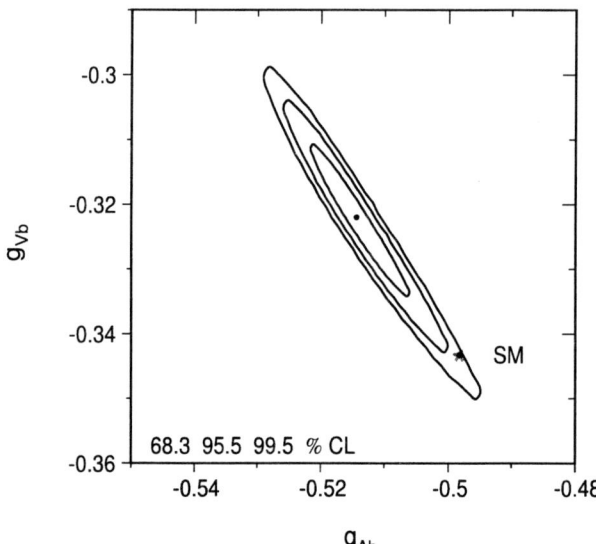

Fig. 7. Comparison of the effective vector and axial-vector coupling constants for the b-quark from the electroweak fit (*contour lines*) and the Standard Model expectation (*star*). From [31]

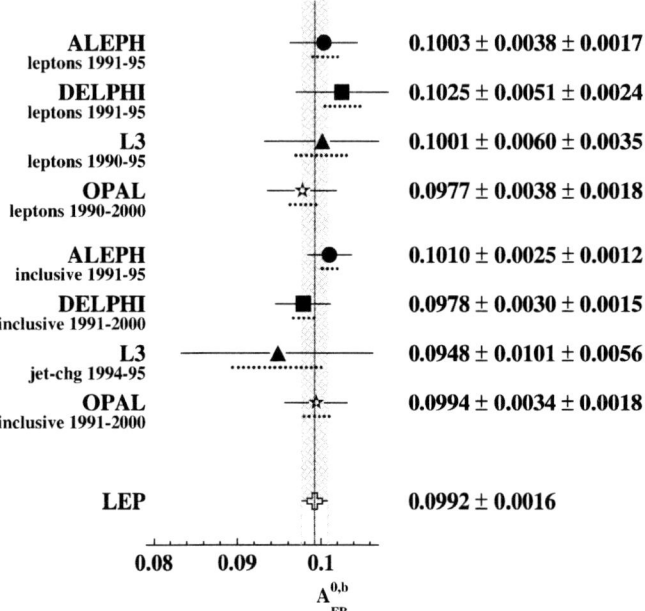

Fig. 6. $A_{FB}^{0,b}$ measurements used in the heavy flavour combination in the electroweak multi-parameter fit. The *dotted lines* indicate the size of the systematic error

1.3 Indirect constraints on the mass of the top quark

The precise electroweak measurements performed at LEP, SLC, NuTeV and the $p\bar{p}$ colliders can be used to check the validity of the Standard Model and within its framework, to infer valuable information about its fundamental parameters. Due to the accuracy of those measurements sensitivity to the mass of the top quark and the Higgs boson through radiative corrections is gained.

All electroweak quantities (mass, width and couplings of the W- and the Z-boson) depend in the Standard Model only on five parameters. At leading order this dependence is reduced to only three parameters, two gauge couplings and the Higgs-field vacuum expectation value. The three best-measured electroweak quantities can be used to determine these three parameters: The electromagnetic coupling constant, α, measured in low-energy experiments [39], the Fermi constant, G_F, determined from the μ lifetime [40], and the mass of the Z-boson, measured in e^+e^- annihilation at LEP and SLC [31]. By defining the electroweak mixing angle θ_W through:

$$\sin^2 \theta_W \equiv 1 - \frac{m_W^2}{m_Z^2} , \qquad (23)$$

the W-boson mass can be expressed as:

$$m_W^2 = \frac{\frac{\pi\alpha}{\sqrt{2}G_F}}{\sin^2\theta_W(1-\Delta r)} , \qquad (24)$$

where Δr contains all the one-loop corrections. Contributions to Δr originate from the top quark by the one-loop diagrams shown in Fig. 8, which contribute to the W and Z masses via:

$$(\Delta r)_{top} \simeq -\frac{3G_F}{8\sqrt{2}\pi^2 \tan^2\theta_W} m_t^2 . \qquad (25)$$

from low energy experiments that are sensitive to the interference between neutral current and electromagnetic amplitudes. Figure 7 shows the comparison of confidence level contour lines of the electroweak fit to the Standard Model in the plane of the vector and axial-vector coupling of the b-quark. Good agreement between the fit and the Standard Model at the level of ≈ 2 standard deviations ($2\sigma\hat{=}95.5\%$ CL) is found. From earlier measurements of $\Gamma_{b\bar{b}}$ and A_{FB} at LEP, SLC, and the low energy experiments (PEP, PETRA and TRISTAN [35–37]), one obtains [38]

$$T_3^L(b) = -0.490^{+0.015}_{-0.012} \Rightarrow T_3^L(b) = -1/2 , \qquad (21)$$

$$T_3^R(b) = -0.028 \pm 0.056 \Rightarrow T_3^R(b) = 0 , \qquad (22)$$

for the third component of the isospin of the b-quark. This implies that the b-quark must have a weak isospin partner, i.e. the top quark with $T_3^L(t) = +1/2$ must exist.

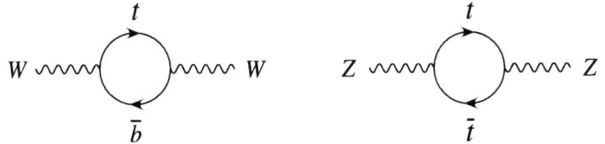

Fig. 8. Virtual top quark loops contributing to the W and Z boson masses

Fig. 9. Virtual Higgs boson loops contributing to the W and Z boson masses

Also the Higgs boson contributes to Δr via the one-loop diagrams, shown in Fig. 9:

$$(\Delta r)_{\text{Higgs}} \simeq \frac{3 G_{\text{F}} m_W^2}{8\sqrt{2}\pi^2}\left(\ln\frac{m_{\text{H}}^2}{m_Z^2}-\frac{5}{6}\right). \qquad (26)$$

While the leading m_t dependence is quadratic, i.e. very strong, the leading m_{H} dependence is only logarithmic, i.e. rather weak. Therefore the inferred constraints on m_{H} are much weaker than those on m_t. This was used to successfully predict the top quark mass from the electroweak precision data before it was discovered by CDF and DØ in 1995 [42, 43]. Neutral current weak interaction data, such as e^+e^- annihilation near the Z pole, νN and eN deep-inelastic scattering, νe elastic scattering, and atomic parity violation can also be used to constrain the top quark mass. Figure 10 shows the χ^2 of the Standard Model electroweak fit to the precision data as a function of the assumed top quark mass for three different choices of the Higgs boson mass [41]. $m_{\text{H}} = 50\,\text{GeV}/c^2$ was the lower limit of the Higgs boson mass from direct searches at LEP1 at the time, $1000\,\text{GeV}/c^2$ is the theoretical upper limit

of the Higgs boson mass, and $300\,\text{GeV}/c^2$ was chosen to be a representative, central value as a logarithmic average between the two extremes. The minimum of the χ^2 curve indicates the best estimate of the top quark mass, the width of the curves gives an estimate of the uncertainty of this determination. The most recent indirect measurements of the top quark mass using the Z-pole data together with the direct measurements of the W-boson mass and total width and several other electroweak quantities yields [44, 45]:

$$m_{\text{top}} = 179.4^{+12.1}_{-9.2}\,\text{GeV}/c^2, \qquad (27)$$

which is in very good agreement with the world average of the direct measurements [46]

$$m_{\text{top}} = 172.7 \pm 2.9\,\text{GeV}/c^2. \qquad (28)$$

The global fit to all electroweak precision data including the world average of the direct top quark mass measurements yields [44, 45]:

$$m_{\text{top}} = 173.3 \pm 2.7\,\text{GeV}/c^2, \qquad (29)$$

while a fit only to the Z-pole data gives [31]:

$$m_{\text{top}} = 172.6^{+13.2}_{-10.2}\,\text{GeV}/c^2. \qquad (30)$$

The successful prediction of the mass of the top quark before its discovery provides confidence in the precision and predictive power of radiative corrections in the Standard Model. Therefore, the Standard Model fit to the electroweak precision data including the direct measurements of the top quark and W-boson mass is used to infer on the mass of the Standard Model Higgs boson. Figure 11 (left) shows the $\Delta\chi^2$ of the latest fit as a function of the Higgs boson mass. The most likely value of the Higgs mass, determined from the minimum of the $\Delta\chi^2$ curve is $91^{+45}_{-32}\,\text{GeV}/c^2$ [44, 45], clearly indicating that the data prefers a light Higgs boson, $m_{\text{H}} < 186\,\text{GeV}/c^2$ [44, 45].

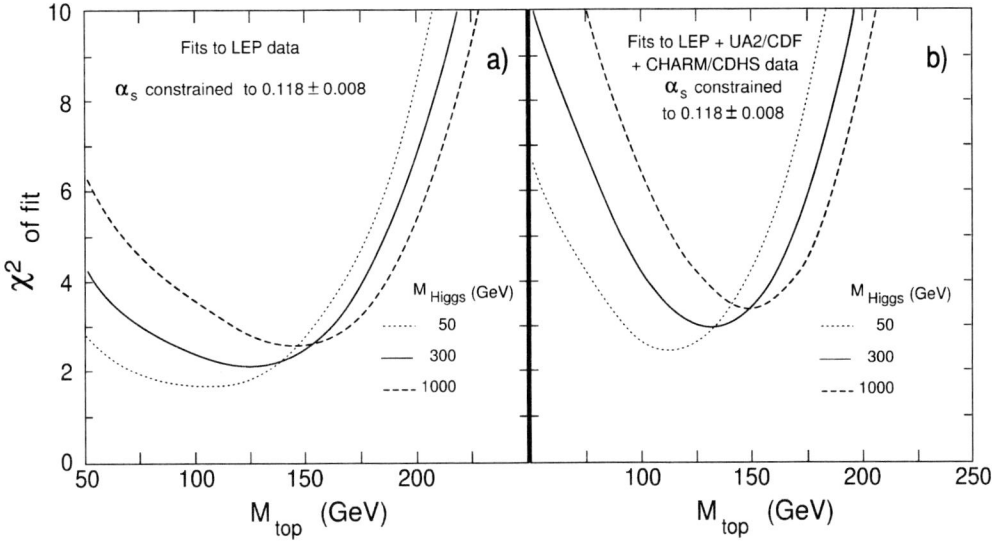

Fig. 10. χ^2 of the Standard Model fit to the electroweak data as a function of the top quark mass using LEP 1 data (*left*) and using LEP 1, hadron collider and neutrino experiment data (*right*) [41]. The dependence on the Higgs boson mass, here chosen to be 50, 300 or $1000\,\text{GeV}/c^2$, is weak, since m_{H} enters only logarithmically in the electroweak fit, whereas m_t enters quadratically

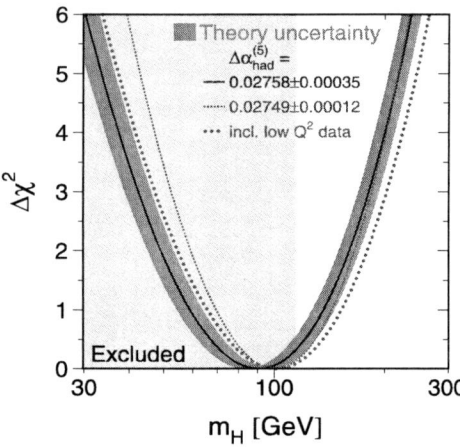 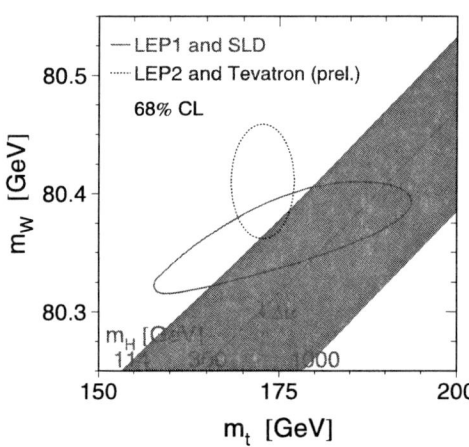

Fig. 11. *Left*: Blueband plot, showing the indirect determination of the Higgs boson mass from all electroweak precision data together with the 95% CL lower limit on the Higgs boson mass from the direct searches [47]. *Right*: Lines of constant Higgs mass on a plot of M_W vs. m_t. The *dotted ellipse* is the 68% CL direct measurement of M_W and m_t. The *solid ellipse* is the 68% CL indirect measurement from precision electroweak data

The preferred value is slightly above the exclusion limit of $114.4\,\mathrm{GeV}/c^2$ from the direct search for the Standard Model Higgs boson at LEP [47].

Figure 11 (right) shows the 68% CL contour in the (m_t, m_W) plane from the global electroweak fit [44,45]. It shows the direct and indirect determination of m_t and m_W. Also displayed are the isolines of Standard Model Higgs boson mass between the lower limit of $114\,\mathrm{GeV}/c^2$ and the theoretical upper limit of $1000\,\mathrm{GeV}/c^2$. As can be seen from the figure, the direct and indirect measurements are in good agreement, showing that the Standard Model is not obviously wrong. On the other hand, the fit to all data has a χ^2 per degree of freedom of 18.6/13, corresponding to a probability of 13.6%. This is mostly due to three anomalous measurements: the b forward-backward asymmetry (A_{FB}^b) measured at LEP, which deviates by 2.8σ, the total hadronic production cross section (σ_{had}^0) at the Z-pole from LEP and the left-right cross section asymmetry (A_{LR}) measured at SLC, both of which deviate from the Standard Model fit value by about 1.5σ. If $\sin^2\theta_W(\nu N)$, measured by the NuTeV collaboration [48], is in addition included in the fit, the measured and fitted value of $\sin^2\theta_W(\nu N)$ differ by 3σ. It seems there is some tension in the fit of the precision electroweak data to the Standard Model.

Measurements of M_W and m_t at the TEVATRON could resolve or exacerbate this tension. Improvements in the precision of the measurement of the top quark or the W-boson mass at the TEVATRON translate into better indirect limits on the Higgs boson mass. This will also be a service to the LHC experiments which optimise their analysis techniques and strategies for the search for the yet elusive Standard Model Higgs boson in the lower mass range, preferred by the Standard Model electroweak fit.

1.4 Historic overview over top quark searches at e^+e^- and pp colliders

In 1977, the b-quark was discovered at Fermilab [49]. As explained in Sect. 1.2, the existence of a weak isospin partner of the b-quark, the top quark, was anticipated and the search for the top quark began. At the e^+e^- colliders

PETRA at DESY [50–63] (1979–84, $\sqrt{s} = 12$–$46.8\,\mathrm{GeV}$), TRISTAN at KEK [64–68] (1986–90, $\sqrt{s} = 61.4\,\mathrm{GeV}$), and SLC at SLAC [69] and LEP at CERN [70–72] (1989–90, $\sqrt{s} = M_Z$) the production of top-antitop bound states (toponium) $e^+e^- \to t\bar{t}$ was searched for. Based on the lack of observation of such states, the experiments increased the lower bound on the top quark mass from $m_t > 23.3\,\mathrm{GeV}/c^2$ at PETRA to $m_t > 30.2\,\mathrm{GeV}/c^2$ at TRISTAN and finally to $m_t > 45.8\,\mathrm{GeV}/c^2$ at SLC and LEP. Provided a minimum amount of data, the sensitivity at e^+e^- colliders is limited by half of the achieved centre-of-mass energy, since the top quarks would have to be pair-produced.

In the 1980s, the development of hadron colliders started with the intersecting storage ring (ISR) [73] at CERN, followed by the $Sp\bar{p}S$ at CERN with $\sqrt{s}$ up to $630\,\mathrm{GeV}$ and the TEVATRON at Fermilab with $\sqrt{s} = 1.8\,\mathrm{TeV}$. The search for the top quark at these hadron colliders was not limited by the available centre-of-mass energy, but by the luminosity and the expected resulting rate of top quark events. The dominant mechanism for the production of top quarks was expected to be the production of W-bosons with the subsequent decay $W \to tb$. This search mode provides sensitivity to the top quark to masses of up to $\approx 77\,\mathrm{GeV}/c^2$, since the W-boson can be produced singly in electroweak interactions at $p\bar{p}$ colliders. For a heavier top quark, the strong $t\bar{t}$ pair production with the subsequent weak decay $t \to Wb$ dominates. After some initial indication for the production of top quark at the $Sp\bar{p}S$ experiments UA1 and UA2 in 1984 with $m_t = 40 \pm 10\,\mathrm{GeV}/c^2$ [74], more data and improved analyses proved this result to be a fluctuation [75]. The experiments set a lower bound on the top quark mass of $m_t > 45\,\mathrm{GeV}/c^2$. With more data, the UA1 and UA2 experiments increased this limit in 1989 to $m_t > 60\,\mathrm{GeV}/c^2$ and $m_t > 69\,\mathrm{GeV}/c^2$, respectively [7,76,77]. In 1988, the central collider detector (CDF) at the $p\bar{p}$ collider TEVATRON at FERMILAB started data taking. Already in 1991, with only $\int \mathcal{L}\,dt = 4.4\,\mathrm{pb}^{-1}$, CDF set limits of $m_t > 77\,\mathrm{GeV}/c^2$ from the e + jets channel and $m_t > 72\,\mathrm{GeV}/c^2$ from the $e\mu$ channel [78–80] for $m_t < m_W$. This limit was already stronger than the one achievable at the $Sp\bar{p}S$ despite the larger luminosity of $\int \mathcal{L}\,dt = 7.5\,\mathrm{pb}^{-1}$ collected by the UA2 experiment due to the higher beam energy at the TEVATRON.

Table 2. History of the search for the top quark at e^+e^- and at hadron colliders. The quoted uncertainties for the top quark mass from the 1995 discovery publications are statistical and systematic uncertainties, respectively

Year	Collider	Particles	References	Limit on m_t
1979–84	PETRA (DESY)	e^+e^-	[50]–[63]	$> 23.3\,\mathrm{GeV}/c^2$
1987–90	TRISTAN (KEK)	e^+e^-	[64]–[68]	$> 30.2\,\mathrm{GeV}/c^2$
1989–90	SLC (SLAC), LEP (CERN)	e^+e^-	[69]–[72]	$> 45.8\,\mathrm{GeV}/c^2$
1984	$Sp\bar{p}S$ (CERN)	$p\bar{p}$	[75]	$> 45.0\,\mathrm{GeV}/c^2$
1990	$Sp\bar{p}S$ (CERN)	$p\bar{p}$	[76, 77]	$> 69\,\mathrm{GeV}/c^2$
1991	TEVATRON (FNAL)	$p\bar{p}$	[78]–[80]	$> 77\,\mathrm{GeV}/c^2$
1992	TEVATRON (FNAL)	$p\bar{p}$	[81, 82]	$> 91\,\mathrm{GeV}/c^2$
1994	TEVATRON (FNAL)	$p\bar{p}$	[84, 85]	$> 131\,\mathrm{GeV}/c^2$
1995	TEVATRON (FNAL)	$p\bar{p}$	[42]	$= 174 \pm 10^{+13}_{-12}\,\mathrm{GeV}/c^2$
			[43]	$= 199^{+19}_{-21} \pm 22\,\mathrm{GeV}/c^2$

By adding more search channels and due to the use of soft-lepton b-tagging, CDF reached in 1992 a top quark mass limit of $m_t > 91\,\mathrm{GeV}/c^2$ [81, 82]. In 1992, the DØ experiment was commissioned and had comparable sensitivity to the top quark as CDF [83]. In 1994, DØ set a limit on the top quark mass of $m_t > 131\,\mathrm{GeV}/c^2$ (later corrected down to $128\,\mathrm{GeV}/c^2$ due to a re-calibration of the DØ luminosity measurement) [84, 85]. Later that year, CDF claimed the first evidence for $t\bar{t}$ production [86, 87] with a measured $t\bar{t}$ production cross section approximately 2.4 times that expected in the Standard Model. Shortly after that, CDF improved the determination of the background normalisation factor, reducing the obtained $t\bar{t}$ cross section and the significance of the claimed signal. A review of the status of searches for the top quark in 1994 with the supposedly

$t\bar{t}$ production cross section $\sigma_{t\bar{t}} = 13.9^{+6.1}_{-4.8}\,\mathrm{pb}$, measured by CDF [86, 87], being significantly higher than the Standard Model expectation of $\sigma_{t\bar{t}} \approx 5\,\mathrm{pb}$ and the DØ results (7 events observed, 3.2 ± 1.1 events expected from background, yielding $\sigma_{t\bar{t}} = 6.5 \pm 4.9\,\mathrm{pb}$ for $m_t = 180\,\mathrm{GeV}/c^2$) being consistent with the Standard Model prediction albeit not very significant yet is given in [83]. Finally, in 1995, both CDF and DØ published the discovery of the top quark in strong $t\bar{t}$ production [42, 43], which marked the beginning of a new era, moving on from the search for the top quark to the studies and measurements of the properties of the top quark. During the exciting time of the searches for and the discovery of the top quark at the TEVATRON, the journalist Kent W. Staley accompanied both collaborations, CDF and DØ, at FERMILAB and describes his scientific and non-scientific experiences in [88].

Table 2 summarises the history of searches for the top quark and Fig. 12 shows the development of limits and measurements on the top quark mass from indirect and direct studies at e^+e^- and hadron colliders. The top quark was discovered with a mass of exactly the value that was predicted from global fits to electroweak precision data.

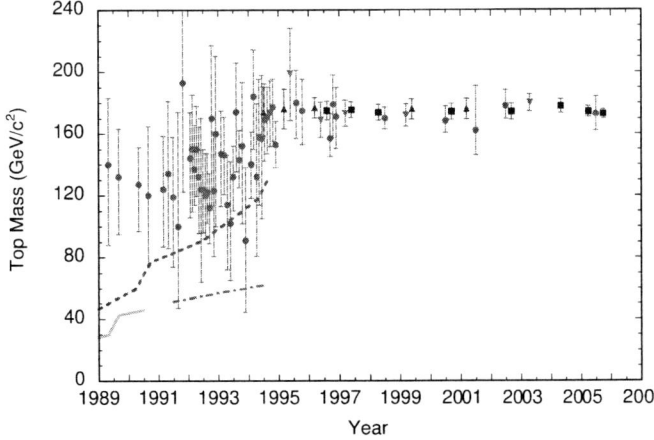

Fig. 12. History of the limits on or measurements of the top quark mass (updated Sept. 1995 by C. Quigg from [89]): (•) Indirect bounds on the top-quark mass from precision electroweak data; (■) world-average direct measurement of the top-quark mass (including preliminary results); (▲) published CDF and (▼) DØ measurements; Lower bounds from $p\bar{p}$ colliders $Sp\bar{p}S$ and the TEVATRON are shown as *dash-dotted* and *dashed lines*, respectively, and lower bounds from e^+e^- colliders (PETRA, TRISTAN, LEP and SLC) are shown as a *solid light grey line*

2 Top quark production and decay at hadron colliders

2.1 Strong pair production of top quarks

The $t\bar{t}$ production at high energy interactions of a $p\bar{p}$ or a pp collision at the TEVATRON or LHC, respectively, is described by perturbative QCD. In this approach, a hard scattering process between two hadrons (proton or antiproton) is the result of an interaction between the quarks and gluons which are the constituents of the incoming hadrons. The incoming hadrons provide broad band beams of partons which possess varying fractions x of the momenta of their parent hadrons. The description of hadron collisions can be separated into a short distance (hard scattering) partonic cross section for the participating partons of type i and j, $\hat{\sigma}^{ij}$, and into long distance pieces

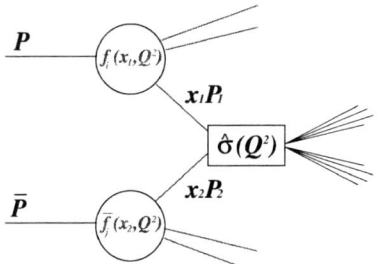

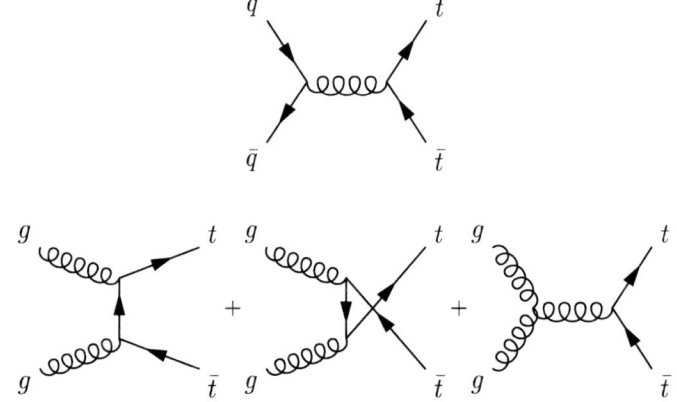

Fig. 13. Parton model description of a hard scattering process using the factorisation approach

which are factored into the parton longitudinal momentum distribution functions (PDFs) $f_i(x_i, \mu_{\mathrm{F}}^2)$. This separation is called factorisation and is schematically shown in Fig. 13.

The separation is set by the factorisation scale μ_{F}^2. The short distance cross section only involves high momentum transfer and is calculable in perturbative QCD. It is insensitive to the physics of low momentum scale. In particular, it does not depend on the hadron wave functions or the type of the incoming hadrons. This factorisation property of the cross section can be proven to all orders in perturbation theory [90]. When higher order terms are included in the perturbative expansion, the dependence on this arbitrary scale μ_{F}^2 gets weaker.

The parton distribution function (PDF), $f_i(x_i, \mu_{\mathrm{F}}^2)$, can be interpreted as the probability density to observe a parton of flavour i and longitudinal momentum fraction x_i in the incoming hadron, when probed at a scale μ_{F}^2. Since the PDFs can not be calculated a priori by perturbative QCD, they are extracted in global QCD fits from deep-inelastic scattering and other data [91–93]. An example parameterisation, obtained by the CTEQ collaboration [94], for two different $Q^2 = \mu_{\mathrm{F}}^2$ scales, is shown in Fig. 14.

In higher order calculations, infinities such as ultraviolet divergences appear. These divergences are removed by a renormalisation procedure, which introduces another artificial scale μ_{R}^2. However, the physical quantities cannot depend on the arbitrary scale, μ_{R}^2, as expressed by the renormalisation group equation [13–15,91]. It is common to choose the same scale $Q^2 = \mu^2$ for both, the factorisation

Fig. 15. Top-quark pair production via the strong interaction at hadron colliders proceeds at lowest order through quark–antiquark annihilation (*top*) and gluon fusion (*bottom*)

scale μ_{F}^2 and the renormalisation scale μ_{R}^2. The convention is used in the following.

The total top quark pair production cross section for hard scattering processes, initiated by a $p\bar{p}$ or a pp collision at a centre-of-mass energy $\sqrt{s}$ can be calculated as [95,96]:

$$\sigma^{t\bar{t}}(\sqrt{s}, m_{\mathrm{t}}) = \sum_{i,j=q,\bar{q},g} \int \mathrm{d}x_i \, \mathrm{d}x_j \, f_i\left(x_i, \mu^2\right) \bar{f}_j\left(x_j, \mu^2\right)$$
$$\times \hat{\sigma}^{ij \to t\bar{t}}\left(\rho, m_{\mathrm{t}}^2, x_i, x_j, \alpha_s(\mu^2), \mu^2\right) . \quad (31)$$

$f_i(x_i, \mu^2)$ and $\bar{f}_j(x_j, \mu^2)$ are the PDFs for the proton and the antiproton, respectively. The summation indices i and j run over all $q\bar{q}$, gg, qg, and $\bar{q}g$ pairs, $\rho = 4m_{\mathrm{t}}^2/\sqrt{\hat{s}}$ and $\hat{s} = x_i x_j s$ is the effective centre-of-mass energy squared for the partonic process. The corresponding lowest order parton model processes are shown in Fig. 15.

Since there has to be at least enough energy to produce a $t\bar{t}$ pair at rest, $\hat{s} \geq 4m_{\mathrm{t}}^2$. Therefore, $x_i x_j = \hat{s}/s \geq 4m_{\mathrm{t}}^2/s$. Since the probability of finding a quark of momentum fraction x in the proton falls off with increasing x (see Fig. 14), the typical value of $x_i x_j$ is near the threshold for $t\bar{t}$ produc-

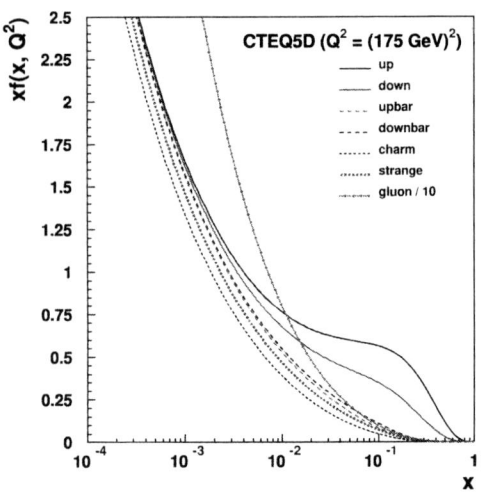

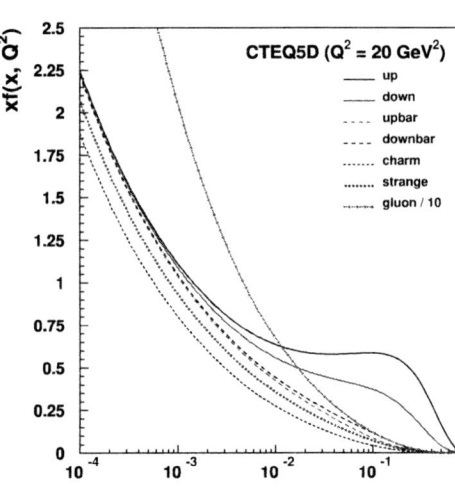

Fig. 14. The quark, antiquark and gluon momentum densities in the proton as a function of the longitudinal proton momentum fraction x at $Q^2 = m_t^2$ (*left*) and at $Q^2 = 20\,\mathrm{GeV}^2$ (*right*) from the CTEQ5D parameterisation [94]

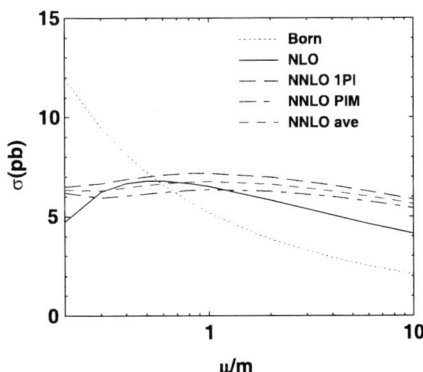
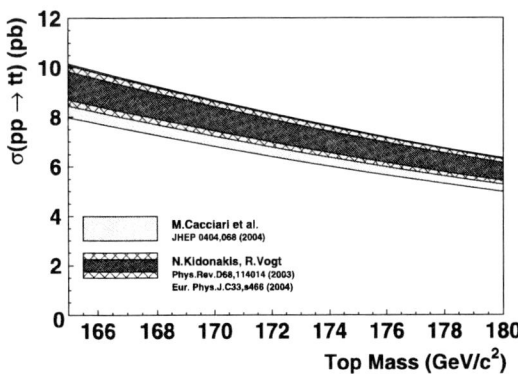

Fig. 16. Left: The scale dependence for $m_t = 175$ GeV of the $t\bar{t}$ cross section at $\sqrt{s} = 1.96$ TeV in $p\bar{p}$ collisions at the TEVATRON. The exact definition of the terms which are considered in the perturbative expansion referred to as "NNLO" can be found in [116]. *Right:* Top quark mass dependence for $\mu = m_t$ of the $t\bar{t}$ cross section at $\sqrt{s} = 1.96$ TeV in $p\bar{p}$ collisions at the TEVATRON. The *error band* for the calculations of Cacciari et al. [114] contains scale and PDF uncertainties. The inner error band for the calculation of Kidonakis and Vogt [116, 118] contains kinematics uncertainties (one-particle inclusive versus pair-invariant mass), while the outer error band also contains PDF uncertainties according to [119]

tion. Setting $x_i \approx x_j \equiv x$ gives:

$$x \approx \frac{2m_t}{\sqrt{s}} \qquad (32)$$
$$= 0.19 \qquad \text{at the TEVATRON in Run I}$$
$$= 0.18 \qquad \text{at the TEVATRON in Run II}$$
$$= 0.025 \qquad \text{at the LHC}$$

as the typical value of x for $t\bar{t}$ production. For the typical values of x at the TEVATRON, the quark distribution functions, in particular the u- and d-valence quark distribution, are much larger than that of the gluon. This explains why quark–antiquark annihilation dominates at the TEVATRON. At Run II, in comparison to Run I, a slightly lower x value is already sufficient to produce a $t\bar{t}$ pair, resulting in a $\approx 30\%$ increase in the $t\bar{t}$ production cross section at Run II compared to Run I. Since the gluon distribution increases more steeply towards low x than the valence- or even the sea-quark distributions, the fraction of gluon–gluon initiated interactions in the total $t\bar{t}$ production increases from 10% in Run I to 15% in Run II. For the same reason, at the LHC, where x-values as small as 0.025 are sufficient for $t\bar{t}$ production, the total $t\bar{t}$ production cross section increases by more than a factor of 100 and is vastly dominated by gluon–gluon fusion. In reality x_i and x_j of the partons in the proton and antiproton do not necessarily have the same value, allowing asymmetric momenta of the incoming partons in $t\bar{t}$. Consequently, in particular at the LHC, low-x gluons contribute a large fraction of the $t\bar{t}$ production cross section. On the other hand, at the LHC $t\bar{t}$ pairs are typically produced above the mass threshold due to the large available centre-of-mass energy.

The top quark cross section was calculated at next-to-leading order in QCD many years ago [97–100]. These calculations were later improved with the resummation to all orders of perturbation theory of classes of large soft logarithms. Large logarithmically enhanced corrections due to soft-gluon radiation are a general feature in the study of the production cross section of high-mass systems near

threshold. Techniques for re-summing these corrections have been developed over the past several years, starting from the case of Drell–Yan (DY) pair production [101, 102] and then applied to heavy quark production in [103–107] or the bottom-quark fragmentation in top-quark decays in [108]. This transfer is possible since these logarithms are universal between electroweak and QCD induced cross sections. To go beyond leading logarithms one has to take into account the complex colour structures of QCD cross section calculations [109, 110]. The soft-gluon resummation for $t\bar{t}$ production at the TEVATRON and the LHC[5] of QCD corrections at next-to-leading logarithm (NLL) accuracy including part of the higher order corrections is performed in [109–117][6].

The introduction of resummation turns out to have only a mild impact on the overall rates (the effects at next-to-leading logarithm (NLL) are typically of the order $\mathcal{O}(5\%)$), but improves the stability of the predictions with respect to changes of the renormalisation or factorisation scale (Fig. 16, left). In theoretical studies of the systematic uncertainties due to parton densities and scale dependence [114], the importance of including the α_s uncertainty

---

[5] Since $t\bar{t}$ pairs are produced at the LHC mostly well above threshold, soft-gluons are a small effect and their resummation a small correction to this small effect. Consequently, the soft-gluon resummation is less important for the LHC than for the TEVATRON.

[6] The available $t\bar{t}$ cross section calculations include the exact NLO corrections and estimate part of the higher order NLLO corrections. Kidonakis and Vogt [116] include estimates, derived from a resummation approach, of part of the higher order corrections at NNLO (2-loop) level, where they consider scale uncertainties and the choice of kinematic variables as systematic uncertainties. Cacciari et al. [114] include estimates, also derived from resummation, of part of the higher order corrections of all orders, where they consider scale uncertainties and uncertainties from the parton distribution functions in their systematic uncertainty.

into the PDF fits in a more systematic fashion is under-scored. On the same footing, the impact of higher order corrections, as well as the treatment of higher twist effects in the fitting of low-Q^2 data, may need some more study before a final tabulation of the PDF uncertainties can be achieved [120]. The PDF uncertainty on the top quark pair production cross section is mostly driven by the poorly known gluon density, whose luminosity in the relevant kinematic range for the TEVATRON varies by up to a factor of 2 within the 1σ PDF range. For the LHC cross section calculations, dominated by the gluon–gluon fusion, this uncertainty is even larger. In recent years, with increasing precision of the measurements of the deep-inelastic scattering cross sections at HERA [121–124], experimental and theoretical groups have focused on the proper evaluation and propagation of uncertainties on the parton distribution functions, starting with [125] and followed by [120,121,126–135]. While the overall top pair production rate at the TEVATRON has a large relative uncertainty of approximately 15% (Fig. 16, right shows the total uncertainty of the $t\bar{t}$ production cross section calculations with gluon resummation [114,116], including scale, kinematics and PDF uncertainties, as a function of the top quark mass), it is important to point out that the ratio of cross sections at $\sqrt{s} = 1.96$ TeV and $\sqrt{s} = 1.8$ TeV is very stable.

Table 3 summarises the $t\bar{t}$ production cross section calculation for Run I and Run II at the TEVATRON and for the LHC. Reference [113] only considers uncertainties from scale variations, resulting in a $\approx 10\%$ uncertainty. Another $\approx 6\%$ come from PDFs and α_s. Reference [116] only considers uncertainties from scale variations, resulting in a $\approx 4\%$ uncertainty. Another $\approx 5\%$ come from PDFs. Reference [114] considers uncertainties from scale variations, PDFs and α_s. At the TEVATRON, for every 1 GeV/c^2 increase in the top quark mass over the interval $170 < m_{\text{top}} < 190$ GeV/c^2, the $t\bar{t}$ cross section decreases by 0.2 pb. The hard scattering cross sections for several processes, including $t\bar{t}$ production, are shown in Fig. 17 as a function of the centre-of-mass energy, covering the energy range for the TEVATRON and the LHC. In addition to having similar event topology to the Standard Model Higgs production, $t\bar{t}$ production also has a similar cross section, many orders of magnitude lower than the W- or Z-production or the inclusive QCD b-production.

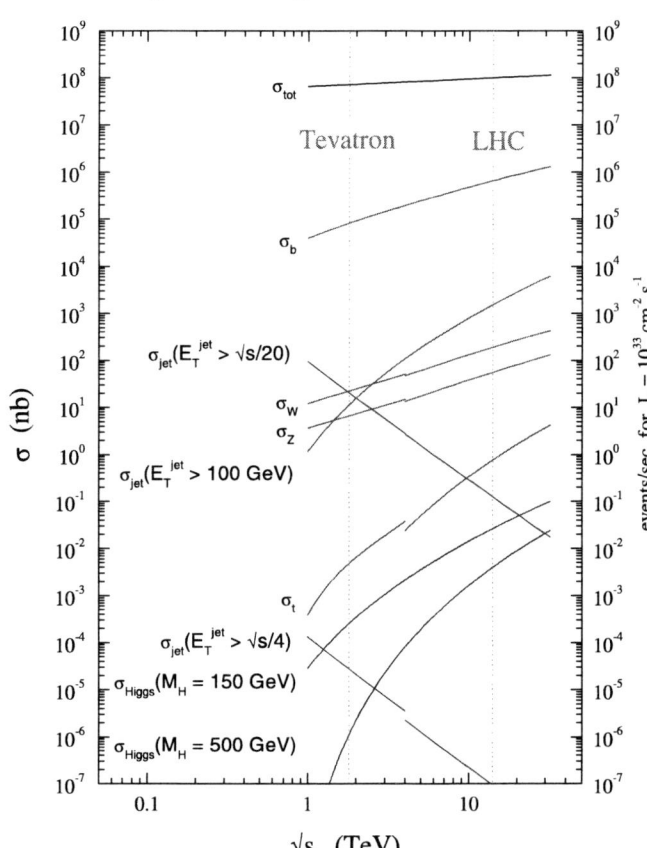

Fig. 17. QCD predictions for hard scattering cross sections at the TEVATRON and the LHC [141]. σ_t stands for the $t\bar{t}$ production cross section. The steps in the *curves* at $\sqrt{s} = 4$ TeV mark the transition from $p\bar{p}$ scattering at the TEVATRON to pp scattering at the LHC

An accurate calculation of the cross section for top quark pair production is a necessary ingredient for the measurement of $|V_{tb}|$ since $t\bar{t}$ production is an important background for the electroweak single-top production. More importantly, this cross section is sensitive to new physics in top quark production and/or decay. A new source of top quarks (such as gluino production, followed by the decay $\tilde{g} \to \tilde{t}t$) would appear as an enhancement

Table 3. Cross section, at next-to-leading order in QCD including gluon resummation corrections, for $t\bar{t}$ production via the strong interaction at the TEVATRON and the LHC for $m_t = 175$ GeV/c^2. Details on the meaning of the quoted uncertainties are given in the text and in references [114,116]. For the $\sqrt{s} = 1.96$ TeV result of reference [116], the quoted error includes the uncertainty from the PDFs according to [119]

	σ_{NLO} (pb)	$q\bar{q} \to t\bar{t}$	$gg \to t\bar{t}$
TEVATRON($\sqrt{s} = 1.8$ TeV, $p\bar{p}$)	$5.19 \pm 13\%$ [114]	90%	10%
	$5.24 \pm 6\%$ [116]	90%	10%
TEVATRON($\sqrt{s} = 1.96$ TeV, $p\bar{p}$)	$6.70 \pm 13\%$ [114]	85%	15%
	$6.77 \pm 9\%$ [116]	85%	15%
LHC ($\sqrt{s} = 14$ TeV, pp)	$833 \pm 15\%$ [113]	10%	90%

of the cross section, and a new decay mode (such as $t \to \tilde{t}\tilde{\chi}^0$) would appear as a suppression. Resonances in $t\bar{t}$ production would also increase the top quark cross section [136–140]. The latest $t\bar{t}$ cross section measurements from the TEVATRON are discussed in Sect. 4.

2.2 Electroweak single top quark production

The best way to study the properties of the Wtb vertex and to directly measure $|V_{tb}|$ at a hadron collider is via the measurement of the electroweak single top quark production, shown in Fig. 18. There are three separate processes: (a) W-gluon fusion or t-channel process [142–144], which is similar to heavy-flavour production via charged-current deep-inelastic scattering, and (b) Wt production [145], and (c) quark–antiquark annihilation or s-channel process [146, 147], which is similar to the Drell–Yan process and also called W^* process. Only (a) and (c) are relevant to the electroweak single top production at the TEVATRON. So far, electroweak single-top quark production has not yet been observed in experiments, but the processes (a) and (c) are both expected to be observed in Run II at the TEVATRON. While the Wt production (b) is expected to be observed at the LHC. All three processes involve the top quark charged current, so their cross sections are proportional to $|V_{tb}|^2 g_W^2(tb)$. Assuming the Standard Model weak $SU(2)$ coupling for a doublet pair of quarks, the electroweak single-top quark production cross section provides direct sensitivity to the CKM matrix element $|V_{tb}|$.

Calculations of fully-differential NLO single-top quark cross sections have been performed in [148–151] and, including NLO top quark decay, in [152–156]. The total

s-channel production cross section has been calculated to next-to-leading order in QCD (for example in [144]), and some of the technology to extend this calculation to next-to-next-to-leading order exists [157, 158]. The total t-channel production cross section has also been calculated to next-to-leading order [159, 160].

Consider the strengths and weaknesses of the separate single-top processes. The s-channel process involves the quark distribution functions, which are better known than the gluon distribution function in the t-channel process and in Wt production. Furthermore, the s-channel process benefits from its similarity to the Drell–Yan process, which can be used as a normalisation. The t-channel process has the advantage that it will be observable at the LHC, while s-channel production will be difficult to observe, due to large backgrounds. The large rate of the t-channel process at the LHC implies that the measurement of $|V_{tb}|$ will have negligible statistical uncertainty.

It is interesting to study the three processes separately, since they have separate backgrounds, their systematic uncertainties for $|V_{tb}|$ are different, and they are sensitive to new physics in different ways. For example, the presence of a heavy W'-boson would result in a decrease of the s-channel signal. Instead, the existence of a flavour-changing neutral current $gu \to t$ would be seen in the t-channel. Discriminants for the three signals are for example: the jet multiplicity (higher for Wt), the presence of more than one jet tagged as a b (this increases the s-channel signal with respect to the t-channel one), the mass distribution of the 2-jet system (which has a peak near the W mass for the Wt signal and not for the other two).

The electroweak single top quark production cross sections, expected in the Standard Model at the TEVATRON and the LHC, are summarised in Table 4. The latest TEVATRON analyses and experimental cross section limits are discussed in Sect. 5.

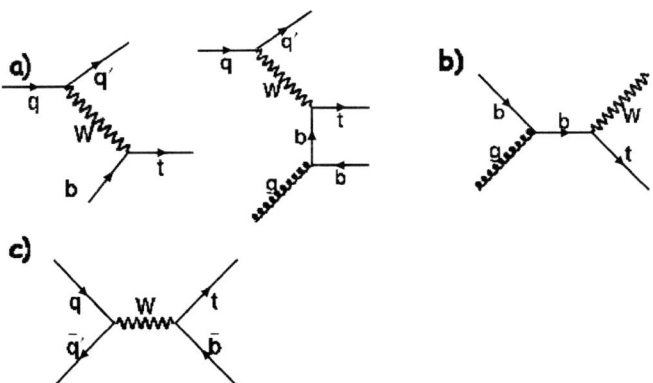

Fig. 18. Feynman diagrams for the electroweak single top quark production processes at the TEVATRON and the LHC: **a** W-gluon fusion or t-channel, **b** Wt production, **c** s-channel or W^* process

2.3 The top quark decay

With a mass above the Wb threshold, the decay width of the top quark is expected to be dominated by the two-body channel $t \to Wb$. Neglecting terms of order m_b^2/m_t^2, α_s^2 and those of order $(\alpha_s/\pi)m_W^2/m_t^2$ in the decay amplitude, the width predicted in the Standard Model is [163]:

$$\Gamma_t = \frac{G_F m_t^3}{8\pi\sqrt{2}}\left(1 - \frac{M_W^2}{m_t^2}\right)^2\left(1 + 2\frac{M_W^2}{m_t^2}\right)$$
$$\times\left[1 - \frac{2\alpha_s}{3\pi}\left(\frac{2\pi^2}{3} - \frac{5}{2}\right)\right]. \tag{33}$$

Table 4. Cross section, at next-to-leading order in QCD, for electroweak single top quark production in the t-channel, Wt production, and the s-channel at the TEVATRON and the LHC for $m_t = 175 \text{ GeV}/c^2$

	t-channel	Wt production	s-channel
TEVATRON($\sqrt{s} = 1.8$ TeV, $p\bar{p}$)	1.98 ± 0.30 pb [149, 161]	≈ 0 pb	0.88 ± 0.14 pb [149, 161]
LHC $\quad(\sqrt{s} = 14$ TeV, $pp)$	245 ± 27 pb [161, 162]	$62.2^{+16.6}_{-3.7}$ pb [145]	10.2 ± 0.7 pb [144, 161]

The G_F Fermi coupling appearing in this equation contains the largest part of the one-loop electroweak radiative corrections, providing an expression accurate to better than 2%. The width increases with mass, changing, for example, from $1.02\,\mathrm{GeV}/c^2$ for $m_t = 160\,\mathrm{GeV}/c^2$ to $1.56\,\mathrm{GeV}/c^2$ for $m_t = 180\,\mathrm{GeV}/c^2$ (using $\alpha_s(M_Z) = 0.118$). With its correspondingly short lifetime of $\approx 0.5 \times 10^{-24}\,\mathrm{s}$, the top quark is expected to decay before top-flavoured hadrons or $t\bar{t}$-quarkonium bound states can form [164]. The order α_s^2 QCD corrections to Γ_t are also available [165, 166], thereby improving the overall theoretical accuracy to better than 1%.

In top decay, the Ws and Wd final states are expected to be suppressed relative to Wb by the square of the CKM matrix elements $|V_{ts}|$ and $|V_{td}|$. The CKM matrix elements involving the top quark have never been measured directly. Assuming unitarity of the three-generation CKM matrix, their values can be estimated to be [167]:

$$|V_{td}| = 0.004\ -0.014\,, \tag{34}$$
$$|V_{ts}| = 0.037\ -0.044\,, \tag{35}$$
$$|V_{tb}| = 0.9990 - 0.9993\,. \tag{36}$$

Thus $|V_{td}|$, $|V_{ts}|$, and $|V_{tb}|$ are known with a precision of 50%, 10%, and 0.02%, respectively. It is briefly described how these CKM matrix elements can be measured:

$|\mathbf{V_{td}}|$ This may be determined indirectly from $B_d^0 - \overline{B_d^0}$ mixing, shown in Fig. 19. The frequency of oscillation, Δm_d, is proportional to $|V_{tb}^* V_{td}|^2$. Measurements give [167]

$$|V_{tb}^* V_{td}| = 0.0079 \pm 0.0015\,, \tag{37}$$

where the uncertainty (20%) is almost entirely from the theoretical uncertainty in the hadronic matrix element. Assuming three generations ($|V_{tb}| \approx 1$), this is a more accurate measurement of $|V_{td}|$ than can be inferred from unitarity (50%).

$|\mathbf{V_{ts}}|$ This may be determined indirectly from $B_s^0 - \overline{B_s^0}$ mixing, which is the same as Fig. 19, but with the d quark replaced by an s quark. The frequency of oscillation, Δm_s, is proportional to $|V_{tb}^* V_{ts}|^2$. The Particle Data Group thus far only quotes a lower limit on the oscillation frequency [167],

$$\Delta m_s > 14.4\,\mathrm{ps}^{-1}\,, \tag{38}$$

but the DØ experiment has released an analysis of semileptonic B_s^0 decays in $1\,\mathrm{fb}^{-1}$ yielding a first two-sided interval of $17 < m_s < 21\,\mathrm{ps}^{-1}$ at 90% CL with a most probable value of $19\,\mathrm{ps}^{-1}$ [168]. The anticipated value from the

range of $|V_{ts}|$ listed above is $\Delta m_s \approx 18\,\mathrm{ps}^{-1}$, just above the current lower bound. This should be observable in Run II at the TEVATRON. However, the theoretical uncertainty is very similar to that of Δm_d, which means that $|V_{ts}|$ can only be extracted with an uncertainty of 20%, which is larger than the uncertainty in the value inferred from unitarity (10%).

$|\mathbf{V_{ts}}|/|\mathbf{V_{td}}|$ The similarity in the hadronic matrix elements involved in Δm_s and Δm_d can be exploited by taking the ratio:

$$\frac{\Delta m_s}{\Delta m_d} = \frac{M_{B_s}}{M_{B_d}} \xi^2 \left| \frac{V_{ts}}{V_{td}} \right|^2\,. \tag{39}$$

The theoretical uncertainty in the ratio of the hadronic matrix elements, ξ^2, is much less than the uncertainty in the hadronic matrix elements themselves. Using the value of $|V_{ts}|$ from unitarity yields an uncertainty in $|V_{td}|$ that is less than the uncertainty obtained from Δm_d alone.

Figure 20 shows the $\bar{\rho}$–$\bar{\eta}$ plane. The radius of the large circles centred at $(1, 0)$ is proportional to $|V_{td}|$. The large annulus is from the measurement of Δm_d, the small annulus that lies inside it is from the ratio $\Delta m_s / \Delta m_d$ and Δm_d combined, using the current lower bound on Δm_s. The measurement of Δm_s at the TEVATRON will reduce the width of this annulus by about a half, making it one of the most precise measurements in the $\bar{\rho}$–$\bar{\eta}$ plane.

$|\mathbf{V_{tb}}|$ Despite the fact that is has never been measured directly, $|V_{tb}|$ is the best known CKM matrix element (0.02%), assuming three generations. It is only interesting to measure if the assumption of three generations is relaxed, in which case $|V_{tb}|$ is almost completely uncon-

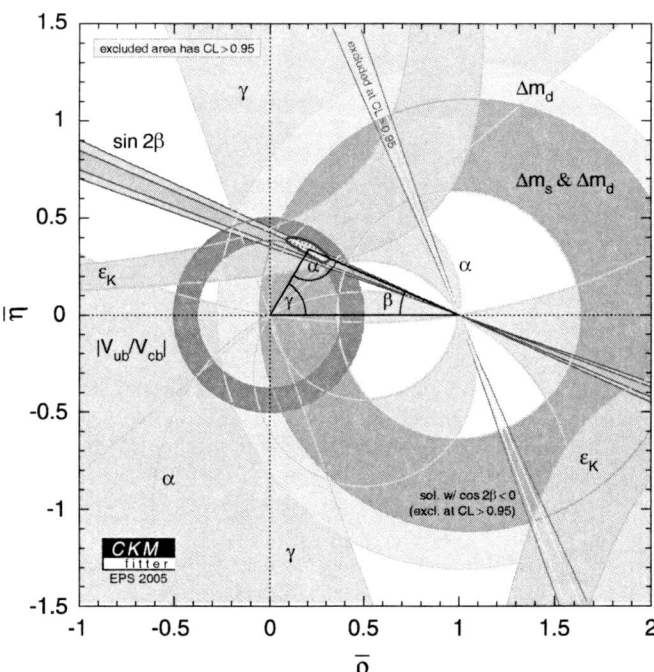

Fig. 20. The $\bar{\rho}$–$\bar{\eta}$ plane, showing constraints from various measurements, as well as the best fit. From [169]

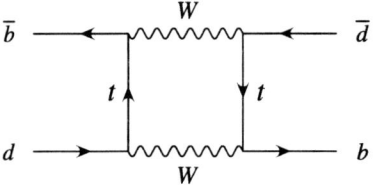

Fig. 19. $B_d^0 - \overline{B_d^0}$ mixing proceeds via a box diagram

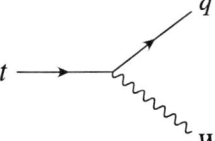

Fig. 21. Top-quark decay to a W-boson and a light quark ($q = d, s, b$)

strained [167],

$$|V_{tb}| = 0.08 - 0.9993 . \qquad (40)$$

In this scenario, $|V_{tb}|$ can be measured directly at the TEVATRON. Considering the top-quark decay, $t \to Wq$, shown in Fig. 21, CDF and DØ measure the fraction of top decays that yield a b-quark (see Sect. 6.2):

$$R = \frac{B(t \to Wb)}{B(t \to Wq)} = \frac{|V_{tb}|^2}{|V_{td}|^2 + |V_{ts}|^2 + |V_{tb}|^2} , \qquad (41)$$

where q denotes any light quark (d, s, b). The last term is the interpretation of this measurement in terms of CKM matrix elements. If one assumes three generations, the denominator of this expression would be unity. Without this assumption, the measurements of this fraction, which come out close to unity, show that $|V_{tb}| \gg |V_{ts}|$ and $|V_{td}|$, but they do not allow conclusions on its absolute magnitude. The latest experimental results on the ratio R from the TEVATRON are discussed in Sect. 6.2.

The way to measure $|V_{tb}|$ directly, with no assumptions about the number of generations, is to measure single top quark production via the weak interaction [170] (Sect. 2.2). The cross sections for three processes are proportional to $|V_{tb}|^2$, and thus provide a direct measurement of this CKM matrix element. The s- and t-channel processes are expected to be observed in Run II. At the TEVATRON, $|V_{tb}|$ is expected to be measured with an uncertainty of about 10% via the $q\bar{q}$ annihilation process (assuming $|V_{tb}|$ near unity), where the uncertainty is statistical. The measurement of $|V_{tb}|$ at the LHC via the W-gluon-fusion process will be limited mostly by the uncertainty in the gluon distribution function: $\Delta|V_{tb}| \sim \Delta g(x)/2$. An uncertainty of 5% requires knowledge of the gluon distribution function to 10%. $|V_{tb}|$ can be extracted from the top width measured from e^+e^- and $\mu^+\mu^-$ colliders operating at the $t\bar{t}$ threshold: $\Delta|V_{tb}| \sim \Delta\Gamma/2$. An uncertainty in the width of less than 30 MeV (2%) may be possible [171–174], yielding an uncertainty in $|V_{tb}|$ of about 1%.

The measurement of $|V_{tb}|$ at a hadron collider requires input from a variety of sources: deep-inelastic scattering (for the parton distribution functions), theory (for precise QCD calculations), and of course the actual experiment. It is a good example of the coordinated effort that is often required to measure a fundamental parameter of the Standard Model.

Given that the top quark decays almost 100% of the time as $t \to Wb$, typical final states for the leading pair-production process can therefore be divided into three classes:

Table 5. Born level theoretical and best measured branching fractions [167] of the real W^+-boson decay, assuming lepton universality. Identical values are calculated and measured for the charge conjugates modes of the W^-

Decay mode	Born level W branching fraction	Measured
$W^+ \to e^+ \nu_e$	1/9	$10.72 \pm 0.16\%$
$W^+ \to \mu^+ \nu_\mu$	1/9	$10.57 \pm 0.22\%$
$W^+ \to \tau^+ \nu_\tau$	1/9	$10.74 \pm 0.27\%$
$W^+ \to \ell^+ \nu_\ell$	3/9	$32.04 \pm 0.36\%$
$W^+ \to u\bar{d}, c\bar{s}$	6/9	$67.96 \pm 0.35\%$

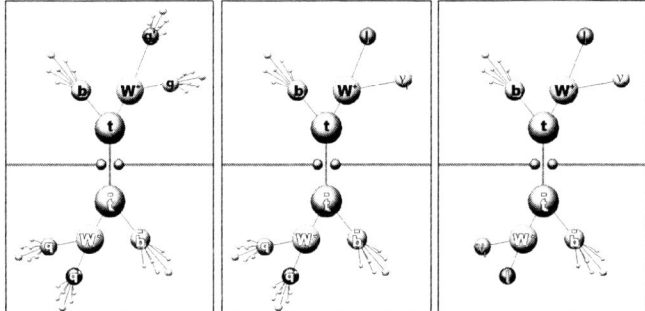

Fig. 22. Schematic diagrams of the three $t\bar{t}$ decay channels: *Left* (A) the alljets channel; *middle* (B) the lepton + jets channel; *right* (C) the dilepton channel

A. $t\bar{t} \to W^+bW^-\bar{b} \to q\bar{q}'bq''\bar{q}'''\bar{b},$ (46.2%)

B. $t\bar{t} \to W^+bW^-\bar{b} \to q\bar{q}'b\ell\bar{\nu}_\ell\bar{b} + \ell\nu_\ell bq\bar{q}'\bar{b},$ (43.5%)

C. $t\bar{t} \to W^+bW^-\bar{b} \to \ell\nu_\ell b\ell'\bar{\nu}_{\ell'}\bar{b},$ (10.3%)

The quarks in the final state evolve into jets of hadrons. A, B, and C are referred to as the all-jets, lepton + jets ($\ell +$ jets), and dilepton ($\ell\ell$) channels, respectively. Because of fermion universality in electroweak interactions[7], in lowest order the W-boson decays 1/3 of the time into an $\ell\nu$ pair and 2/3 of the time into a $q\bar{q}$ pair (see Table 5). The resulting decay branching ratios at Born level for the $t\bar{t}$ decay are shown in Fig. 23. The relative contribution of the three channels A, B, C, including hadronic corrections, are given in parentheses above. The event topologies of the three channels are shown in Fig. 22 schematically. While ℓ in the above processes refers to e, μ, or τ, most of the results to date rely on the e and μ channels. Therefore, in what follows, ℓ will be used to refer to e or μ, unless noted otherwise.

---

[7] The W-boson can decay to pairs of leptons from all three generations and to pairs of quarks from the first and the second generation, each coming in three different colour states. The sum of the masses of the quarks in the third generation exceeds the mass of the W-boson, so that such a decay is strongly suppressed. Therefore, the W-boson can decay to $3 + 2 \cdot 3 = 9$ different fermions pairs with equal rate, yielding a branching ratio of 1/9 for each.

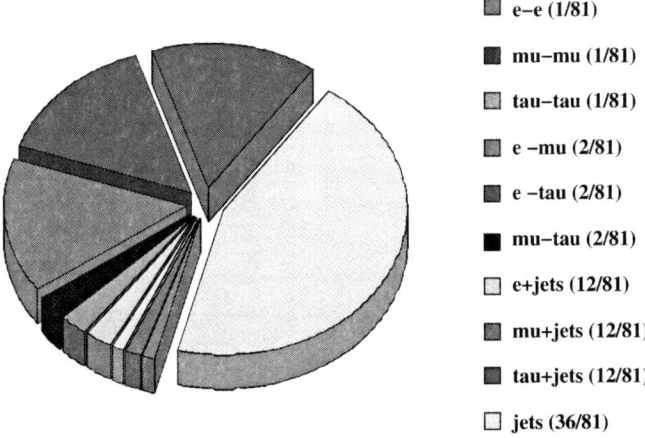

■ e–e (1/81)

■ mu–mu (1/81)

▨ tau–tau (1/81)

■ e –mu (2/81)

■ e –tau (2/81)

■ mu–tau (2/81)

☐ e+jets (12/81)

▨ mu+jets (12/81)

■ tau+jets (12/81)

☐ jets (36/81)

Fig. 23. Pie chart of the branching ratios of the different $t\bar{t}$ decay channels at Born level

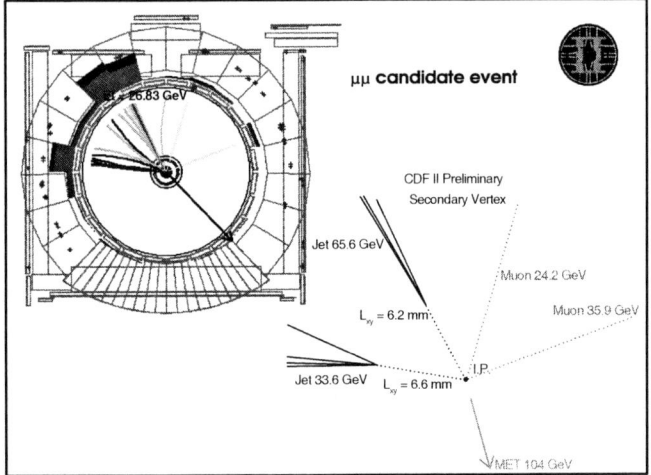

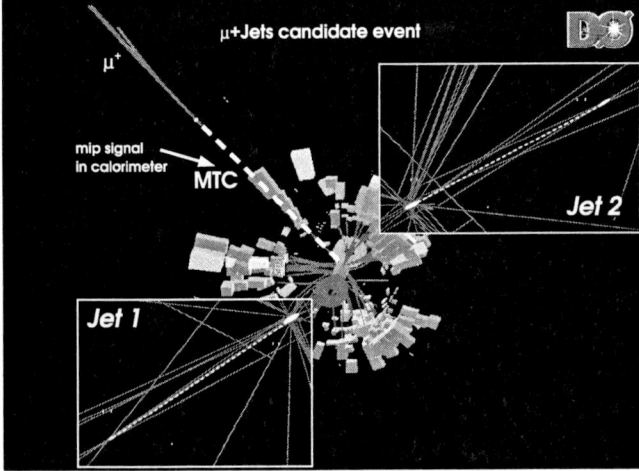

Fig. 24. *Top:* $\mu\mu\nu\nu bb$ example event recorded by CDF. *Bottom:* μ + jets example event recorded by DØ

The initial and final-state quarks can radiate (or emit) gluons that can be detected as additional jets. The number of jets reconstructed in the detectors depends on the decay kinematics as well as on the algorithm for reconstructing jets used by the analysis. The transverse momenta of neu-

trinos are reconstructed from the imbalance in transverse momentum measured in each event (missing E_{T}).

The observation of $t\bar{t}$ pairs has been reported in all of the above decay classes. The production and decay properties of the top quark extracted from the three decay classes are consistent within their experimental uncertainty. In particular, the $t \to Wb$ decay mode is supported through the reconstruction of the $W \to jj$ invariant mass in events with two identified b-jets in the $\ell\nu_{\ell}b\bar{b}jj$ final state [175]. Also the CDF and DØ measurements of the top quark mass in lepton + jets events, where the jet energy scale is calibrated *in situ* using the invariant mass of the hadronically decaying W boson [176, 177], support this decay mode.

Figure 24 shows example event displays in the $\mu\mu$ channel recorded by CDF (top) and in the μ + jets channel recorded by DØ (bottom). In both events, two of the jets show distinct secondary vertices well separated from the primary event vertex. In the DØ μ + jets example event, the muon leaves a clear MIP (= minimum ionising particle) signal in the calorimeter. In Run I such a signature was used for the muon identification via muon tracking in the calorimeter (= MTC). In Run II, such identification criteria are being worked out, but not yet used in most analyses presented in this review.

2.4 Top quark properties

2.4.1 Top quark mass

The mass of the top quark is larger than that of any other quark. Furthermore, the top quark mass is measured with better relative precision (1.7%) than any other quark, as shown in Fig. 25. Given the experimental technique used to extract the top mass, these mass values should be taken as representing the top *pole mass*. The top pole mass, like any quark mass, is defined up to an intrinsic ambiguity of order $\Lambda_{\mathrm{QCD}} \sim 200$ MeV [178].

The desired precision of the top quark mass is generally derived from the relation of the masses of the W-boson, the top quark and the Higgs boson, shown in Fig. 11 (right) [179]. Once a Higgs boson is discovered, even a crude knowledge of

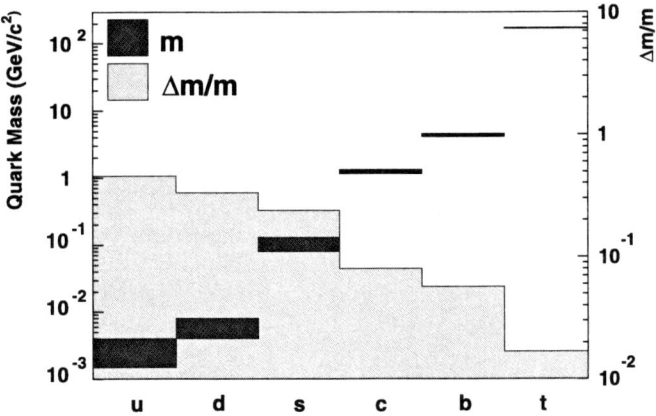

Fig. 25. Quark masses and their absolute and relative uncertainties, indicated by the vertical size of the error bands

its mass will define a narrow line in Fig. 11, since precision electroweak measurements are sensitive only to the logarithm of the Higgs boson mass. An uncertainty in M_W of $20\,\mathrm{MeV}/c^2$, expected to be in reach at the TEVATRON in Run II, projected onto a line of constant Higgs mass corresponds to an uncertainty of $3\,\mathrm{GeV}/c^2$ in the top quark mass. Thus, a precision of $\Delta m_\mathrm{t} \approx 3\,\mathrm{GeV}/c^2$ is desired in order to make maximal use of the precision measurement of M_W for consistency tests of the Standard Model. However, the achieved precision in m_t is already now better than that initial goal. The combined measurement of the top quark mass from Run I yields $178.0 \pm 4.3\,\mathrm{GeV}/c^2$ [180]. The most recent combination of top quark mass measurements by the TEVATRON Electroweak/Top Working group, including preliminary CDF and DØ measurements from Run II, yields $m_\mathrm{t} = 172.7 \pm 2.9\,\mathrm{GeV}/c^2$ [46][8]. The prospects for the precision of the top quark mass measurements at the TEVATRON have recently been revised to better than $2\,\mathrm{GeV}/c^2$ per experiment with the full Run II data set. At the LHC, a final precision of the top quark mass measurement of 1–$2\,\mathrm{GeV}/c^2$ is expected. The latest measurements of the top quark mass at the TEVATRON are discussed in Sect. 7.

At a future linear e^+e^- collider, the expected precision of a measurement of the top quark mass from a cross section scan at the $t\bar{t}$ production threshold is $\Delta m_\mathrm{t} = 20$–$100\,\mathrm{MeV}/c^2$ [171,182,183].

2.4.2 Electric charge of the top quark

Like most of its fundamental quantum numbers, the electric charge of the top quark, q_top, has not been measured so far. The electric charge of the top quark is easily accessible in e^+e^- production by measurements of the ratio $R = \frac{\sigma(e^+e^- \to \mathrm{hadrons})}{\sigma(e^+e^- \to \mu^+\mu^-)}$ through the top quark production threshold. However, this region of energy is not yet available at e^+e^- colliders. Thus, alternative interpretations for the particle that is believed to be the charge $2/3$ isospin partner of the b-quark are not ruled out. For example, since the correlations of the b-quarks and the W-bosons in $p\bar{p} \to t\bar{t} \to W^+W^-b\bar{b}$ events are not determined by CDF or DØ, it is conceivable that the "t quark" observed at the TEVATRON is an exotic quark, Q_4, with charge $-4/3$ with decays via $Q_4 \to W^- b$. This interpretation is consistent with current precision electroweak data. In order to determine the charge of the top quark, one can either measure the charge of its decay products, in particular of the b-jet via jet charge techniques, or investigate photon radiation in $t\bar{t}$ events [184]. The latter method actually measures a combination of the electromagnetic coupling strength and the charge quantum number. Combining the results of the two methods will thus make it possible to determine both quantities.

At the TEVATRON, $q\bar{q}$ annihilation dominates the $t\bar{t}$ production and photon radiation off the incoming quarks

constitutes an irreducible background which limits the sensitivity to q_top. In contrast, at the LHC, gluon fusion dominates, and the $t\bar{t}\gamma$ cross section scales approximately with q_top^2.

At the TEVATRON, with an integrated luminosity of 1–$2\,\mathrm{fb}^{-1}$, one will be able to exclude at 95% CL the possibility that an exotic quark Q_4 with charge $-4/3$ and not the Standard Model top quark was found in Run I. At the LHC with $10\,\mathrm{fb}^{-1}$ obtained at $10^{33}\,\mathrm{cm}^{-2}\,\mathrm{s}^{-1}$, it is expected to be possible to measure the electric charge of the top quark with an accuracy of 10%. For comparison, at a linear collider with $\sqrt{s} = 500\,\mathrm{GeV}$ and $\int \mathcal{L}\,\mathrm{d}t = 200\,\mathrm{fb}^{-1}$, one expects that q_top can be measured with a precision of about 10% [172,185].

The present status and used techniques for the measurements of the top quark electric charge are discussed in Sect. 7.2.

2.4.3 Helicity of the W-boson in top-quark decay

The Standard Model dictates that the top quark has the same vector-minus-axial-vector (V-A) charged-current weak interaction $\left(-i\frac{g}{\sqrt{2}} V_{tb} \gamma^\mu \frac{1}{2}(1 - \gamma_5)\right)$ as all the other fermions. It is easy to see that this implies that the W-boson in top quark decay cannot be right-handed, i.e. have positive helicity. The argument is sketched in Fig. 26. In the idealised limit of a massless b-quark, the V-A current dictates that the b-quark in top decay is always left-handed. If the W-boson were right-handed, then the component of total angular momentum along the decay axis would be $+3/2$ (there is no component of orbital angular momentum along this axis). But the initial top quark has spin angular momentum $\pm 1/2$ along this axis, so this decay is forbidden by conservation of angular momentum. The status of the experimental searches for a right-handed W-boson in top quark decay is summarised in Sect. 6.4.

The top quark may decay to a left-handed (negative helicity) or a longitudinal (zero helicity) W-boson. Its coupling to a longitudinal W-boson is similar to its Yukawa coupling, which is enhanced with respect to the weak coupling. Therefore the top quark prefers to decay to a longitudinal W-boson, with a branching ratio

$$B(t \to W_0 b) = \frac{m_\mathrm{t}^2}{m_\mathrm{t}^2 + 2M_W^2} \approx 0.70\,. \qquad (42)$$

CDF and DØ measure this branching ratio (Sect. 6.4) and find it to be consistent with the Standard Model expectation, even though with very limited statistics. This measurement will improve further during Run II. The present status of the measurements of the helicity of the W-boson in top quark decay are described in Sect. 6.4.

---

[8] An update of the combined top quark mass measurement by the TEVATRON Electroweak/Top Working group yields $m_\mathrm{t} = 171.4 \pm 2.1\,\mathrm{GeV}/c^2$ [181]. This update arrived after the editorial deadline of this review and could therefore not be included in more detail.

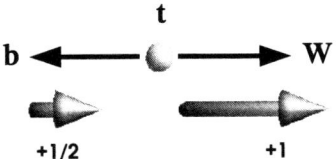

Fig. 26. Illustration that the top quark cannot decay to a right-handed (positive helicity) W-boson

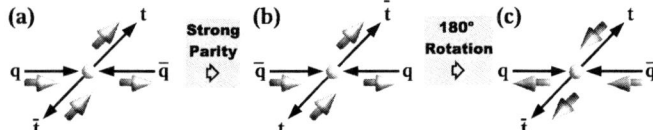

Fig. 27. Parity symmetry of the strong interaction and rotational symmetry are used to show that an ensemble of top quarks is produced unpolarised by the strong interaction in (unpolarised) $p\bar{p}$ collisions. Higher order effects such as QCD final state interactions and mixed QCD/weak interactions, however, can produce small polarisations perpendicular to or in the scattering plane

2.4.4 Spin correlation in strong $t\bar{t}$ production

One of the unique features of the top quark is that on average the top quark decays before there is time for its spin to be depolarised by the strong interaction [164]. Thus, the top quark polarisation[9] is directly observable via the angular distribution of its decay products. This means that it should be possible to measure observables that are sensitive to the top quark spin.

It is well known that top quarks can be polarised at an e^+e^- collider by polarising the electron beam[10], and that this is a useful tool to study the weak decay properties of the top quark. There is an analogue of this tool at hadron colliders.

Although the top and antitop quarks are produced essentially unpolarised[11] [188–191] in (unpolarised) hadron collisions (Fig. 27), the spins of the t and $\bar{t}$ are correlated [192–197], as shown in Fig. 28. In $t\bar{t}$ production by $q\bar{q}$ annihilation the correlation can be 100% with respect to a suitably chosen axis. The spins are also correlated in unpolarised e^+e^- collisions (LO [198], NLO [187]). The spin correlation can be used to study the $t\bar{t}$ production mechanisms, which result in the spin correlation, as well as the weak decay properties of the top quark by observing the angular correlations between the decay products of the t and $\bar{t}$. The spin correlation is expected to be observed in Run II at the TEVATRON.

The origin of the spin correlation in $t\bar{t}$ production is as follows:

For QCD processes close to the production threshold, the $t\bar{t}$ system is dominantly produced in a $^3S_1$ state for $q\bar{q}$ annihilation (Fig. 28b), or in a $^1S_0$ state for gluon–gluon fusion (Fig. 28c) [199]. Hence, in the first case, the top

---

[9] The spin of an individual top quark cannot be measured, only the spin polarisation of an ensemble of top quarks.

[10] Top quarks are naturally polarised to a small degree (-20% to -40%) via the weak interaction in unpolarised e^+e^- collisions (at threshold [186], above threshold [187]). Using polarised beams, the top quark polarisation is dramatically enhanced.

[11] Top and antitop quarks receive a small (2%) polarisation perpendicular to the scattering plane via QCD final state interactions [188–190]. An additional, very small contribution of top/antitop quark polarisation is received from mixed QCD/weak interactions in the scattering plane [191].

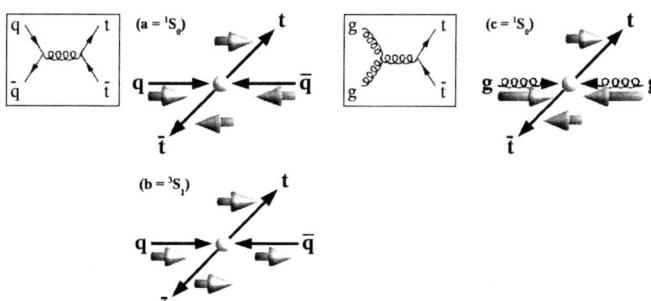

Fig. 28. Schematic of the $t\bar{t}$ spin correlation in the $q\bar{q}$ annihilation (*left*) and gg annihilation (*right*). The parton momenta are shown as *thin arrows*, the parton spins as *big arrows*. In $q\bar{q}$ annihilation the cross section for opposite-helicity $t\bar{t}$ production (**b**) is larger than that for same-helicity production (**a**). Configurations with reversed spin directions are not shown explicitly, but always meant to be included implicitly. The spin configurations shown are strictly valid only at the $t\bar{t}$ production threshold. Above threshold orbital angular momentum effects need to be considered in addition

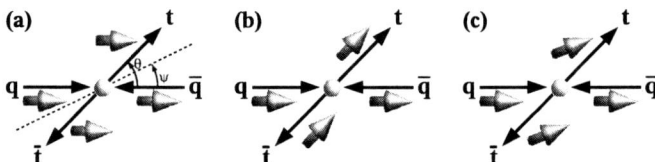

Fig. 29. In $t\bar{t}$ production via $q\bar{q}$ annihilation the spins of the top quark and antiquark are 100% correlated when measured along an axis that makes an angle ψ with respect to the beam axis, where $\tan \psi = \beta^2 \sin \theta \cos \theta / (1 - \beta^2 \sin^2 \theta)$: **a** near threshold, **b** far above threshold, **c** intermediate energies

and the antitop tend to have parallel spins, i.e. opposite helicities, while in the second case the spins tend to be antiparallel, i.e. the same helicities. Since the $q\bar{q}$ annihilation dominates the $t\bar{t}$ production at the TEVATRON while gg annihilation dominates the $t\bar{t}$ production at the LHC, the spin correlation coefficient κ (43) is expected to have opposite sign at both colliders (see Table 6). The absolute sign of the spin correlation coefficient depends on the convention of its definition (for example (43)), which varies in the literature.

At energies large compared to the top mass, chirality conservation implies that the t and $\bar{t}$ are produced with opposite helicities ("helicity basis"). At the other extreme, the t and $\bar{t}$ are produced with zero orbital momentum at threshold, so spin is conserved. Since the colliding quark and antiquark have opposite spins (due to chirality conservation), the t and $\bar{t}$ have opposite spins along the beam axis ("beam-line basis" [195], "beam basis" [200, 201]). Remarkably, for $q\bar{q}$ annihilation there exists a basis which interpolates at all energies between these two extremes ("diagonal basis"), such that the t and $\bar{t}$ spins are always opposite [198] (Fig. 29).

In single-top production at hadron colliders, the spin of the top quark is 100% left-handed polarised along the di-

Table 6. Coefficient κ to leading and next-to-leading order in α_s for the helicity basis, the beam basis and the off-diagonal basis for the TEVATRON in Run II (left) and for the LHC (right) from [200]. The numbers in brackets are taken from [201]. They are not directly comparable as they have been calculated with different parton distribution functions, but they clearly show that the spin correlations at the LHC are very small in the beam and the off-diagonal axis

| | | $p\bar{p}$ at $\sqrt{s} = 1.96$ TeV | | | pp at $\sqrt{s} = 14$ TeV | | |
		Dilepton	Lepton-Jet	All-Jet	Dilepton	Lepton-Jet	All-Jet
$\kappa_{\text{heli.}}$	LO	-0.471	-0.240	-0.123	0.319	0.163	0.083
	NLO	-0.352	-0.168	-0.080	0.326	0.158	0.076
κ_{beam}	LO	0.928	0.474	0.242	(-0.005)		
	NLO	0.777	0.370	0.176	(-0.072)		
$\kappa_{\text{off-diag.}}$	LO	0.937	0.478	0.244	(-0.027)		
	NLO	0.782	0.372	0.177	(-0.089)		

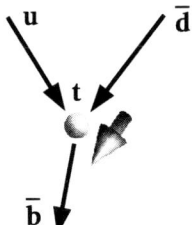

Fig. 30. In single top production, the top quark is 100% polarised along the direction of motion of the d quark, in the top quark rest frame

rection of motion of the d quark[12], in the top quark rest frame, since they involve the weak interaction (Fig. 30). Given the large number of top quark pairs and single top quarks that will be produced at the TEVATRON and the LHC, the spin correlation and the single top polarisation should be powerful tools to analyse the properties of the top quark.

In the dilepton channel, defining θ_+ (θ_-) as the angle between the direction of flight of the lepton ℓ^+ (ℓ^-) in the t ($\bar{t}$) rest frame and arbitrarily chosen directions, the spin correlation can be expressed as [193, 195, 202]:

$$\frac{1}{\sigma}\frac{\mathrm{d}^2\sigma}{\mathrm{d}(\cos\theta_+)\,\mathrm{d}(\cos\theta_-)} = \frac{1 - \kappa\cos\theta_+\cos\theta_-}{4}, \quad (43)$$

where the correlation coefficient κ describes the degree of correlation resulting from the production dynamics as well as the spin analysis factors for the reactions $t \to a \ldots$ and $\bar{t} \to b \ldots$ [203, 204], present prior to imposition of selection criteria or effects of detector resolutions. In the lepton + jets and the all-jets channel, κ can be defined analogously. Different choices of quantisation axes as arbitrary directions in the definition of θ_+ and θ_- yield different values for the correlation coefficient κ. Table 6 summarises the values of κ expected in the Standard Model at leading and at next-to-leading order in α_s at the TEVATRON and at the LHC [200, 201] in all three $t\bar{t}$ decay channels for different choices of quantisation axes.

---

[12] This consideration is only strictly valid in the Born-approximation of the 2-to-2 process ($ub \to dt$ or $u\bar{d} \to \bar{b}t$) with massless quarks. Initial state gluon radiation changes the centre-of-mass system of the initial partons and hence the helicity of the massive top quark.

At the TEVATRON, the dilepton spin correlations are large in the beam and the off-diagonal axis. There appears to be practically no difference between these two choices as far as the sensitivity to QCD-induced spin correlations is concerned [200, 201]. Yet, the beam axis might be simpler to implement in the analysis of experimental data. The QCD corrections are about -10%. At the LHC, the beam and off-diagonal bases are not very good choices due to the dominance of $gg \to t\bar{t}$. Here, the helicity basis is a good choice, and the QCD corrections are small. Another set of observables for measuring the correlation of the $t\bar{t}$ spins at the LHC, which are expected to have only relatively small experimental errors (see Sect. 6.1), are the opening angle distributions predicted within QCD in [200]. For the LHC a basis exists [205] which yields a larger effect than the helicity basis. Since the contributions from the gg and the $q\bar{q}$ initial state to κ enter with a different sign, the measurement of the $t\bar{t}$ spin correlations offers the possibility to constrain the PDFs. Furthermore, a measurement of spin correlations would provide a lower bound on $|V_{tb}|$ without assuming the existence of three quark generations [196].

The present status of experimental studies of $t\bar{t}$ spin correlation at the TEVATRON is described in Sect. 6.1.

2.4.5 Asymmetry in strong $t\bar{t}$ production

Another interesting aspect of the strong production of $t\bar{t}$ pairs is an asymmetry in the rapidity-distribution of the t and $\bar{t}$ quarks [206–208]. This effect arises at next-to-leading order, and leads to a forward-backward asymmetry of about 5% in $t\bar{t}$ production at the TEVATRON. With the present experimental precision, this effect is not yet visible. Therefore, no studies of the asymmetry in strong $t\bar{t}$ production are presently pursued at the TEVATRON.

2.4.6 Rare top quark decays

Rare top decays in the Standard Model tend to be very rare, outside the range of the TEVATRON. The flavour changing neutral current decays $t \to Zq$ and $t \to \gamma q$ have negligible branching ratios in the Standard Model [209]. Deviations from this expectation are searched for at the

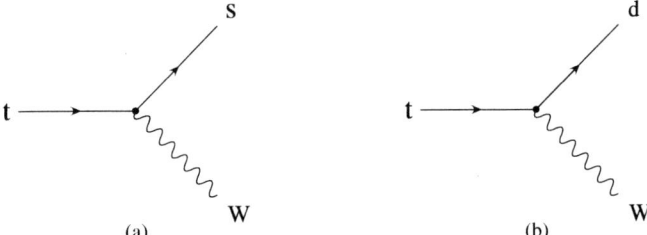

Fig. 31. Rare top quark decays: **a** $t \to Ws$ and **b** $t \to Wd$

TEVATRON (Sect. 6.5) and will be searched for with much higher sensitivity at the LHC.

The least rare of the rare top quark decays in the Standard Model are the CKM suppressed decays $t \to Ws$ and $t \to Wd$, shown in Fig. 31. These decays are interesting because they allow a direct measurement of the CKM matrix elements V_{ts} and V_{td}. Assuming three quark generations, the branching ratios are predicted to be [167]:

$$B(t \to Ws) \approx 0.1\%, \tag{44}$$
$$B(t \to Wd) \approx 0.01\%, \tag{45}$$

which are small, but not zero. Since there will be about 10 000 raw $t\bar{t}$ pairs produced at the TEVATRON in Run-II with 2 fb^{-1} and about 8 million $t\bar{t}$ events at the LHC in one year of running at luminosities of 10^{33} cm^{-2} s^{-1}, events of this type will be present in the data. However, there is at present no generally accepted strategy for identifying these events.

2.4.7 Top quark Yukawa coupling

Yukawa coupling is the Higgs coupling to fermions and thus relates the fermionic matter content of the Standard Model to the source of mass generation, the Higgs sector [20–22]. In the Standard Model, the Yukawa coupling to the top quark, $y_t = \sqrt{2}m_t/v$ (where $v \approx 246$ GeV is the vacuum expectation value), is very close to unity. This theoretically interesting value leads to numerous speculations that new physics might be accessed via top quark physics [210]. The Yukawa coupling will be measured in the associated $t\bar{t}H$ production at the LHC.

Indirect determinations of y_t represent an independent and complementary approach to the direct measurement of y_t via $t\bar{t}H$ production at the LHC or even a linear collider, which of course provides the highest accuracy [172, 173]. At a linear collider with $\sqrt{s} = 500$ GeV, the top quark Yukawa coupling is expected to be measured directly with 33% precision with a possible improvement to 10% when using polarised beams [211], at $\sqrt{s} = 800$ GeV even a 5% measurement appears possible in unpolarised e^+e^- collisions due to the increased $t\bar{t}H$ production cross section [172]. In order to obtain indirect constraints on the top quark Yukawa coupling y_t from electroweak precision observables a high precision on the top quark mass m_t is important [212]. The top coupling enters the Standard Model prediction of electroweak precision observables starting at

$\mathcal{O}(\alpha\alpha_t)$ [213, 214]. Indirect bounds on this coupling can be obtained if one assumes that the usual relation between the Yukawa coupling and the top quark mass, $y_t = \sqrt{2}m_t/v$, is modified.

Assuming a precision $\Delta m_t = 2$ GeV/c^2, an indirect determination of y_t with an accuracy of only about 80% can be obtained from the electroweak precision observables measured at a Linear Collider with GigaZ option. A precision of $\Delta m_t = 0.1$ GeV/c^2, on the other hand, leads to an accuracy of the indirect determination of y_t of about 40%, which is competitive with the indirect constraints from the $t\bar{t}$ threshold [171] with a precision of 35%.

2.5 Modelling of top quark and background events

The extraction of top-quark properties from TEVATRON and the LHC data rely on good understanding of the production and decay mechanisms of the top quark, as well as of the background processes. For the background, the jets are expected to have a steeply falling E_T spectrum, to have an angular distribution peaked at small angles with respect to the beam, and to contain b- and c-quarks at the few percent level. On the contrary, for the top signal, the b fraction is expected to be $\approx 100\%$ and the jets rather energetic, since they come from the decay of a massive object. It is therefore possible to improve the S/B ratio by requiring the presence of a b quark, or by selecting very energetic and central kinematic configurations, or both.

Background estimates can be checked using control data samples with fewer jets, where there is little top contamination (0 or 1 jet for dilepton channels, 1 or 2 jets for lepton + jets channels, and, ≤ 4 jets or multi-jets ignoring b-tagging for the all-jets channel). Wherever possible, estimates of the background rate and shape in relevant kinematic distributions are performed in data, since the leading-order (LO) Monte Carlo simulations are subject to large theoretical uncertainties, in particular in their normalisation.

Next-to-leading order Monte Carlo programs have recently become available for both signal ($t\bar{t}$, single-top so far only s- and t-channel) and background processes [215–220], but for the backgrounds the jet multiplicities required in $t\bar{t}$ analyses are not yet available. To date, only leading-order Monte Carlo programs have been used in the analyses. Theoretical estimates of the background processes (W- or Z-bosons+jets and dibosons+jets) using LO calculations have large uncertainties. While this limitation affects estimates of the overall production rates, it is believed that the LO determination of event kinematics and of the fraction of W + multi-jet events that contain b- or c-quarks are relatively accurate [221, 222].

The simulation of $p\bar{p}$ or pp interactions in Monte Carlo programs makes use of the factorisation (Sect. 2.1) in a short distance hard scattering interaction, calculable in perturbative QCD, and the long range physics, including the parton momentum distributions and further soft physics interactions, referred to as the underlying event. This separation introduces the artificial factorisation scale

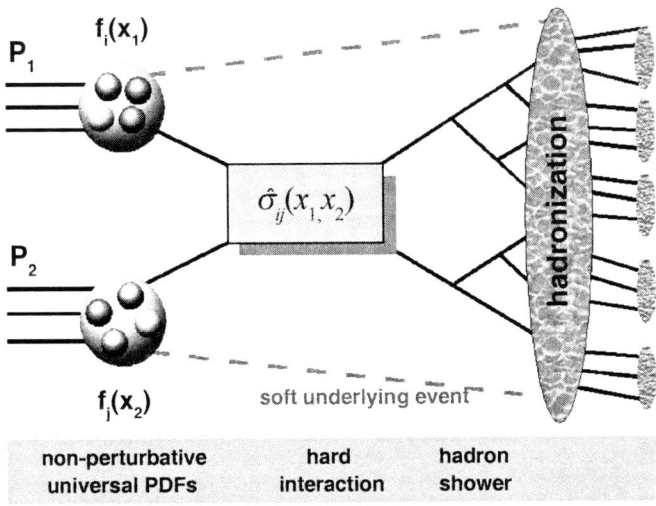

Fig. 32. Sketch of a $p\bar{p}$ or pp interaction

Q^2. Additional effects such as multiple proton interactions and pile-up can occur in the detector.

Figure 32 shows a sketch of the $p\bar{p}$ or pp interactions. The full chain of the simulation is briefly described in the following:

The hard scatter interaction is described by calculating the leading order matrix element using PYTHIA [223] or ALP-GEN [224]. The set of parton distribution functions used is CTEQ5L [94], and CTEQ6.1M [132]. The latter is derived in NLO, which is strictly speaking not adequate to be used with a leading order matrix element. However, proper PDF uncertainties are at present only available for NLO PDFs, and numerically the change in the $t\bar{t}$ cross section at the TEVATRON is found to be small.

The underlying event is comprised of a hard component and of a soft component. The hard component describes the particles that arise from initial and final state radiation and from the outgoing hard scatter partons. The soft component consists of beam–beam remnants and multiple parton interactions.

- The beam–beam remnant describes the outgoing partons of the $p\bar{p}$ or pp interaction, which do not participate in the hard scattering process. The colour connection between these spectator partons and the two partons from the hard scattering is the origin of this soft interaction and is hard to model.
- Multiple parton interactions describe the possibility that a hard scattering event also contains "semi-hard" interactions between the remaining partons from a given $p\bar{p}$ or pp pair. There is a colour connection between the "semi-hard" and the hard scattering partons, and in addition a dependence on the p_T of the hard scattering process.

The transverse region defined as the phase space around the plane orthogonal to the jet with highest E_T in the event, is sensitive to the underlying event. A data to Monte Carlo comparison of the average charged particle density and p_T distribution in the transverse region provides

a measurement of the underlying event and allows the tuning of its Monte Carlo modelling, dubbed "Tune-A" [225].

PYTHIA 6.202 and JETSET [223], including multiple parton interactions, are used to model the underlying event. PYTHIA models the soft component of the underlying event with colour string fragmentation. Recently, also HERWIG [226, 227] was modified to provide a simulation of the underlying event including multiple parton interactions using JIMMY [228, 229].

Multiple proton interactions can occur when more than one $p\bar{p}$ or pp interaction takes place in the colliding bunches of hadrons. The multiple proton interactions are simulated by superimposing minimum bias data events to the event. Minimum bias events are defined as events which show a minimum activity in the detector, i.e. not being triggered by a high p_T lepton, jet or $\not{E}_T$. At the TEVATRON, the number of added events is taken from a Poisson distribution with a mean between 0.5 and 0.8 events. This number is luminosity dependent and will be significantly larger at the LHC.

Pile-up describes overlapping $p\bar{p}$ or pp interactions from consecutive bunch crossings in the detector, which are reconstructed in one event. The pile-up can be simulated or modelled by adding randomly recorded data events (called zero bias).

Hadronisation: The collections of partons must then be hadronised into colourless mesons and baryons. Different approaches are used by the event generators. The Lund model implemented in PYTHIA [223] splits gluons into $q\bar{q}$ pairs and turns them into hadrons via the string fragmentation model. HERWIG [226, 227] forms colourless clusters from quarks and gluons with low invariant mass, which are turned into hadrons (cluster fragmentation).

The Detector Simulation at the TEVATRON is based on GEANT 3 [230], while the LHC experiments describe their detector geometry using GEANT 4 [231]. The generation of large samples of Monte Carlo events can be necessary for example for studies of systematic uncertainties or the construction of template distributions from Monte Carlo samples generated with different generation parameters such as a varied top quark mass. Since the generation of such samples is very CPU intensive, the TEVATRON as well as the LHC experiments use in addition to the 'full simulation' of the detector geometry and material distribution also fast simulations, which are based on parameterisations of their detector response. These fast simulations are tuned to the full detector simulation. They are followed by the signal digitisation and the reconstruction software.

2.5.1 Event generators

Several Monte Carlo programs are available to calculate tree level matrix elements or to generate full scattering events including parton showering and hadronisation. A brief summary is given in the following, while a more detailed overview can be found in [232].

Full event simulation packages. These packages provide a full event simulation including the hard process generation, showering and hadronisation with subsequent decays of the unstable hadrons.

HERWIG [226, 227] contains a wide range of Standard Model, Higgs and supersymmetric processes. It uses the parton-shower approach for initial- and final-state QCD radiation, including colour coherence effects and azimuthal correlations both within and between jets. HERWIG is particularly sophisticated in its treatment of the subsequent decay of unstable resonances, including full spin correlations for most processes.
Processes included: $t\bar{t}$, single-top (t- and s-channels), $t\bar{t}H$, $Zt\bar{t}$, $gb \to tH^+$.

PYTHIA [223] is a general-purpose generator for hadronic events in pp, $p\bar{p}$, e^+e^- and ep collisions. It contains a subprocess library and generation machinery, initial- and final-state parton showers, underlying event, hadronisation and decays, and analysis tools.
Processes included: $t\bar{t}$, single-top (t- and s-channels), $t\bar{t}H$, $gb \to tH^+$, no spin correlations.

ISAJET [233] is a general-purpose generator for hadronic events. ISAJET is based on perturbative QCD plus phenomenological models for parton and beam jet fragmentation.
Processes included: $t\bar{t}$, no spin correlations.

SHERPA [234] is a new multi-purpose event generator a powerful matrix element generator AMEGIC++.
Processes included: Standard Model, MSSM and an ADD model of large extra dimensions.

Tree level matrix element generators. Such packages generate the hard processes kinematic quantities, such as masses and momenta, the spin, the colour connection, and the flavour of initial- and final-state partons. Then such information is stored in the "Les Houches" format and is passed to full event simulation generators, such as PYTHIA or HERWIG.

ALPGEN [224] is designed for the generation of the Standard Model processes in hadronic collisions, with emphasis on final states with large jet multiplicities. It is based on the exact LO evaluation of partonic matrix elements, as well as top quark and gauge boson decays with helicity correlations. The code generates events in both a weighted and unweighted mode.
Processes included: $t\bar{t}+$ up to 6 jets, single-top: tq, tb, tW, tbW (no extra jets), $t\bar{t}t\bar{t}+$ up to 4 jets, $t\bar{t}b\bar{b}+$ up to 4 jets, $t\bar{t}H+$ up to 4 jets, $W/Zt\bar{t}+$ up to 4 jets.

COMPHEP [235] computes squared Feynman diagrams symbolically and calculates numerically the corresponding total and differential cross sections. The event output is provided in the "Les Houches" format.
Processes included: $t\bar{t}$, single-top, $t\bar{t}b\bar{b}$, $W/Zt\bar{t}$, spin correlations are included.

MADEVENT [236] is a multi-purpose, tree-level event generator, which is powered by the matrix element generator MADGRAPH [237]. MADGRAPH automat-

ically generates the amplitudes for all relevant subprocesses and produces the mappings for the integration over the phase space.
Processes included: $t\bar{t}+$ up to 3 jets, single-top, $t\bar{t}b\bar{b}+$ up to 1 jet, $t\bar{t}H+$ up to 2 jets.

MC@NLO [215–220] combines a Monte Carlo event generator with exact NLO calculations of rates for QCD processes at hadron colliders.
Processes included: $t\bar{t}$, single-top (s- and t-channel).

ACERMC [238, 239] is dedicated to the generation of Standard Model background processes in pp collisions at the LHC.
Processes included: $t\bar{t}$, single-top, $t\bar{t}t\bar{t}$, $t\bar{t}b\bar{b}$, $W/Zt\bar{t}$, spin correlations are included.

SINGLETOP [240] is a generator based on the COMPHEP package.
Processes included: t-channel single-top production ($2 \to 2$ and $2 \to 3$), spin correlations are included.

TOPREX [241] provides a simulation of several important processes in hadronic collisions, which are not implemented in PYTHIA. Several top-quark decay channels are included: the Standard Model channel ($t \to qW^+$, $q = d, s, b$), b-quark and charged Higgs ($t \to bH^+$) and the channels with flavour changing neutral current (FCNC): $t \to u(c)V$, $V = g, \gamma, Z$. The implemented matrix elements take into account spin polarisation of the top quark.
Processes included: $gg(q\bar{q}) \to t\bar{t}$, single-top ($t$-, s- and tW-channel), $q\bar{q}' \to H^{\pm *} \to t\bar{b}$, $q\bar{q} \to W^*/Z^*Q\bar{Q}$, with $W^*/Z^* \to f\bar{f}$ and $Q = c, b, t$, $gu(c) \to t \to bW$ (due to FCNC).

MCFM [221, 242] includes the matrix elements at NLO and incorporates full spin correlations.
Processes included: $t\bar{t}$, single-top (t- and s-channel), $t\bar{t}H$, $W/Zt\bar{t}$.

ZTOP [149, 161] includes the full NLO corrections to single-top production (t- and s-channel).

ONETOP [153–155] includes the full NLO corrections to single-top production (t- and s-channel) and to the top quark decay.

2.5.2 $t\bar{t}$ signal simulation

In CDF, the production and decay of the $t\bar{t}$ signal is simulated using PYTHIA [223] for the hard scatter process, followed by HERWIG for the hadronisation step. DØ uses ALPGEN [224], which includes the complete $2 \to 6$ Born level matrix elements, followed by PYTHIA for the simulation of the underlying event and the hadronisation. This procedure takes advantage of the full spin correlation information for top quarks that is provided in ALPGEN. In almost all analyses, the top quark mass is set to $175\,\mathrm{GeV/c^2}$. EVTGEN [243], known to successfully describe the spin correlations between the decay particles, is used to provide the branching fractions and lifetimes for the following b-quark states: B^0, B^+, B_s^0, B_c^+, and Λ_b. In earlier CDF analyses, the program QQ [244] is used for that purpose. The decay of Taus in the final state is simulated using TAUOLA [245, 246].

2.5.3 $W/Z+$ jets background simulation

The dominant background for the $t\bar{t}$ analyses in the lepton + jets and the dilepton channel is the $W+$ jets and $Z+$ jets or diboson background. The all-jets channel is dominated by instrumental background from multijet events, which is estimated from control data samples.

In DØ, the $W/Z+$ jets background is simulated using ALPGEN followed by PYTHIA, while CDF uses HERWIG to simulate the hadronisation. Example Feynman diagrams of the inclusive $W/Z+$ jets processes considered are shown in Fig. 33. In the b-tag analyses, each data sample ($\mu+$ jets and $e+$ jets) is subdivided into disjoint event samples with 1, 2, 3, and ≥ 4 jets in the final states, and each sample represents an individual counting experiment. In the example topology of the $W+4$-jets analysis, the following flavour combinations are generated separately: $Wjjjj$, $Wcjjj$, $Wc\bar{c}J$, and $Wb\bar{b}J$, where j is any flavour of u, d, s, g and J is any flavour of u, d, s, g, c. Similar combinations are generated separately for the 1, 2, and 3-jet case.

While the overall production rate of the LO Monte Carlo program ALPGEN does not describe the data very well, the event kinematics of the $W/Z+$ jets samples do describe the data reasonably well. Also, the ratio of b-jets

to light flavour jets changes very little upon the inclusion of NLO radiative corrections and appears to be predicted very well by LO perturbation theory [221, 222]. This is demonstrated in Fig. 34, which compares the shape of the ratio of b-jets to light flavour jets in LO and NLO calculations as a function of the di-jet mass. Good agreement between the LO and the NLO results is found for the $W+2$ jets (left) and the $Z+2$ jets (right). First studies in $W/Z+4$ jets production indicate that for higher jet multiplicities the NLO corrections to the shape of the distribution might be somewhat larger [222]. Further studies are clearly needed in this area. Also, CDF and DØ are expected to provide measurements of the $W/Z+$ jets cross section as a function of the jet multiplicity with increased precision soon, allowing for more stringent tests of the perturbative QCD calculations at LO and NLO.

Diboson (WW, WZ, and ZZ) backgrounds are modelled in CDF using PYTHIA [223] and ALPGEN + HERWIG Monte Carlos [224, 226, 227], while DØ uses ALPGEN and PYTHIA. Both experiments normalise their diboson samples to the theoretical total NLO cross section: 13.3 pb for WW, 4.0 pb for WZ, and 1.5 pb for ZZ [247]. The normalisation uncertainties are determined different using Monte Carlo calculations for the same process.

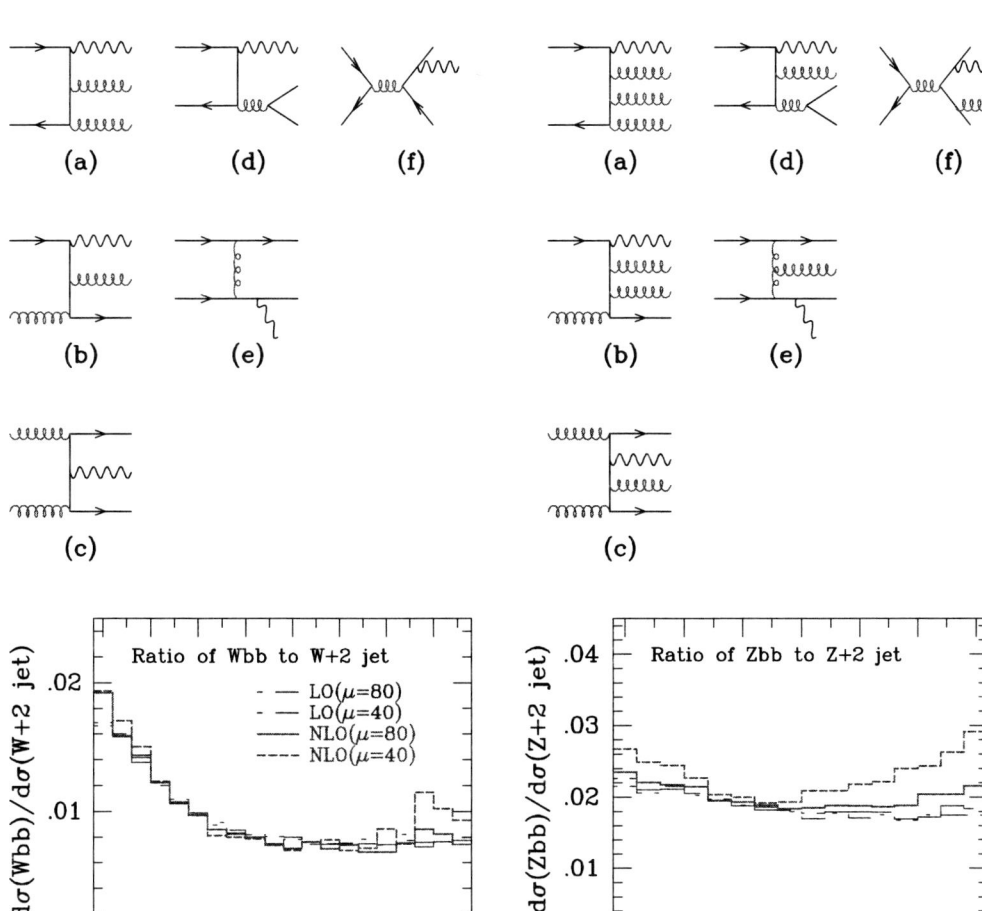

Fig. 33. Example diagrams for the process parton + parton $\rightarrow W/Z+2$ partons (*left*) and parton + parton $\rightarrow W/Z +3$ partons (*right*). The vector boson is denoted by a wavy line

Fig. 34. Ratio of $W/Z+2b$-jets to $W/Z+2$ jet events in LO and NLO at two different factorisation scales. From [221]

2.6 Jet-parton matching

In the recent years, the consistent combination of the leading order parton level calculation, performed for example by ALPGEN, with the partonic shower Monte Carlo programs PYTHIA and JETSET or HERWIG is a topic that has received much attention [248–250]. In this merging process the problem of possible double counting of configurations with different number of hard partons at the matrix element level but similar final state occurs:

The parton shower Monte Carlo programs serve to model the higher order corrections to the leading order matrix element calculation in all orders of α_s. Two sources of double counting are identified. (i) Consider the final state of $W + n$ jets. This final state can be produced by a $W + n$ parton matrix element calculation with a parton shower step that transforms each parton into a jet. Another possibility is a $W + (n-1)$ parton matrix element calculation with a parton shower step that generates an additional jet, so that n jets are reconstructed in the detector. In general, $W + (n-m)$ partons with m additional jets from the parton showering and $0 \leq m \leq n$ can give the same final state in the detector. Figure 35 shows the example of $W + 3$ jets events with $m = 0$ (left) or $m = 1$ (right) additional jets from the parton shower step. (ii) The second source of double counting can occur when events differ in the number of hard scatter partons, but some jets are too soft or too forward to be reconstructed in the detector or selected, yielding identical final states.

A matching of partons, produced by the matrix element calculation, to reconstructed jets is performed in order to eliminate the double counting. This matching procedure also reduces the sensitivity of the parton-level cross section, predicted by the fixed-order matrix element generation (ALPGEN or PYTHIA), to the parton generation cuts. Two matching procedures have been proposed:

CKKW matching. The multijet matrix elements are merged with the shower development by reweighting the matrix elements weights with Sudakov form factors and vetoing shower emissions in regions of phase space already covered by the parton level configurations [251, 252]. This

matching scheme is implemented in the Monte Carlo program SHERPA [253–255].

MLM matching. Matrix element partons are matched to parton jets [248, 256]. Events are rejected if there are extra jets which fail to match to the light partons generated at the matrix element level or if there are missing jets. In the special case of heavy flavour partons, the strict matching criteria are relaxed because the two partons may be merged into one jet due to the parton mass.

Although it minimises double counting of generated events, this procedure introduces a new type of systematic uncertainty which depends on the matching criteria and the jet definition. The jet-parton matching procedures are presently used by CDF and DØ in some of the b-tagging analyses. Further concerted effort by theorists and experimentalists will be needed to study the matching procedures, their effects on kinematic distributions of the corresponding Monte Carlo samples and the resulting systematic uncertainties. In particular for the LHC these techniques will play an important role due to the very large rate of $t\bar{t}$ events and the high rate of $W/Z +$ jet events with large jet multiplicity.

3 Accelerator and detectors

In this chapter experimental aspects of the top quark production and detection are discussed. First the TEVATRON collider and the CDF and DØ experiments are described, followed by a discussion on the identification algorithms of physics objects such as electrons, muons, tau, neutrinos, jets and b-jets by the CDF and DØ collaboration.

3.1 The TEVATRON accelerator

The $p\bar{p}$ collider TEVATRON [257, 258] at FERMILAB in Batavia, Illinois, near Chicago is the world's highest energy particle accelerator, with a centre-of-mass energy of $\sqrt{s} = 1.96$ TeV. It is at present the only collider with sufficient energy to produce top quarks. During the data-taking period from 1992–1996 (Run-I), the TEVATRON experiments CDF and DØ each collected about $125 \, \text{pb}^{-1}$ of $p\bar{p}$ collisions data at a centre-of-mass energy of 1.8 TeV, leading to the discovery of the top quark, a measurement of its mass, a precision measurement of the mass of the W-boson, detailed analyses of gauge boson couplings, studies of jet production and vastly improved limits on the production of new phenomena, such as leptoquarks and supersymmetric particles, among many other accomplishments. The new data-taking period (Run-II) started in March 2001 and is expected to deliver between $4 \, \text{fb}^{-1}$ and $9 \, \text{fb}^{-1}$ by the year 2009. Since most of the analyses and measurements discussed in this document have been performed at the TEVATRON Run-II, only that experimental setup is discussed. For details on the experimental environment in Run-I see for example [3].

Figures 36 and 37 show an aerial view of the FERMILAB accelerator complex. Negative hydrogen ions are

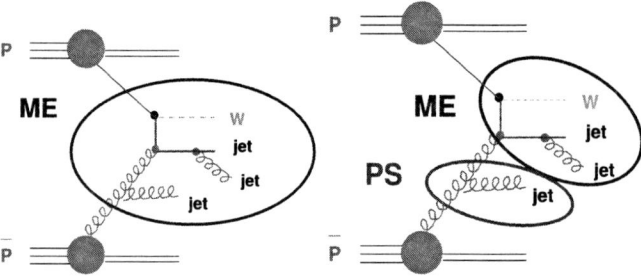

Fig. 35. *Left*: $W + 3$ parton process calculated by the matrix element (ME) and no additional jets from the parton shower (PS). *Right*: $W + 2$ parton process calculated by the matrix element and one additional jet generated by the parton shower. Both processes lead to the same final state

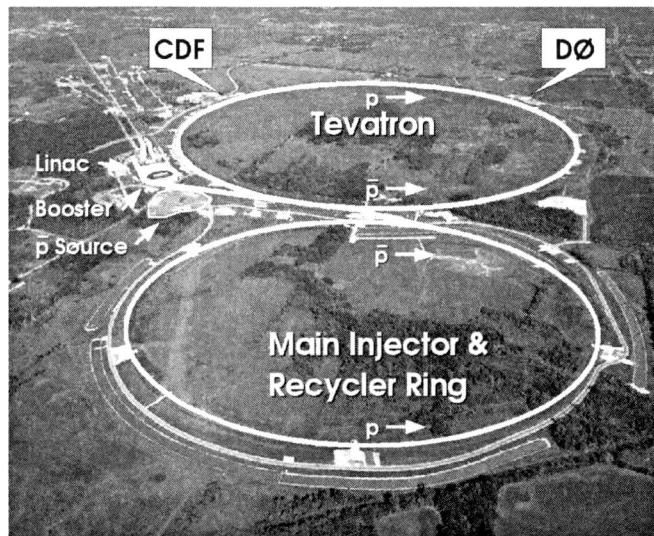

Fig. 36. Aerial view of the FERMILAB accelerator complex

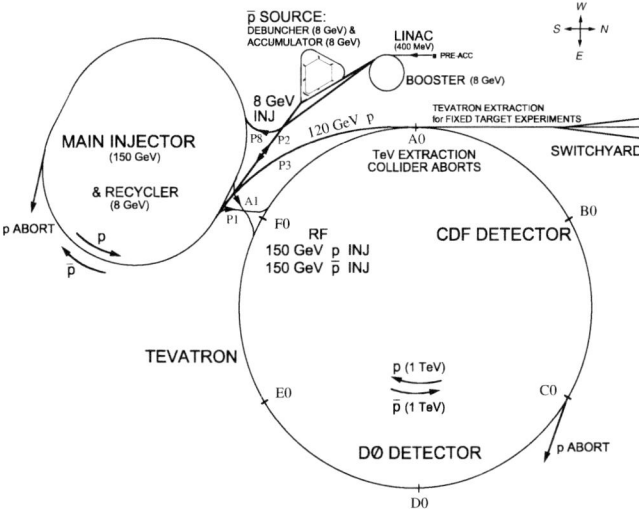

Fig. 37. Schematic view of the pre-accelerator chain and the TEVATRON

first accelerated to 750 keV by an electrostatic Cockcroft–Walton accelerator and then injected into a linear accelerator which boosts their energy to 400 MeV. These ions are stripped of their electrons as they pass through a sheet of graphite and are injected into the Booster. This 75 m radius synchrotron accelerates the protons to 8 GeV. From there they are injected into the main injector, where they are further accelerated to 120 or 150 GeV, depending on their destination. The 150 GeV protons are ejected into the TEVATRON. Antiprotons used in the collisions are collected from the interaction products of a portion of the 120 GeV proton beam incident on a Nickel–Copper target. They are collected from the production target using a lithium lens, momentum-selected around 8 GeV. The antiprotons are cooled and debunched in the large aperture accelerators debuncher and accumulator, two rounded triangular-shaped concentric storage rings with radii of about 75–90 m, using multiple stochastic cooling [259] systems. When enough antiprotons have been accumulated (stacked), they are extracted into the Main Injector, accelerated to 150 GeV and injected into the TEVATRON.

The TEVATRON is a synchrotron made from superconducting magnet coils and warm ion magnets. In collider mode, the TEVATRON is filled with 36 bunches of protons and antiprotons, arranged in three bunch trains with long abort gaps, circulating in opposite directions and separated by 396 ns bunch spacing. The protons and antiprotons are accelerated to their final energy of 980 GeV before colliding at the centre of the CDF and DØ detectors. The beams are typically kept colliding for ≈ 24 h, after which the beam intensity is too low and the spread too high, so that the beams are dumped and the machine is refilled. The dumping and refilling process typically takes about 2.5 h. The length of each bunch is ≈ 38 cm, resulting from the accelerator RF system.

After the TEVATRON upgrade in 1996–2001, the collider was commissioned for Run-II starting in May 2001. The Run-I record for the instantaneous luminosity of 2×10^{31} cm^{-2} s^{-1} was surpassed in the spring of 2002. Since then, the TEVATRON performance is continuously increasing. Figure 38 shows the time development of the instantaneous and of the integrated luminosity. Since the fall of 2004, when the recycler was commissioned, peak luminosities of 100×10^{30} cm^{-2} s^{-1} = $100\,\mu$b^{-1} s^{-1} are routinely achieved. The TEVATRON delivered over 1 fb^{-1}, CDF and DØ, of which 200–350 pb^{-1} have been studied by each experiment in the analyses presented in this report. Figure 38 also shows the delivered luminosity per fiscal year up to 2005 compared to the conservative 'baseline' and the optimistic 'design' scenario of luminosity development as predicted in the summer of 2003. In 2005 the

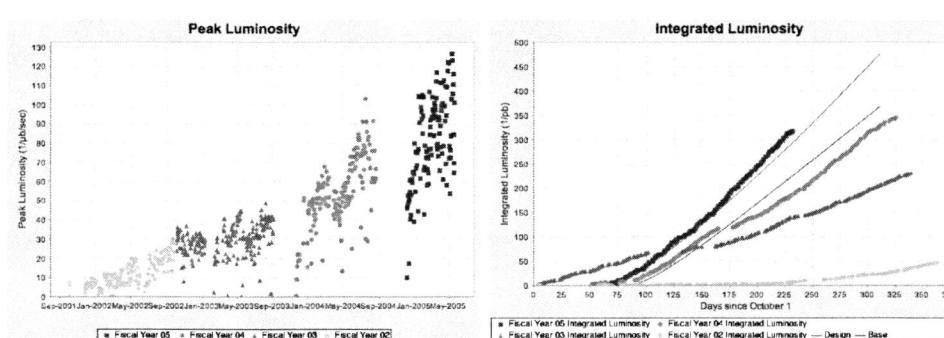

Fig. 38. *Left*: Time evolution of the TEVATRON instantaneous luminosity. The Run-I record of 2×10^{31} cm^{-2} s^{-1} = $20\,\mu$b^{-1} s^{-1} was broken in spring 2002; *right*: Integrated luminosity delivered to CDF and DØ by the TEVATRON (from [260])

TEVATRON delivered more integrated luminosity than expected in the optimistic 'design' scenario.

Since the summer of 2003 several measures have been or will be taken to increase the TEVATRON luminosity further [261–264]: by the end of 2004, the number of protons on the antiproton production has been increased by the use of slip stacking (antiprotons loaded from both accumulator and recycler), increasing the number of protons on the antiproton target from 5×10^{12} to more than 8×10^{12}, along with an upgrade of the target itself in order to be able to handle higher beam intensity. The antiproton collection efficiency has been increased by an increase in the gradient of the antiproton collection lens as well as an increase in the aperture of the antiproton collection transfer line and the Debuncher ring. Furthermore, the antiproton stacking and storing capabilities will be increased by increasing the flux capability of the Accumulator stacktail stochastic cooling system, and by using the Recycler as a second antiproton storage ring, with both stochastic [259] and electron cooling [265]. The additional electron cooling is being commissioned in the summer and fall 2005. In addition, the TEVATRON itself will be upgraded to be able to handle higher intensity bunches. Improvements to the helix separation and smoothness as well as an active compensation for beam–beam tune shift will be implemented. The expected luminosity projection, resulting from this TEVATRON upgrade programme, is summarised in Table 7. After the completion of the above mentioned TEVATRON upgrade in about 2006/2007, the TEVATRON is expected to continue operation until the end of 2009. By that time, an integrated luminosity of 4.5 to 8.6 fb^{-1} will presumably be delivered to each experiment.

Table 8 summarises the most important collider parameters, comparing the $Sp\bar{p}S$ at CERN, the TEVATRON

Table 7. Expected integrated luminosity projection for the TEVATRON Run-II in fb^{-1} per experiment (from [261–264])

year	2003	2004	2005	2006	2007	2008	2009
baseline	0.28	0.59	0.98	1.48	2.11	3.25	4.41
design	0.30	0.68	1.36	2.24	3.78	6.15	8.57

at FERMILAB and the upcoming LHC at CERN. For the first two colliders, the factor limiting the luminosity is the number of antiprotons which can be produced and stored at high energies. Note that the TEVATRON stores about 1×10^{30} antiprotons per beam, i.e. 'only' twice as many as were stored in the $Sp\bar{p}S$. Nevertheless the instantaneous luminosity at the TEVATRON is much higher than at the $Sp\bar{p}S$ thanks to the improved magnet technology, resulting in a smaller transverse beam size and beta function β^* at the interaction point:

$$\mathcal{L} = \frac{f N_{\mathrm{p}}(B N_{\mathrm{p}})}{2\pi(\sigma_{\mathrm{p}}^2 + \sigma_{\bar{\mathrm{p}}}^2)} F(\sigma_z / \beta^*), \qquad (46)$$

where f is the bunch revolution frequency, N_{p} the number of protons per bunch, $N_{\bar{\mathrm{p}}}$ the number of antiprotons per bunch, B the number of bunches, σ_{p} and $\sigma_{\bar{\mathrm{p}}}$ the transverse size of the proton and antiproton beams, respectively. The luminosity also depends on the beam shape form factor F which in turn is a function of the longitudinal bunch length σ_z and the beam amplitude function β^*, which is the ratio of the transverse beam size σ_x and the corresponding direction with respect to the beam axis, σ_x' at the interaction point. To achieve high luminosities, one wants β^* to be as small as possible.

Table 8. Summary of design parameters for the hadron colliders $Sp\bar{p}S$ at CERN, the TEVATRON at FERMILAB and the upcoming LHC at CERN (from [167, 266])

	$Sp\bar{p}S$	TEVATRON	LHC
Physics start	1981	1987	2007
Particles	$\bar{p}p$	$\bar{p}p$	pp
cm energy (TeV)	0.62	1.96	14
Lumi (10^{30} cm^{-2} s^{-1})	6	50–100	0.1–1.0×10^4
Lumi (fb^{-1} year^{-1})	0.05	0.5	100
Bunch spacing (ns)	3800	396	25
Transverse emittance ($10^{-9} \pi \mathrm{rad} - \mathrm{m}$)	p: 9 $\bar{p}$: 5	p: 4.3 $\bar{p}$: 2.7	0.5
Bunch length σ_z (cm)	20	38	7.5
β^*, ampl. function at interaction point (m)	0.6 (H) 0.15 (V)	0.35	0.5 - 0.55
Particles per bunch (10^{10})	p: 15 $\bar{p}$: 8	p: 24 $\bar{p}$: 3	p: 11.5
Max. no $\bar{p}$ in accumulator	1.2×10^{12}	2.6×10^{12}	–
Bunches	$6 + 6$	$36 + 36$	$2835 + 2835$
Circumference (km)	6.9	6.28	26.7
No. dipoles	232	774	1232 (main dipoles)
Magnet type	warm	cold, warm iron	cold, cold iron
Peak magnetic field (T)	1.4	4.4	8.3

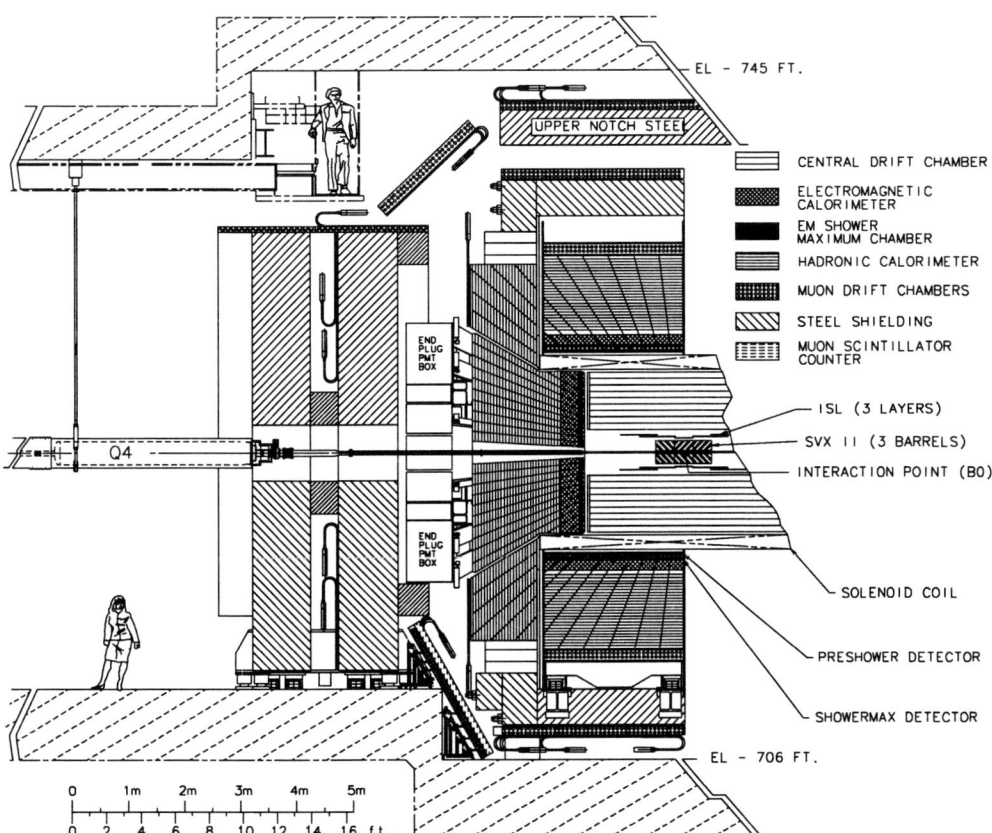

CENTRAL DRIFT CHAMBER

ELECTROMAGNETIC
CALORIMETER

EM SHOWER
MAXIMUM CHAMBER

HADRONIC CALORIMETER

MUON DRIFT CHAMBERS

STEEL SHIELDING

MUON SCINTILLATOR
COUNTER

ISL (3 LAYERS)

SVX II (3 BARRELS)

INTERACTION POINT (B0)

SOLENOID COIL

PRESHOWER DETECTOR

SHOWERMAX DETECTOR

Fig. 39. Elevation view of the CDF detector in Run-II

3.2 The CDF and DØ detectors

The CDF and DØ detectors [267–270] are large omni-purpose detectors. They have been designed for the identi-fication of the particles in the final states of the $p\bar{p}$ collisions and precision measurement of their four-momenta. To serve this purpose, both experiments consist of three ma-jor subsystems. At the core of the detector, a magnetised tracking system records precisely the angles and trans-verse momenta of charged particles. A hermetic, finely grained calorimeter measures the energy of electromag-netic and hadronic showers, and a muon system detects and measures the momenta of escaping muons. Both ex-periments have undergone substantial upgrades [271–274] after the end of Run-I and have been recommissioned in Run-II [275–282].

Both experiments use a right-handed coordinate sys-tem, centred on the detector, with the z-axis along the pro-ton beam direction. The y-axis is vertical, and the x-axis points towards the centre of the accelerator ring, defining the transverse plane. In addition, the polar angle θ, the azimuthal angle ϕ, and the pseudorapidity[13] η, defined as $\eta \equiv -\ln(\tan\theta/2)$, are frequently used. The separation of

two physics objects is typically expressed by their distance in the (η, ϕ) plane, i.e. $\Delta R = \sqrt{(\Delta\eta)^2 + (\Delta\phi)^2}$, which is a Lorentz-invariant quantity with respect to boosts along the z-axis. Depending on the choice of the origin of the co-ordinate system, the coordinates are referred to as *physics* coordinates, when the origin is the reconstructed events vertex (ϕ and η), while *detector* coordinates (ϕ_{det} and η_{det}) are calculated with respect to the centre of the detector.

3.2.1 CDF

The CDF Run-II detector [271, 272], in operation since 2001, is an azimuthally and forward-background sym-metric apparatus designed to study $p\bar{p}$ collisions at the TEVATRON. It is a general purpose solenoidal detec-tor which combines precision charged particle tracking with fast projective calorimetry and fine grained muon detection. The detector is shown in an elevation view in Fig. 39. The CDF detector consists of three main func-tional sections going radially outwards from the beam-line. The tracking system is used for particle charge and three-momentum measurements. It is contained in a super-conducting solenoid, 1.5 m in radius and 4.8 m in length, which generates a 1.4 T magnetic field parallel to the beam axis. The solenoid is surrounded by the scintillator-based calorimeter system with separate electromagnetic and hadronic measurements, which covers the region $|\eta| \leq 3$. Outside the calorimeters, layers of steel absorb the re-maining hadrons leaving only muons, which are detected

---

[13] The pseudorapidity η is obtained as an approximation of the rapidity $y = \frac{1}{2}\left(\frac{E+p_z}{E-p_z}\right)$ when the particle mass is ignored, i.e. for $p \gg m$. Rapidity intervals Δy are Lorentz-invariant. In inclusive QCD the number of particles produced per rapidity interval is a flat plateau reaching out to $(y_{\text{cm}})_{\text{max}} = \pm\frac{1}{2}\ln\left(\frac{s}{m^2}\right)$ in the centre-of-mass frame.

by the outermost muon detectors. The CDF Run-I and Run-0 detector, which was operated between 1987 and 1996, is described elsewhere [267–269]. Major differences for Run-II include: the replacement of the central tracking system, the replacement of a gas sampling calorimeter in the plug-forward region with a scintillating tile calorimeter, preshower detectors, extension of the muon coverage, a time-of-flight (TOF) detector and upgrades of trigger, readout electronics, and the data acquisition systems. The main features of the Run-II detector systems are summarised below.

Luminosity monitor. The beam luminosity is determined by using low pressure gas Cherenkov counters located in the $3.7 < |\eta| < 4.7$ region which measure the average number of inelastic $p\bar{p}$ collisions per bunch crossing [275]. The detector consists of 48 thin, long, conical counters, arranged around the beam pipe in three concentric layers of 16 counters, each oriented with their small end pointing to the centre of the interaction region and readout via photomultiplier tubes (PMTs) at their larger end. The cones in the outer two layers are about 180 cm long, the inner ones about 110 cm. Prompt particles from $p\bar{p}$ interactions traverse the full length of the counter and generate a large amplitude PMT signal, ~ 100 photoelectrons, while beam-halo particles traverse the counter at larger angles with shorter path length, yielding a much smaller signal. Since the counters effectively measure the actual number of primary particles, the Cherenkov monitor does not saturate at high TEVATRON luminosity.

The central tracking system. The tracking system consists of a silicon microstrip system [276] and of an open-cell wire drift chamber (COT) [277] that surrounds the silicon detector. The silicon microstrip detector (Fig. 40) consists of seven layers (eight layers for $1.0 < |\eta| < 2.0$) in a barrel geometry that extends from a radius of 1.5 cm from the beam line to $r = 28$ cm, at a length from 90 cm to nearly two meters. The layer closest to the beam pipe is a radiation-hard, single sided axial strip detector called Layer 00 which employs LHC designs for sensors supporting high-bias voltages. This enables good signal-to-noise performance even after extreme radiation doses. The layer 00 silicon has an implant pitch of 25 μm and a readout pitch of 50 μm. The remaining seven layers are radiation-hard, double sided detectors. The next five layers after Layer 00 at radii from 2.45 to 10.6 cm comprise the SVXII system consisting of 300 μm thick, n-type, double sided sensors. The SVXII uses 90-degree and small angle stereo sensors for the n-strips from the innermost to outermost SVXII layers in the pattern $(90°, 90°, -1.2°, 90°, +1.2°)$. The p-strips on the non-stereo side run in the axial direction, spaced in $r\phi$ by 60–65 μm. The two outer layers comprise the intermediate silicon layer (ISL) system. The ISL consists of two symmetric silicon layers in the forward and backward region ($|\eta| \geq 1.1$) located at radii of $R \simeq 20$ cm and $R \simeq 29$ cm, respectively, and one in the central region ($|\eta| < 1.1$) at $R \simeq 23$ cm. It provides one space point in the central region which improves the matching between the SVXII tracks and COT tracks and its fine granularity helps to resolve ambiguities in dense

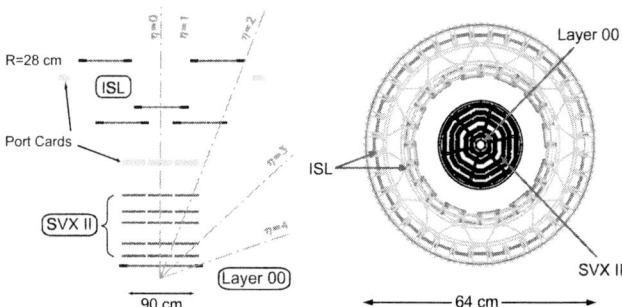

Fig. 40. *Left*: A side view of half of the CDF silicon system. The z coordinate is strongly compressed. *Right*: An endview of the CDF silicon tracker showing the SVXII cooling bulkheads and ISL support structures

track environments. This entire silicon tracking system allows track reconstruction in three dimensions. The impact parameter resolution of the combination of SVXII and ISL is 40 μm including a 30 μm contribution from the transverse width of the beam-line. The z_0 resolution of the SVXII and ISL is 70 μm.

The central outer tracker (COT, Fig. 39) is a 3.1 m long cylindrical drift chamber outside the silicon microstrip detector which covers the radial range from 40 to 137 cm and provides 96 measurement layers, organised into alternating axial and $\pm 2°$ stereo superlayers. Supercells of 12 sense wires each are tilted by $35°$ with respect to the radial direction in order to compensate for the Lorentz angle and the drifting charge particles in the magnetic field. The COT provides coverage for $|\eta| \leq 1$. The hit position resolution is approximately 140 μm and the momentum resolution $\sigma(p_T)/p_T^2 = 0.0015\,(\text{GeV}/c)^{-1}$. The COT provides in addition particle identification information based on the dE/dx energy loss.

The time-of-flight detector. A time-of-flight (TOF) detector [278], based on 3 m long plastic scintillator bars and fine mesh photomultipliers, attached to both ends of each bar, is installed in a few centimetres clearance just outside the COT. The TOF resolution is ≈ 100 ps and it provides at least two standard deviation separation in reconstructed particle mass[14] between $K^\pm$ and $\pi^\pm$ for momenta $p < 1.6\,\text{GeV}/c$. The TOF is mainly used for heavy flavour physics and for searches for new phenomena, such as stable heavy particle production in CDF.

The calorimeter system. Segmented electromagnetic and hadronic sampling calorimeters, arranged in projective towers, surround the tracking system and measure the energy flow of interacting particles in the pseudorapidity range $|\eta| < 3.64$. The central calorimeters (and the endwall hadronic calorimeter) cover the pseudorapidity range $|\eta| < 1.1\,(1.3)$. The central electromagnetic calorimeter [279] (CEM, see Fig. 41) uses lead sheets interspersed with polystyrene scintillator as the active medium and em-

---

[14] $m = \frac{p}{c}\sqrt{\left(\frac{ct}{L}\right)^2 - 1}$, where p is the particle momentum from the tracker, t is the time of flight from the TOF, L is the path length, and c is the speed of light.

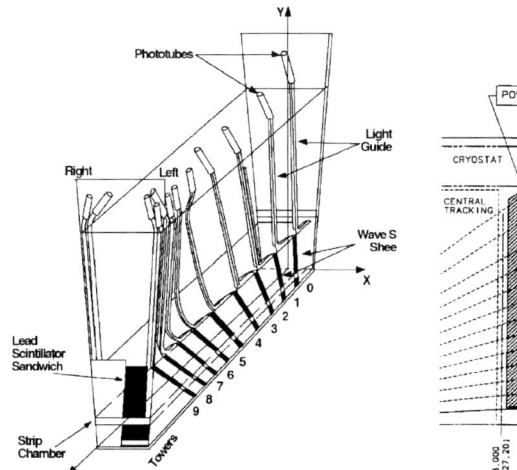 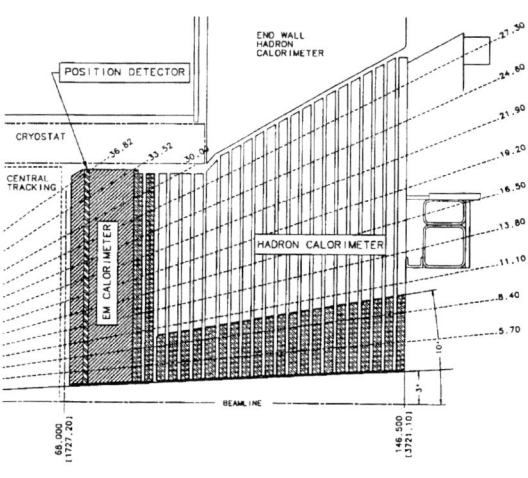

Fig. 41. *Left*: One wedge of the CDF central electromagnetic calorimeter. The ten $(\Delta\phi, \Delta\eta) = (15°, 0.11)$ projective towers are shown. *Right*: Cross section of the CDF plug calorimeter

ploys phototube readout. The CEM thickness corresponds to $18X_0$ and its energy resolution is $13.5\%/\sqrt{E} \oplus 2\%$. The central hadronic calorimeter [280] (CHA) uses steel absorber interspersed with acrylic scintillator as the active medium. It is about $4.5\lambda_I$ thick and its energy resolution is $75\%/\sqrt{E} \oplus 3\%$. The central calorimeters are divided into 24 wedges, each extending about 250 cm along the beam axis on either side of $z = 0$. The calorimeter towers cover a range of $\Delta\eta = 0.11$. To enable a more precise measurement of the transverse shower profile, a proportional strip and wire chamber, called the central electromagnetic shower counter (CES), is embedded in each tower of the central calorimeter at the shower maximum in $5.9X_0$. In addition to the CES, the central pre-radiator detector (CPR), composed of proportional chambers, is placed between the solenoid and the CEM. Both, the CES and CPR help in distinguishing electrons from hadrons.

The plug calorimeters (Fig. 41), divided into projective towers in 12 concentric η regions, cover the pseudorapidity region $1.1 < |\eta| < 3.64$. They are sampling scintillator calorimeters which are read out with plastic fibres and phototubes. The plug electromagnetic calorimeter is $21X_0$ thick and has an energy resolution of $16\%/\sqrt{E} \oplus 1\%$ [281]. As in the central calorimeter, a shower maximum detector (PES) is also embedded in the plug EM section. The plug hadronic calorimeter is $7\lambda_I$ thick and has an energy resolution of $74\%/\sqrt{E} \oplus 4\%$.

The muon system. The muon system has been significantly upgraded for Run-II, in particular to complete the coverage in the central region. The muon system resides beyond the calorimetry. Three muon detectors are used for most top physics analyses: The central muon detector (CMU), the central muon upgrade (CMP), and the central muon extension chambers (CMX). The CMU consists of four layers of planar drift chambers and detects muons with $p_T > 1.4$ GeV/c which penetrate the five absorption lengths of calorimeter steel. The additional four layers of planar CMP drift chambers instrument 0.6 m of steel ($\approx 3.4\lambda_I$) outside the magnet return yoke and detect muons with $p_T > 2.0$ GeV/c. The CMU and CMP chambers each provide coverage in the pseudorapidity range $|\eta| < 0.6$. The CMX, covering $0.6 < |\eta| < 1.0$, now have a full 2π azimuthal coverage. In addition, the intermediate muon detectors (IMU) are covering the region $1.0 < |\eta| < 1.5$. Figure 42 shows the coverage of each sub-detector in the (η, ϕ) coordinates. It should be noted that the CMU and the CMP coverage do not exactly overlap. The CMU is located outside the Central Hadronic Calorimeter ($\approx 5\lambda_I$) at a radius of 3.47 m from the beam. The CMS is an arch-shaped detector built around the plug calorimeter. The muon system relies on proportional wire chambers to provide coarse tracking information, and scintillation counters for triggering. The three detectors are designed with the same four-layer configuration of drift chambers. Wires in the first and third layer are slightly offset in ϕ with respect to the second and fourth layer, in order to remove the ϕ ambiguity in the track reconstruction. The z-position of the track is obtained by comparing the pulse heights at each end of the sense wires. The resolution in the (r, z) plane is 1.2 mm. Tracks measured in at least 3 of the 4 layers form a track segment, called a *stub*.

The trigger system. The trigger and data acquisition systems are designed to accommodate the high rates and large data volume of Run-II. The trigger system is comprised of three levels and is able to function with a 132 ns bunch separation while keeping dead time as short as possible. The trigger architecture is shown in Fig. 42. In the Level-1 trigger, the information for all detectors is buffered in a 42-event deep synchronous pipeline and stored for 5.5 μs. During this time the received data is analysed by three parallel synchronous streams, analysing calorimetry, the muon system, and the extremely fast tracker, XFT. The calorimeter triggers are formed by applying thresholds to energy deposits in calorimeter trigger towers with a segmentation of $\Delta\eta \times \Delta\phi = 0.2 \times 15°$. The thresholds are applied to individual triggers (object triggers) as well as to the sum of energies from all towers (global triggers). The muon trigger looks for stubs in the muon chambers. The XFT reconstructs tracks in the transverse plane of the COT, and an extrapolation unit matches these tracks to the calorimeter and muon chambers. Based on preliminary

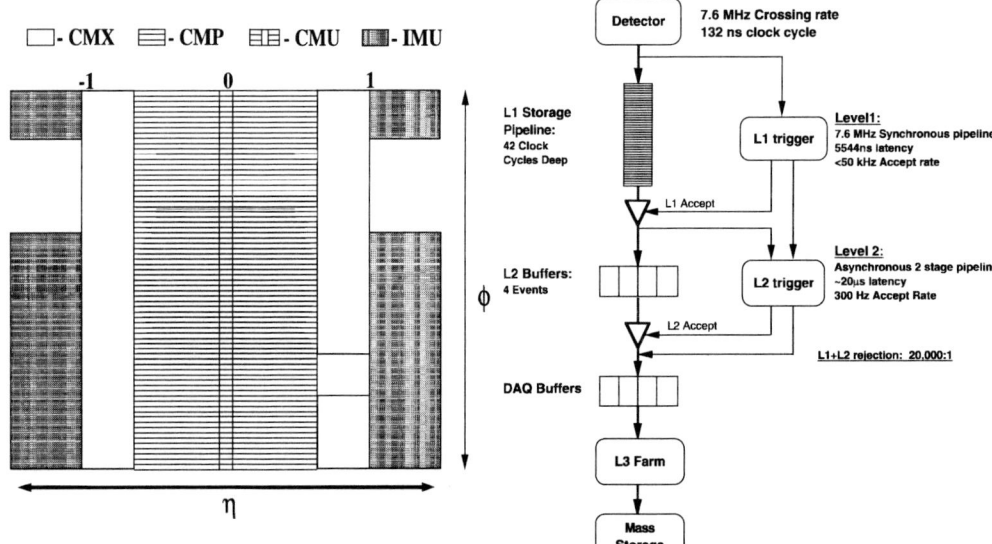

Fig. 42. *Left*: The (η, ϕ) muon coverage of the CDF detector; *Right*: Functional block diagram of the CDF trigger system and data flow

information from tracking, calorimetry, and muon system, the output of the first level of the trigger is used to limit the rate for accepted events to $\approx 18\,\mathrm{kHz}$ at the luminosity range of $3\text{--}7 \times 10^{31}\,\mathrm{cm}^{-2}\,\mathrm{s}^{-1}$.

Events satisfying the Level-1 trigger requirements are downloaded into one of four asynchronous event buffers and processed via programmable Level-2 hardware processors. While Level-2 analyses the events, the buffer cannot be used for additional Level-1 accepts. If all four buffers are full, then the experiment starts to incur deadtime. To keep the deadtime at an acceptable level of 10% and maintain to 50 kHz Level-1 rate, the Level-2 latency is set to $20\,\mu\mathrm{s}$ by using pipelines in two stages, each taking approximately $10\,\mu\mathrm{s}$. The first phase is an event building stage, where clusters and jets are formed and tracks are matched to electromagnetic clusters. On the second pipelined stage, the results of the first phase are collected in the Level-2 processor memory and compared to the Level-2 trigger requirements. About one hundred different Level-2 triggers can be formed. Exploiting the more refined Level-2 information and additional tracking information from the silicon detector, the accept rate is reduced further to $\approx 300\,\mathrm{Hz}\text{--}1\,\mathrm{kHz}$.

At the third and final level of the trigger, events are transferred via a network switch to event builder CPU nodes, where they are assembled from their fragments, and passed to the Level-3 farm of parallel processor nodes. Taking advantage of the full detector information and improved resolution, they analyse and classify each event and apply the Level-3 filter mechanisms. The rate of accepted events, written to permanent storage, is $\approx 75\,\mathrm{Hz}$, with an average event size of 250 kB, corresponding to up to 20 MB/s total output rate.

3.2.2 DØ

To take advantage of the improvements in the TEVATRON and to enhance the physics reach of the experiment, the DØ detector has been significantly upgraded. The detector consists of three major subsystems:

central tracking detectors, uranium/liquid-argon calorimeters, and a muon spectrometer. The central tracking system has been completely replaced. The old Run I system lacked a magnetic field and suffered from radiation damage, and improved tracking technologies are now available. The new system includes a silicon microstrip tracker and a scintillating-fibre tracker located within a 2 T solenoidal magnet. The silicon microstrip tracker is able to identify displaced vertices for b-quark tagging. The magnetic field enables a measurement of the energy to momentum ratio (E/p) for electron identification and calorimeter calibration, opens new capabilities for τ identification and hadron spectroscopy, and allows precision muon momentum measurements. Between the solenoidal magnet and the central calorimeter and in front of the forward calorimeters, preshower detectors have been added for improved electron identification. In the forward muon system, proportional drift chambers have been replaced by mini drift tubes and trigger scintillation counters have been added for improved triggering. Also a forward proton detector has been added for the study of diffractive physics. Figure 43 shows the cross sectional view of the upgraded Run-II DØ detector, as installed in the collision hall and viewed from inside the TEVATRON ring. The forward proton detector is not shown. The large reduction in the bunch spacing required the improvement of the readout electronics and the implementation of pipelining for the front-end signals from the tracking, calorimeter, and muon systems. The calorimeter preamplifiers and signal-shaping electronics have been replaced, as have all of the electronics for the muon system. The trigger system has been significantly upgraded, providing three full trigger levels to cope with the higher collision rate and new hardware to identify displaced secondary vertices for b-quark tagging. Muon triggering has been enhanced by the addition of scintillation counters in the central and forward regions.

A significant improvement to the detector's performance results from the removal of the old main ring beam pipe from the calorimeters. During Run-I, the main ring

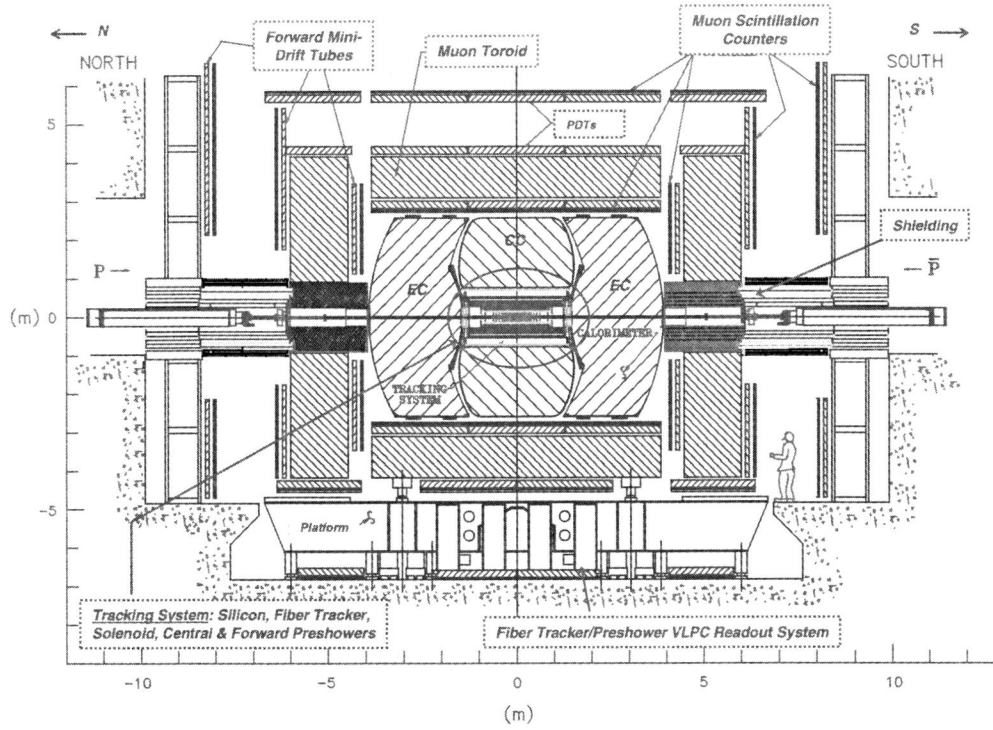

Fig. 43. The DØ detector in cross sectional view

was used to accelerate protons for antiproton production while the TEVATRON operated in collider mode. Losses from the main ring produced spurious energy deposits in the calorimeters and muon system, and most triggers were not accepted while main ring protons passed through the detector. Removal of the main ring increased the lifetime of the detector by approximately 10%, depending on the trigger.

In the following the design and performance of the upgraded DØ detector is described for the various subsystems. A more detailed description of DØ detector in Run-II can be found in [282].

Luminosity monitor. The primary purpose of the Luminosity Monitor (LM) is to make an accurate determination of the TEVATRON luminosity at the DØ interaction region by the detection of inelastic $p\bar{p}$ collisions with a dedicated detector. The LM also serves to measure beam halo rates, to make fast measurements of the z-coordinate of the interaction vertex, and to identify the beam crossing in multiple $p\bar{p}$ interactions.

The luminosity $\mathcal{L}$ is determined from the average number of inelastic collisions per beam crossing $\overline{N}_{\mathrm{LM}}$ measured by the LM: $\mathcal{L} = \frac{f\overline{N}_{\mathrm{LM}}}{\sigma_{\mathrm{LM}}}$ where f is the beam crossing frequency and σ_{LM} is the effective cross section for the LM that takes into account the acceptance and efficiency of the LM detector [283]. Since $\overline{N}_{\mathrm{LM}}$ is typically greater than one, it is important to account for multiple $p\bar{p}$ collisions in a single beam crossing. This is done by counting the fraction of beam crossings with no collisions and using Poisson statistics to determine $\overline{N}_{\mathrm{LM}}$. In this measurement, $p\bar{p}$ interactions and beam halo background are distinguished by the time of flight difference between the

forward and the backward detector, located at $\mp 140\,\mathrm{cm}$, respectively.

The LM detector consists of two arrays of twenty-four plastic scintillation counters with photomultiplier readout. The arrays are located in front of the endcap calorimeters at $z = \pm 140\,\mathrm{cm}$, and occupy the region between the beam pipe and the forward preshower detector. The counters are 15 cm long and cover the pseudo-rapidity range $2.7 < \eta_{\mathrm{det}} < 4.4$.

The central tracking system. Excellent tracking in the central region is necessary for studies of top quark, electroweak, and b physics and to search for new phenomena, including the Higgs boson. The central tracking system consists of the silicon microstrip tracker (SMT) and the central fibre tracker (CFT) surrounded by a 2 T solenoidal magnetic field parallel to the beam axis. The two tracking detectors locate the primary interaction vertex with a resolution of about 35 μm along the beam-line. They can tag b-quark jets with an impact parameter resolution of better than 15 μm in $r - \phi$ for particles with transverse momentum $p_{\mathrm{T}} > 10\,\mathrm{GeV/c}$ at $|\eta| = 0$. The high resolution of the vertex position allows good measurement of lepton p_{T}, jet transverse energy (E_{T}), and missing transverse energy E_{T}. Calibration of the electromagnetic calorimeter using E/p for electrons is now possible. Both the SMT and CFT provide tracking information to the trigger. A schematic view of a quarter of the tracking system, embedded in the solenoid and the calorimeter, is shown in Fig. 46.

Charged particles passing through the 300 μm thick wafers of n-type silicon which comprises the SMT produce pairs of electrons and holes. The ionised charge is collected by strips of p-type or n^+-type silicon strips, whose minute construction (mostly between $\sim 50\,\mu$m and $\sim 80\,\mu$m pitch,

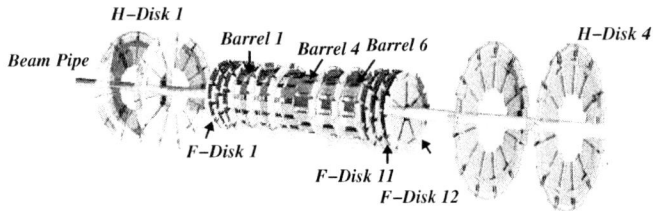

Fig. 44. The disk/barrel design of the DØ silicon microstrip detector

some sensors have $\sim 150\,\mu$m pitch) provide for the measurement of the ionisation. The SMT provides both tracking and vertexing over nearly the full η coverage of the calorimeter and muon systems. The length of the interaction region ($\sigma \approx 25$ cm) sets the length scale of the device. With a long interaction region, it is a challenge to deploy detectors such that the tracks are generally perpendicular to detector surfaces for all η. This led to the design of barrel modules interspersed with disks in the centre and assemblies of disks in the forward regions. The barrel detectors primarily measure the $r - \phi$ coordinate and the disk detectors measure $r - z$ as well as $r - \phi$. Thus vertices for high η particles are reconstructed in three dimensions by the disks, and vertices of particles at small values of η are measured in the barrels and central fibre tracker. An isometric view of the SMT is shown in Fig. 44. The detector has six barrels in the central region. Each has four silicon readout layers, each layer having two staggered and overlapping sub-layers. The outer barrels have single sided and double sided $2°$ stereo ladders. The four inner barrels have double sided $90°$ stereo and double sided $2°$ stereo ladders.

Each barrel is capped at high $|z|$ with a disk of twelve double sided wedge detectors, called an "F-disk". In the far forward and backward regions, a unit consisting of three F-disks and two large-diameter "H-disks" provides tracking at high $|\eta_{det}| < 3.0$. The F-disks are made of 24 pairs of single sided detectors glued back to back. The axial hit resolution is on the order of $10\,\mu$m, the z hit resolution is $35\,\mu$m for $90°$ stereo and $450\,\mu$m for $2°$ stereo ladders.

The central fibre tracker consists of $835\,\mu$m diameter scintillating fibres mounted on eight concentric support cylinders and occupies the radial space from 20 to 52 cm from the centre of the beam pipe (Fig. 45). The two innermost cylinders are 1.66 m long, the outer six cylinders are 2.52 m long. Each cylinder supports one doublet layer of fibres oriented along the beam direction and a second doublet layer at a stereo angle of alternating $\pm 3°$. The two layers of fibres are offset by half a fibre width to provide improved coverage. The small fibre diameter gives the CFT a cluster resolution of about $100\,\mu$m per doublet layer. Light production in the fibres is a multi-step process. When a charged particle traverses one of the fibres, the scintillator emits light at $\lambda = 340$ nm through a rapid fluorescence decay. A wave-shifting dye absorbs the light well at $\lambda = 340$ nm and emits at $\lambda = 530$ nm. The light is then transmitted by total internal reflexion to the end of the scintillating fibres, where the light is transfered through an optical connection to clear fibre waveguides of identical diameter which are 7.8 to 11.9 m long. The light is only observed from one end of each scintillating fibre. The opposite end of each of the scintillating fibres is mirrored by sputtering with an aluminium coating that provides a reflectivity of 85 to 90%. The clear fibre waveguides carry the scintilla-

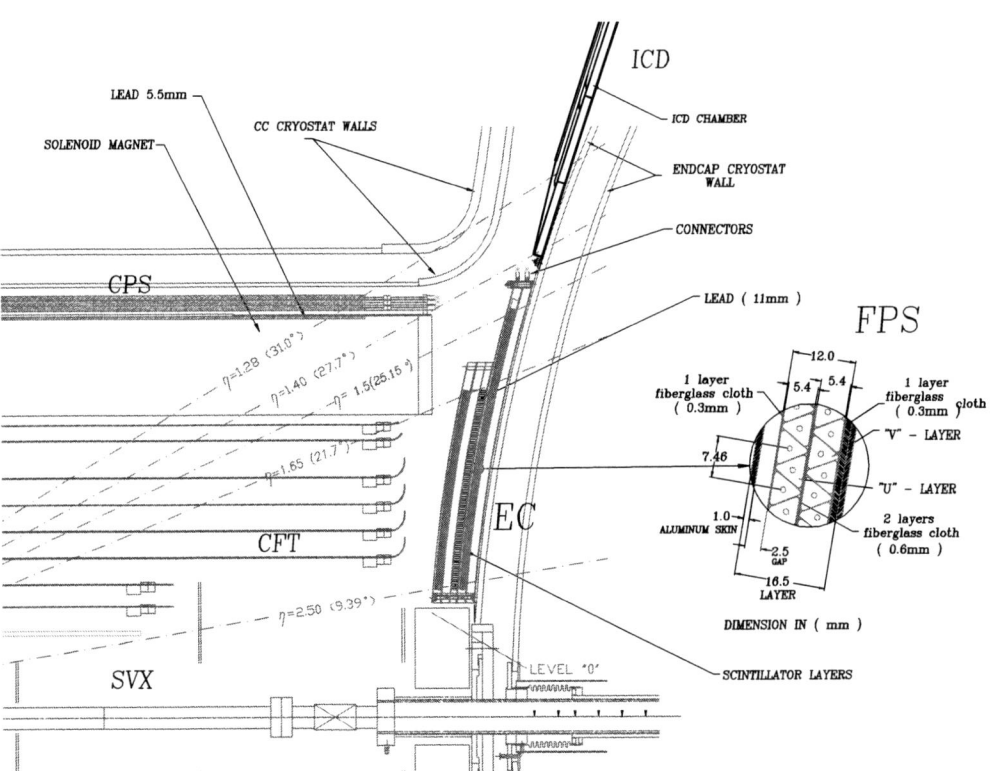

Fig. 45. Cross sectional view of a quarter of the DØ tracking and preshower system

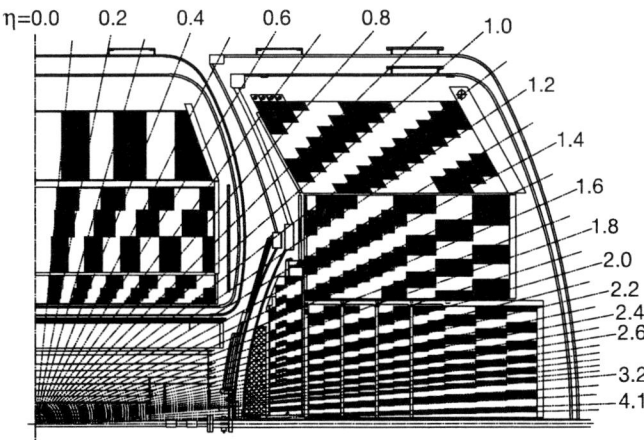

Fig. 46. Cross sectional view of the DØ calorimeter showing the transverse and longitudinal segmentation pattern. The *shading pattern* indicates cells for the signal readout. The *lines* depict the pseudorapidity intervals in steps of 0.2

tion light to visible light photon counters (VLPCs) which convert it into an electronic pulse which is readout. The visible light photon counters are situated in a liquid Helium cryostat and operate at a temperature of 9 K. They detect photons with a quantum efficiency of 85% and provide charge of about 30 to 60 k electrons per photon. A minimum ionising particle creates an average of eight photoelectrons per layer, depending on the angle between the scintillating fibre and particle trajectory.

Hits from both tracking detectors are combined to reconstruct tracks. The momentum resolution of the tracker for minimal ionising particles can be parameterised as:

$$\sigma(p^{-1}) = \frac{\sqrt{(S\sqrt{\cosh\eta})^2 + (Cp_{\mathrm{T}})^2}}{p}, \qquad (47)$$

where p is the particle momentum and η is the pseudorapidity. S accounts for the multiple scattering term and C represents the resolution term. A study of $Z \to \mu^+\mu^-$ events has found $S = 0.015$ and $C = 0.0018$.

The calorimeter and preshower system. The calorimeter system is designed to provide the energy measurement for and assist in the identification of electrons, photons, taus and jets and establish the transverse energy balance in an event. The device is also sensitive to MIPs (minimum ionising particles) and therefore can serve to identify muons. The calorimeter itself (i.e. the modules) is unchanged with respect to Run-I. However, there is significantly more material in front of the calorimeter ($2 \lesssim X_0 \lesssim 4$, depending on the η) and the trigger and readout electronics is rebuilt.

The calorimeter is divided into three parts: the central calorimeter (CC, $|\eta_{\mathrm{det}}| < 1$) and the two end calorimeters (ECs), extending the coverage to $|\eta_{\mathrm{det}}| \approx 4$ (Fig. 46). They consist of an inner electromagnetic (EM) section, a fine hadronic (FH) section, and a coarse hadronic (CH) section. The absorber in the EM and FH sections is depleted Uranium; in the CH section, it is a mixture of stainless steel and Copper. The material and exact geometry of the ab-

sorber plates in the different regions is varied, in order to achieve approximate compensation $e/h \approx 1$, and amounts to approximately 6.4 interaction lengths of Uranium and Copper. The active medium in all cases is liquid argon. The EM sections of the calorimeters are about 21 radiation lengths deep, and are read out in four longitudinal segments (layers). The transverse segmentation is pseudo-projective[15], with a cell size of $\Delta\eta \times \Delta\phi = 0.1 \times 0.1$. In the third layer of the EM calorimeter, near the shower maximum, the segmentation is twice as fine in each direction, with a cell size of $\Delta\eta \times \Delta\phi = 0.05 \times 0.05$. The energy resolution is about $15\%/\sqrt{E} \oplus 0.4\%$ for electromagnetic showers and $50\%/\sqrt{E}$ for single hadrons[16]. The resolution is substantially worse, however, in the transition regions between the CC and the ECs ($0.8 < |\eta_{\mathrm{det}}| < 1.4$), due to the presence of a large amount of uninstrumented material. Some of the energy that would otherwise be lost is collected in extra argon gaps mounted on the ends of the calorimeter modules ("massless gaps") and in scintillator tiles mounted between the CC and EC cryostats (intercryostat detectors, or ICDs). For Run-II, preshower detectors have been installed in front of the central and forward calorimeters. They aid in electron identification and improve their energy measurement. The central preshower detector (CPS, $|\eta_{\mathrm{det}}| < 1.3$) consists of three concentric cylindrical layers of triangular scintillator strips (axial and stereo $\pm 23°$), located between the solenoid and the central calorimeter and a $1X_0$-thick lead-radiator on its inner side. The Forward Preshower detectors (FPS, $1.5 < |\eta_{\mathrm{det}}| < 2.5$) are mounted on the spherical heads of the end calorimeter cryostats. They are made of two layers (stereo $\pm 22.5°$) at different z-position, separated by a $2X_0$-thick lead-stainless-steel absorber. The preshower detectors are readout via wavelength-shifting fibres and visible photon counters.

The muon system. As charged particles, which do not cause electromagnetic or hadronic showers, the muons originating from a $p\bar{p}$ collision penetrate the tracking system and the calorimeter essentially unperturbed. The DØ muon detection system, placed around the calorimeter and depicted in Fig. 47, serves to identify and trigger on these muons and measure their momenta and their charge. For that purpose, the upgraded DØ detector uses the original central muon system proportional drift tubes (PDTs) with radial position resolution of approximately 3 mm and toroidal magnets with an internal field of 1.8 T, partially new central scintillation counters and a completely new forward system. The central muon system provides coverage for $|\eta| < 1.0$. The new forward muon system extends muon detection to $|\eta| \approx 2.0$, uses radiation hard and high segmentation mini

---

[15] Although each cell is non-projective, they form towers which are.

[16] The exact numbers for the Run-II setup are being studied using $J/\Psi, \Upsilon$ and Z events with a leptonic decay to two electrons. Final numbers are not yet available, but preliminary studies indicate an increased energy resolution mainly due to the increased tracker material in front of the calorimeter and reduced charge integration time, a result from shortening the bunch crossing time from $2\,\mu s$ to 396 ns.

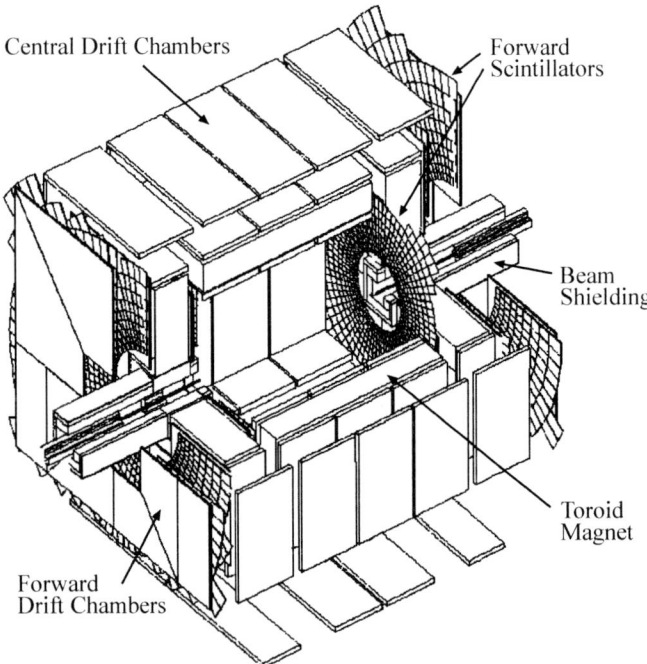

Central Drift Chambers

Forward Scintillators

Beam Shielding

Toroid Magnet

Forward Drift Chambers

Fig. 47. A cut-away view of the DØ muon system

drift tubes (MDTs) with a better coordinate resolution ($\approx$ 0.7 mm) rather than PDTs, and includes trigger scintillation counters and beam pipe shielding.

During Run-I, a set of scintillation counters, the cosmic cap, was installed on the top and upper side of the outer layer of central muon PDTs. This coverage has been extended to the lower sides and bottom of the detector, to form the cosmic bottom. These trigger scintillation counters are fast enough to allow the association of a muon in a PDT with the appropriate bunch crossing and to reduce the cosmic ray background. Additional scintillation counters, the $A\phi$ counters, have been installed on the PDTs mounted between the calorimeter and the toroidal magnet. The $A\phi$ counters provide a fast detector for triggering on and identification of muons and for rejecting out-of-time background events. The scintillation counters are used for triggering; the wire chambers are used for precise coordinate measurements as well as for triggering. Both types of detectors contribute to background rejection: the scintillator with timing information and the wire chambers with track segments. Toroidal magnets and special shielding complete the muon system. Each sub-system has three layers called A, B, and C. The A layer is innermost and located between the calorimeter and the iron of the toroid magnet. B and C layers are located outside the iron.

The most probable value for the energy loss of a muon in the calorimeter is 1.6 GeV, and about 1.7 GeV in the iron. The momentum measurement is corrected for this energy loss.

The momentum resolution for muons, as measured by the muon system in comparison to central tracker measurements in events with ω, ϕ, J/Ψ, Ψ', Υ, or $Z \to \mu\mu$, was found to be $\sigma(p_T)/p_T = 10\%$ for low-momentum muons and 50% for muons with $p_T > 50$ GeV. The overall muon

momentum resolution, including information from the inner tracker is defined by the central tracking system for muons with momentum up to approximately 100 GeV, the muon system improves the resolution only for higher momentum muons.

The forward proton detector. The forward proton detector (FPD) is a series of momentum spectrometers that make use of accelerator magnets in conjunction with position detectors along the beam line in order to determine the kinematic variables t and ξ of the scattered p and $\bar{p}$, where $|t|$ is the squared four-momentum transfer of the scattered proton or antiproton, and $\xi = 1 - x_p$, where x_p is the fractional longitudinal momentum of the scattered particle with respect to the incoming proton. The FPD covers the region $0 \leq t \leq 4.5$ GeV^2 and is of particular importance for DØ's diffractive physics programme.

The trigger system. With the increased luminosity and higher interactions rate delivered by the upgraded TEVATRON, a significantly enhanced trigger is necessary to select the interesting physics events to be recorded. Three distinct levels form this new trigger system with each succeeding level examining fewer events but in greater detail and with more complexity. The first stage (Level 1 or L1) comprises a collection of hardware trigger elements that provide a trigger accept rate of 2 kHz. The pipelined readout makes a trigger decision within 4.2 μs, using field programmable gate arrays (FPGAs). The calorimeter trigger towers of size $\Delta\eta \times \Delta\phi = 0.2 \times 0.2$ provide L1 input up to $|\eta_{det}| < 3.2$. In the second stage (Level 2 or L2), hardware engines and embedded microprocessors associated with specific sub-detectors provide information to a global processor to construct a trigger decision based on individual objects as well as object correlations. The L2 system reduces the trigger rate by a factor of about two and has an accept rate of approximately 1 kHz. Candidates passed by L1 and L2 are sent to a farm of Level 3 (L3) microprocessors; sophisticated algorithms reduce the rate to about 50 Hz and these events are recorded for offline reconstruction.

Further DØ upgrades. Run-IIa at the TEVATRON will deliver ~ 1.5–2 fb^{-1} of integrated luminosity with peak luminosity nearly 1.0×10^{32} cm^{-2} s^{-1} by spring 2006. The DØ detector and trigger are performing very well, however aging of the inner silicon tracker and occupancy-related trigger rate issues will become areas of concern by the end of Run-IIa. The plans for Run-IIb [284, 285], beginning in the summer of 2006, are to achieve peak and integrated luminosities of 2.8×10^{32} cm^{-2} s^{-1} and 8 fb^{-1}, respectively. During a TEVATRON shutdown after Run-IIa, the DØ experiment will complete significant detector and trigger upgrades to deal with the consequences of such an intense beam environment. In particular, a new radiation-hard inner silicon Layer 0 (L0) will be installed on the beam pipe at a radius $R = 1.6$ cm, which will help to recover losses in tracking and b-tagging efficiency that result from dead regions in the inner layer of the Run-IIa SMT. The improved tracking and vertexing resolution resulting from the additional L0 hits near the interaction point corresponds to a 15% improvement over the Run-IIa b-tagging efficiency.

The most ambitious trigger system upgrade is the replacement of the entire L1 calorimeter trigger (L1CAL). The new L1CAL uses digital filtering to improve resolution on the measurement of transverse energy and a sliding window algorithm to perform better clustering. These techniques will dramatically sharpen the trigger threshold turn-on curves and bring much of the current L2 rejection up to L1.

Several upgrades to the data acquisition and online computing systems will increase DØ's capacity to record more high-quality data. The most significant of these projects will be the addition of 96 Linux nodes to the L3 computing farm. The expansion will effectively double the L3 processing power which will confer the ability to efficiently process the more complex high luminosity Run-IIb events and double the L3 output to 100 Hz. Most of the rate benefit is foreseen to be used for triggering and recording of an increased b-physics data sample.

3.3 Particle identification

This section describes the algorithms used by CDF and DØ for the identification and reconstruction of the physics objects in the $t\bar{t}$ or single-top final state such as jets, electrons, muons, taus, missing transverse energy for the neutrinos, and the b-tagging. Wherever possible, also their performance and calibration precision on Run-II is summarised. The corresponding algorithms for the LHC experiments ATLAS and CMS are very similar, but at present only tested on Monte Carlo simulation. They are therefore not described here. More information on those algorithms at the LHC can be found in [286–288].

3.3.1 Quarks, gluons and jets

In $p\bar{p}$ collision, interactions with quarks and gluons in the final state occur at very high rate. These particles hadronise immediately after production, creating a multitude of baryons and mesons or their decay particles which subsequently traverse the detector in the approximate direction of the initial parton and hit the calorimeter. The jet algorithm associates adjacent energy depositions in the calorimeter with the initial parton and forms corresponding jets.

In CDF, jets used in the top quark physics analyses are reconstructed from calorimeter towers using the JETCLU cone algorithm [289]. With a radius of[17] $R = \sqrt{\Delta\phi^2 + \Delta\eta^2} = 0.4$, where the E_T of each tower is calculated with respect to the z coordinate of the event (event vertex from the tracking system). The calorimeter towers belonging to a good electron candidate are not used by the jet clustering algorithm. Due to the construction of the calorimeter, CDF does not observe any noise signal from the calorimeter itself, only from the readout electronics.

At DØ, jets are reconstructed using the *improved legacy cone algorithm*, which was designed following the recommendations of the Run-II QCD workshop [290]. Calorimeter towers are composed from cells (excluding those in the coarse hadronic layer) which share the same pseudorapidity and azimuthal angle. Towers exceeding $E_T > 0.5$ GeV are chosen as seeds, and preliminary jet candidates are identified using a simple cone algorithm with $R = 0.5$. As algorithms operating without seeds show better performance but are computationally too expensive, a compromise is found by considering E_T-weighted centres between pairs of cone jets ('midpoints') as candidates as well. A sophisticated split and merge procedure resolves overlapping cones, and all remaining candidates which fulfil $E_T^{\text{reco}} > 8$ GeV are considered as reconstructed jets.

Calorimeter cells are subject to Gaussian noise from the Uranium as well as from the readout electronics which exceeds the zero suppression threshold: typically, 1000–3000 cells are affected in each event. If such cells are assigned to a jet, the jet energy resolution of real jets is degraded, and fake jets can occur. Therefore, DØ employs the T42 algorithm [291] to improve the interpretation of the calorimeter measurement at the cell level: isolated cells are considered noise if they do not appear to be 'signal-like'. A cell is considered 'signal-like' if its energy is positive and $+4\sigma$ above a threshold, or if it is $+2.5\sigma$ above the threshold but has a neighbouring cell which exceeds the threshold by $+4\sigma$. The T42 algorithm rejects about 30%–60% of all cells in the event, in good agreement with the noise expectation. Towers are subsequently built only from cells not identified as noise.

Reconstructed cone jets must fulfil the following additional quality requirements:

- $0.05 < f_{\text{em}} < 0.95$, where f_{em} is the fraction of jet energy deposited in the electromagnetic section of the calorimeter. Isolated electromagnetic particles are rejected.
- $f_{\text{CH}} < 0.4$, where f_{CH} is the fraction of jet energy deposited in the coarse hadronic section of the calorimeter. Jets which have been formed mainly from cells in this noisy calorimeter section are removed.
- $f_{\text{hot}} < 10$, where f_{hot} is the energy ratio of the highest and the next-to-highest calorimeter cell assigned to the jet. A large value of f_{hot} indicates that the jet is clustered around a hot cell (mostly abnormal electronic noise).
- $n90 > 1$, where $n90$ is the number of calorimeter towers containing 90% of the jet energy. A small $n90$ indicates that the jet is clustered around a hot cell.
- Confirmation of the jet by the Level 1 trigger readout chain. Fake jets surviving all other quality criteria appear mostly at the reconstruction stage, but are not seen in the trigger readout. This electronic noise is due to coherent noise in the precision readout chain and can be efficiently rejected by requiring coincidence between the reconstructed jet and Level 1 trigger signals.
- $p_T > 20$ GeV, after jet energy scale correction.
- $|\eta| < 2.5$

Electrons and photons which pass the cut on f_{EM} and exceed the reconstruction threshold of $E_T > 8$ GeV appear in the list of reconstructed jets. All such jets which are

---

[17] $R = \sqrt{\Delta\eta^2 + \Delta\phi^2}$ is the cone size in (η, ϕ) space.

matched within $\Delta R(\text{jet}, \text{EM}) < 0.5$ to an electromagnetic object are removed from the list, if $p_T^{\text{jet}} > 15$ GeV.

3.3.2 Jet energy scale

The primary goal of the CDF and DØ jet energy corrections is to determine the energy correction to scale the measured energy of the jet energy back to the energy of the final state particle level jet (Fig. 48). Additionally, there are corrections to associate the measured jet energy to the parent parton energy, so that direct comparison to the theory can be made. Currently, the jet energy scale is the major source of uncertainty in the top quark mass measurement and inclusive jet cross section.

The CDF jet energy corrections are divided into different levels to accommodate different effects that can distort the measured jet energy, such as, response of the calorimeter to different particles, non-linearity response of the calorimeter to the particle energies, un-instrumented regions of the detector, spectator interactions, and energy radiated outside the jet clustering algorithm. Depending on the physics analyses, a subset of these corrections can be applied.

The CDF detector has been upgraded for Run-II. All systems, except the central calorimeter and the muon system, were replaced. The data acquisition electronic and simulation and reconstruction software was re-written. For the central calorimeter, the ADC integration gate was reduced

from 600 ns to 132 ns, clipping the tails of the signal. In addition, the material in front of the calorimeter increased due to the new tracking system. Both of these effects reduce the observed energy in the calorimeter. A comparison of the p_T difference in $\gamma + \text{jet}$ events in Run-II with Run-I data shows that the Run-II jet energy scale is -2.8 ± 0.4 (stat.) $\pm$ 0.8 (syst.)% lower than in Run-I, consistent with the drop expected from extra material and the shorter integration gate. The Run-II calorimeter simulation has been tuned to the single particle response measured in Run-II $p\bar{p}$ collisions at low momenta ($p < 20$ GeV) and test beam measurement at higher momenta ($p > 20$ GeV). This tuning takes care of the above-mentioned changes in the detector at least at low momenta. In the central calorimeter, an uncertainty about 50% smaller than the initial CDF Run-II estimate is achieved, which is slightly better than the final Run-I estimate. As a result of having a better CDF detector simulation, the Run-II jet energy scale uncertainties in the non-central regions have been decreased by up to a factor of 5.

Absolute jet energy scale. The jet energy measured in the calorimeter needs to be corrected for any non-linearity and energy loss in the un-instrumented regions of each calorimeter. Since there are no high statistics calibration processes at high E_T, this correction is extracted from Monte Carlo. The simulation of the calorimeter needs to accurately describe the response to single particles (pions, protons, neutrons, etc.). The Monte Carlo fragmentation needs to describe the particle spectra and densities of the data for all jet E_T. CDF measures the fragmentation and single particle response in data and tunes the Monte Carlo to describe it. The correction is obtained mapping the total E_T of the hadron-level jet to the E_T of the calorimeter-level jet. The hadron-level jet consists of particles within a cone of the same size as and within $\Delta R < 0.4$ of the calorimeter-level jet. The main systematic uncertainties on the absolute scale are obtained by propagating the uncertainties on the single particle response (E/p) and the fragmentation. Smaller contributions are included from the comparison of data and Monte Carlo simulation of the calorimeter response close to tower boundaries in azimuth, and from the stability of the calorimeter calibration with time. Figure 49 (left) shows the correction on the absolute jet energy scale and the corresponding uncertainty as a function of jet-p_T.

Relative jet energy scale vs. η. Since the central calorimeters are better calibrated and understood, they are used to

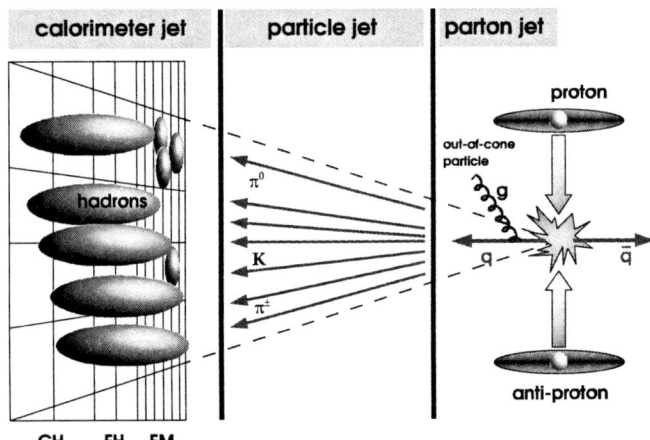

Fig. 48. Schematic of the jet energy scale corrections, taking energy measurements on the calorimeter level to the particle or the parton level

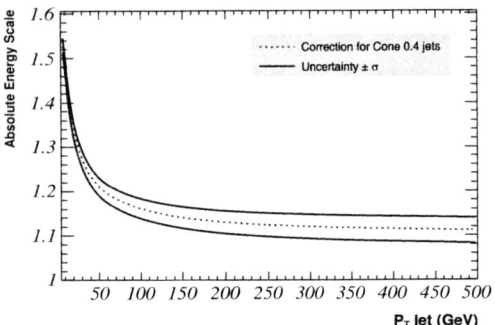

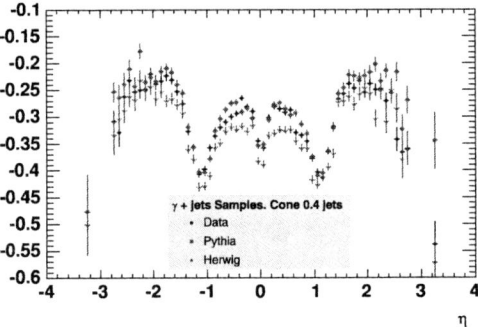

Fig. 49. *Left*: Correction on the absolute jet energy scale and corresponding uncertainty as a function of jet-p_T. *Right*: Relative jet energy scale correction as function of jet η as obtained in $\gamma + \text{jet}$ events before correction

apply a relative scale correction to the forward calorimeters. This correction is obtained using PYTHIA [223] and data di-jet events. The transverse energy of the two jets in a $2 \to 2$ process should be equal. This property is used to scale jets outside the $0.2 < |\eta| < 0.6$ region to jets inside the region. This region is chosen since it is far away from the cracks or non-instrumented regions. This results in a correction as a function of pseudo-rapidity η (Fig. 49 right) and E_T. After corrections, the response of the calorimeter is almost flat with respect to the pseudo-rapidity η. The selection requirements and fitting procedure are varied and the deviation of the calorimeter response versus η from a straight line is taken as a systematic uncertainty. The difference between data and PYTHIA is already accounted for by the uncertainties from other studies of fragmentation and out-of-cone energy, so it does not need to be included again as a systematic here. Good agreement of the relative response of the calorimeter between PYTHIA di-jet production and data is found. Such a good response is not observed with HERWIG di-jet, the origin of these discrepancies is under study.

Multiple interactions. The energy from different (multiple) $p\bar{p}$ interactions during the same bunch crossing falls inside the jet cluster, increasing the energy of the measured jet. The 'multiple interactions' correction subtracts this contribution on average. The correction is derived from minimum bias data and is parameterised as a function of the number of vertices in the event. The systematic uncertainty from this correction is 15%. The sources of uncertainties are the differences observed with different topologies and the luminosity dependence.

Underlying event. The 'underlying event' is defined as the energy associated with the spectator partons in a hard collision event. Depending on the details of the particu-

lar analysis, this energy needs to be subtracted from the particle-level jet energy. The underlying event energy is measured from minimum bias data requiring events with only one vertex. The uncertainty on the underlying event correction is 30%.

Out-of-cone correction. The 'Out-of-Cone correction' corrects the particle-level energy for leakage of radiation outside the clustering cone used in the jet definition, taking the "jet energy" back to the "parent parton energy". The energy flow between cones of size 0.4 and 1.3 is measured. Since the Monte Carlo must describe the jet shape of the data, the systematic is again taken from the difference between data and Monte Carlo for different topologies.

Total systematic uncertainties. The total systematic uncertainties in the central calorimeter ($0.2 < |\eta| < 0.6$) are shown in Fig. 50. For non-central jets, the total uncertainty is obtained by adding in quadrature the relative (η-dependent) and the central uncertainties. The central uncertainties ($0.2 < |\eta| < 0.6$) are of the same order as in Run-I. The CDF simulation has greatly improved since Run-I, therefore in the non-central regions the Run-II uncertainties are smaller by up to a factor of 4 in some regions. At low p_T, the main contribution is from the out-of-cone uncertainty, while at high p_T it is from the absolute jet energy scale. Reducing the uncertainty at low p_T requires a better understanding of the differences between data and Monte Carlo in samples like $\gamma + $ jet (QCD-Compton). A better CDF simulation and larger statistics to determine the uncertainties should reduce the uncertainties at high p_T.

In DØ, the raw energy of a reconstructed jet is given by the sum of energies deposited in the calorimeter cells associated with the jet by the cone algorithm. Several mechanisms cause this energy estimate to deviate from the energy of the initial parton (Fig. 48):

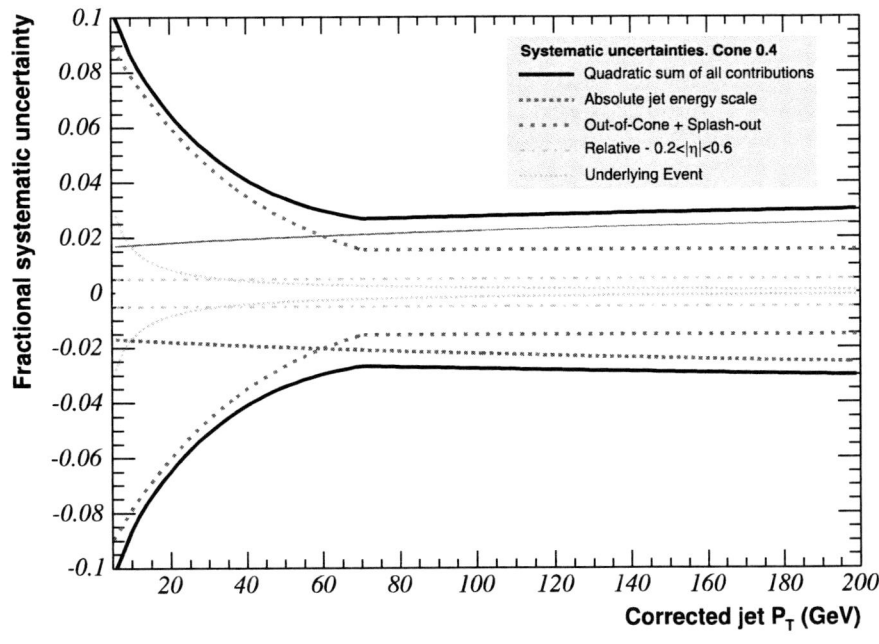

Fig. 50. Total fractional systematic uncertainty on the jet energy calibration as a function of the corrected jet p_T. The contribution from the different sources is also shown as *different line types*

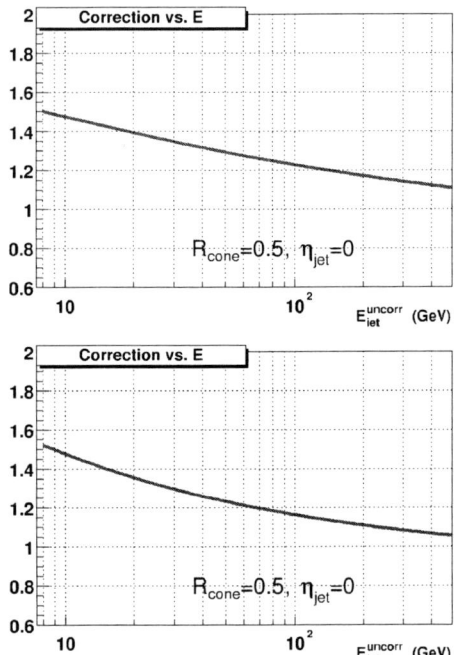
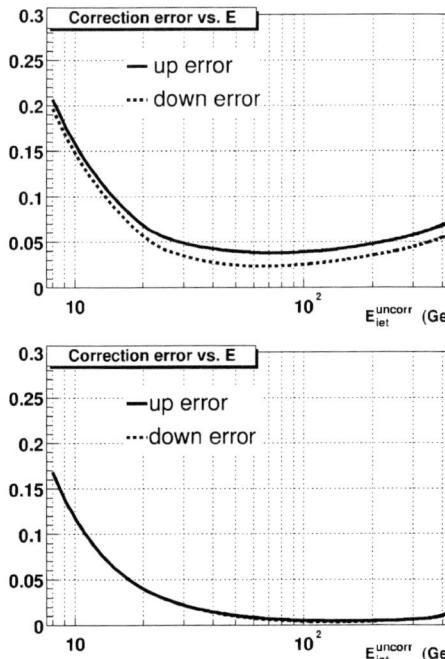

Fig. 51. Jet energy scale (*left*) and corresponding relative uncertainty (*right*) as a function of the jet E_T. *Top*: for data; *bottom*: for Monte Carlo

– Energy offset **O**: Energy in the clustered cells which is due to noise, underlying event, multiple parton interactions, energy pile-up from multiple proton interactions at high instantaneous luminosity and Uranium noise lead to a global offset of jet energies. **O** is determined from energy densities in minimum bias events.

– Calorimeter response **R**: Jets consist of different particles (mostly photons, pions, kaons, (anti-)protons and neutrons), for which the calorimeter response is different. Furthermore, the calorimeter responds slightly non-linearly to particle energies. R is determined with $\gamma + \text{jets}$ ('QCD-Compton') events requiring transverse momentum balance. The photon scale is measured independently from $Z \to ee$ events with high precision.

– Showering corrections **S**: A fraction of the parton energy is deposited outside of the finite-size jet cone. S is obtained from jet energy density profiles.

Consequently, the corrected particle energy $E_{\text{jet}}^{\text{corr}}$ before interaction with the calorimeter is obtained from the reconstructed jet energy $E_{\text{jet}}^{\text{reco}}$ as

$$E_{\text{jet}}^{\text{corr}} = \frac{E_{\text{jet}}^{\text{reco}} - O}{R \times S} . \tag{48}$$

Note that $E_{\text{jet}}^{\text{corr}}$ is not the parton energy: the parton may radiate additional quarks or gluons before hadronisation, which may or may not end up in the jet cone. The correction of the jet energy down to the parton-level, for example for the measurement of the top quark mass, is achieved in the derivation of transfer functions (see the $\ell + \text{jets}$ analyses in Sect. 7.1.2). The jet energy scales for data and Monte Carlo jets are shown in Fig. 51, along with the corresponding uncertainties.

The response measurement is performed for the central and forward calorimeters individually. In a second iteration, with finer binning in η, more subtle effects of the jet

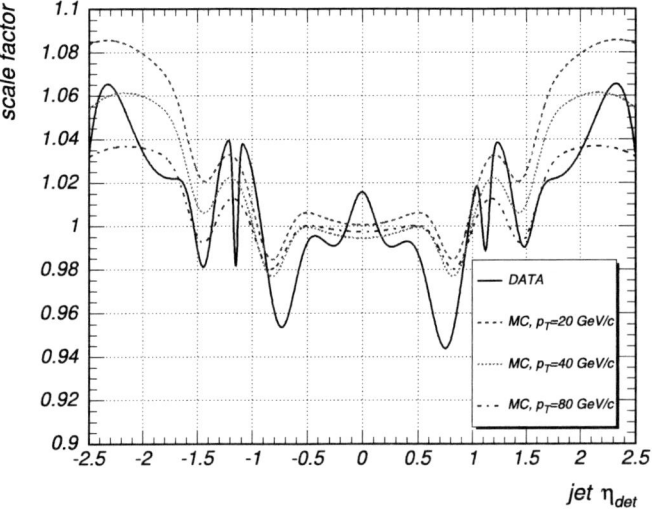

Fig. 52. η-dependent jet energy corrections $\Delta S^{\text{DATA}}/\Delta S^{\text{MC}}$ after jet energy scale correction for data and Monte Carlo jets. The corrections applied to Monte Carlo jets depend on jet E_T as well

energy correction as a function of η are resolved. For the determination of this correction, the scale is applied to the jets in a $\gamma + \text{jets}$ sample, and the variable

$$\Delta S = \frac{p_T^{\text{jet}} - p_T^\gamma}{p_T^\gamma} , \tag{49}$$

reveals additional structure of the jet energy scale as a function of the pseudorapidity. These 'η-dependent corrections', shown as the data to Monte Carlo scale factor $\Delta S^{\text{DATA}}/\Delta S^{\text{MC}}$ in Fig. 52, are applied to jets in Monte Carlo and data separately and are also propagated to the missing transverse energy.

3.3.3 Electrons

In CDF, electron candidates are required to have a COT track with $p_T > 9 \, \text{GeV}/c$ that extrapolates to a cluster of energy with $E_T > 20 \, \text{GeV}$ formed by up to three adjacent towers in pseudorapidity in the central electromagnetic calorimeter. The electron energy is corrected by less than 5% for the non-uniform response across each calorimeter tower by using the CES measurement of the shower position. The shower position is required to be away from the calorimeter tower boundaries to ensure high quality discrimination between electrons and charged hadrons. This fiducial volume for electrons covers 84% of the solid angle in the central $|\eta| < 1$ region. The selection requirements are defined below:

- The ratio of hadronic energy in the cluster, E_{had}, to the electromagnetic energy in the cluster, E_{em}, has to be $\leq (0.055 + 0.00045) \times E(\text{GeV})$
- χ^2 comparison of the lateral shower profile in the calorimeter cluster with that measured from test beam electrons, $L_{\text{shr}} \leq 0.2$.
- χ^2 comparison of the CES shower profiles with those of test beam electrons in the z view, $\chi^2_{\text{strip}} \leq 10.0$.
- Distance between the position of the extrapolated track and the CES shower profiles measured in the $r - \phi$ and z views, δx and δz. The limits on δx are asymmetric and signed by the electric charge Q to allow for energy deposition from bremsstrahlung photons emitted as the electron or positron passes through the detector material ($|\delta z| \leq 3 \, \text{cm}$, $Q \delta z \geq -3.0$, $\leq 1.5 \, \text{cm}$). Central electrons are matched to a track from the central outer tracker (COT). For forward electrons ($|\eta| > 1.2$), this track association uses a calorimeter-seeded silicon tracking algorithm.
- Ratio of cluster energy to track momentum $E/p \leq 2.0$ or $p_T > 50 \, \text{GeV}/c$.
- Isolation, I, defined as the ratio between any additional transverse energy in a cone of radius $R = \sqrt{\Delta \eta^2 + \Delta \phi^2} = 0.4$ around the cluster and the transverse energy of the cluster, $I \leq 0.1$ (for 'tight' electrons).
- A conversion veto is imposed.

In DØ, electrons are identified as narrow clusters in the electromagnetic portion of the calorimeter system. Such an EM cluster is defined by one seed tower selected on the basis of its energy content, and the set of towers within a cone of radius $R = \sqrt{(\Delta \eta)^2 + (\Delta \phi)^2} = 0.2$ around it. An electron has to satisfy the following criteria:

- $f_{\text{EM}} = E_{\text{EM}} / E_{\text{tot}} > 0.9$, i.e. the cluster is required to have 90% of its total energy in the electromagnetic layers, while its energy is determined from all calorimeter systems.
- $f_{\text{iso}} = \frac{E_{\text{tot}}(R < 0.4) - E_{\text{EM}}(R < 0.2)}{E_{\text{EM}}(R < 0.2)} < 0.15$, i.e. the electron has to be isolated. No calorimetric activity in a cone of radius $R = 0.4$ is allowed.
- The longitudinal and lateral development of the shower induced by an electron throughout the layers of the electromagnetic calorimeter is distinct from the properties of showers induced by other particles; each candidate is compared with average distributions from the simulation and test beam measurements and assigned a χ^2 as a measure of electron compatibility: The χ^2 fit is based on seven degrees of freedom, and electron candidates must fulfil $\chi^2 < 75$ (7 degrees of freedom).
- To further suppress the overwhelming background from jet production, candidates are matched to a track in the central tracking system which points to the reconstructed EM cluster in the calorimeter: $|\Delta \phi(\text{EM, track})| < 0.05$, $|\Delta \eta(\text{EM, track})| < 0.05$.
- The major remaining background of photons from π^0 decays, overlapping with a track from a nearby charged particle, is efficiently rejected by an electron likelihood, which is referred to as EM-likelihood. It is based on eight variables, including calorimeter shower shape and track isolation variables. Reference distributions of each variable are obtained from $Z \to ee$ data events for the signal case and from a fake-enriched multi-jet (EM+jet back-to-back) data sample for the background case. Electron candidates are required to have an EM-likelihood of greater than 0.85.

The electron energy resolution and scale are measured in $Z \to ee$ data events. The Monte Carlo simulation is adjusted accordingly.

3.3.4 Muons

In CDF, muon candidates are required to have a COT track with $p_T > 20 \, \text{GeV}/c$ that extrapolates to a track segment in the muon chambers. The muon COT track curvature, and thus the muon transverse momentum, is corrected in order to remove a small azimuthal dependence from residual detector alignment effects. The selection requirements used to separate muons from products of hadrons that interact in the calorimeters and from cosmic rays are defined below:

- Energy deposition in the electromagnetic and hadronic calorimeter expected to be characteristic of minimum ionising particles, $E_{\text{EM}} \leq \max(2, 2 + 0.0115(p - 100)) \, \text{GeV}$ and $E_{\text{had}} \leq \max(6, 6 + 0.0280(p - 100)) \, \text{GeV}$.
- Distance between the extrapolated track and the track segment in the muon chamber, $\Delta x \leq 3.0 \, \text{cm}$ (CMU), $\leq 5.0 \, \text{cm}$ (CMP), $\leq 6.0 \, \text{cm}$ (CMX). A track matched to a segment in the CMU muon chambers is required to have a matched track segment in the CMP chambers as well, and vice versa.
- Cosmic ray muons that pass through the detector close to the beam-line may be reconstructed as a pair of charged particles. The timing capabilities of the COT are used to reject events where one of the tracks from a charged particle appears to travel toward instead of away from the centre of the detector.
- Isolation, I, defined as the ratio between any additional transverse energy in a cone of radius $R = 0.4$ around the track direction and the muon transverse momentum, $I \leq 0.1$.

In DØ, muons are identified in the muon chambers by matching segments on either side of the toroid. The following criteria are required:

– at least two wire hits in the A layer
– at least one scintillator hit in the A layer
– at least two wire hits in the B or C layers
– at least one scintillator hit in the B or C layers (except for central muons with less than four BC wire hits).

A muon signal established according to these criteria is referred to as a 'local muon'. Due to the presence of the toroid magnet, the momentum of the muon can be determined from the muon detector information alone. However, the momentum of the muon is measured with significantly better precision with the tracking detectors, if the local muon can be matched to its corresponding track. Consequently, only muons which can be matched to a central track are considered. The local muon track is extrapolated back to the point of closest approach to the beam and its parameters are compared to nearby charged particle tracks. The local momentum measurement in the muon chambers is disregarded entirely in favour of the tracking information. The central track matched to the muon is required to fulfil the following additional quality criteria:

– $\chi^2_{\text{trak}}/\text{NDF} < 4$ for the track fit
– separation from the primary vertex along the beam axis $\Delta z(\mu, \text{PV}) < 1$ cm
– dca significance $dca/\sigma(dca) < 3$, to reject muons from semi-leptonic heavy flavour decays.

The momentum resolution degrades significantly for tracks without hits in the high precision silicon tracking detectors. As the above dca significance cut relates the muon to the hard scatter interaction in the event, some of the resolution can be recovered by constraining the muon track to the primary vertex. Track parameters are refit accordingly for muon tracks without SMT hits. The muon momentum scale and resolution is measured in $Z \to \mu\mu$ data events. The Monte Carlo simulation is adjusted accordingly.

Muons from leptonic W decays are expected to be isolated from jets and thus any nearby calorimeter or tracking activity. The main source of misidentified $W \to \mu\nu$ decays are muons originating from semi-leptonic heavy flavour decays: if the hadronic signature of the b-quark is not reconstructed as a calorimeter jet, the muon appears to be isolated by mistake. In addition, these muons tend to have lower transverse momentum p_{T}^{μ} than the ones in W decay. Consequently, the following two variables are defined to discriminate between isolated and non-isolated muons:

– **Rat11** $\equiv \text{Halo}(0.1, 0.4)/p_{\text{T}}^{\mu}$, where $\text{Halo}(0.1, 0.4)$ is the E_{T} sum of calorimeter clusters in a hollow cone around the muon direction ranging from $\Delta R = 0.1$ to 0.4. Only the clusters in the electromagnetic and fine hadronic calorimeter layers are considered, where coarse hadronic signals are excluded due to their high noise level. Only muons with Rat11 < 0.08 are accepted.
– **Rattrk** $\equiv \text{TrkCone}(0.5)/p_{\text{T}}^{\mu}$, where $\text{TrkCone}(0.5)$ is the p_{T} sum of all tracks within a cone of radius $\Delta R = 0.5$ around the muon direction. The track matched to the muon itself is excluded from the sum. Only muon with Rattrk < 0.06 are accepted.

3.3.5 τ leptons

In Run-II of the TEVATRON, tau leptons play an important role in electroweak measurements, studies of top quark properties and, in particular, in searches for new phenomena involving the third fermion generation (Higgs and supersymmetry). Tau reconstruction (and triggering) at hadron colliders remains a notoriously difficult task in terms of distinguishing interesting events from background dominated by multi-jet production with its enormous cross section.

In CDF, tau candidates are reconstructed by matching narrow calorimeter clusters with tracks. The calorimeter cluster is required to have E_{T} of the seed tower above 6 GeV. All adjacent towers with transverse energy above 1 GeV are included in the cluster. The sum of transverse energies of the towers is used as the transverse energy of the tau candidate, $E_{\text{T}}^{\text{cal}}$. Only clusters consisting of 6 or less towers are used for tau reconstruction.

Reconstructed tracks that point to the calorimeter cluster are associated with the tau candidate. The highest p_{T} track associated with the tau is called the *seed* track and defines the axes of the signal and isolation cones. The signal cone is defined as a cone with opening angle α around the seed track, taking into account collimation of hadronic tau jets with increasing energy, providing high signal efficiency.

Tau decay modes are often classified by the number of prongs. At CDF, prongs are defined as tracks inside the signal cone of a tau candidate with transverse momentum $p_{\text{T}} > 1$ GeV/c. Tracks are required to pass certain quality requirements, and to have a z-vertex compatible with the one of the seed track: $|z_0^{\text{trk}} - z_0^{\text{seed}}| < 5$ cm. Next, the π^0 information is added. Clusters in the CES chambers are called a π^0 candidate if no COT track of $p_{\text{T}} > 1$ GeV/c is found nearby. Similar to the track case, also for the π^0's a cone of size α_{π^0} around the seed track is defined and all π^0 candidates inside the cone with $E_{\text{T}} > 1$ GeV are associated with the tau candidate. The momentum of the tau is defined as the sum of massless four-vectors of all tracks and π^0's associated with the tau candidate: $p^{\tau} = \sum_{\Delta\Theta < \alpha_{\text{trk}}} p^{\text{trk}} + \sum_{\Delta\Theta < \alpha_{\pi^0}} p^{\pi^0}$. Several variables useful for discriminating between real taus and background fakes are defined using track and π^0 information. The visible mass of a tau candidate, $m_{\text{trk}+\pi^0}^{\tau}$, is defined as the invariant mass of the tau momentum four-vector p^{τ}. The track mass of a tau candidate, m_{trk}^{τ}, is defined as the invariant mass of the track-only part of the tau momentum four-vector, p^{τ}. The charge of a tau candidate is defined as a sum of charges of the prongs associated with it: $Q^{\tau} = \sum_{\tau \text{tracks}} Q_{\text{trk}}$. For discriminating hadronic taus from electrons, a useful variable ξ is defined as $\xi = E_{\text{T}}^{\text{had}} / \sum_{\tau \text{tracks}} p_{\text{T}}^{\text{trk}}$, where $E_{\text{T}}^{\text{had}}$ is the transverse energy of the tau candidate calculated using only hadronic deposition in the calorimeter. For electrons, the ξ value is typically small, allowing substantial suppression of electron background. Furthermore, two kinds of track isolation variables are defined: (i) The scalar sum of the momenta of all tracks inside a cone of 30° around the seed track but outside the signal cone in 3D space:

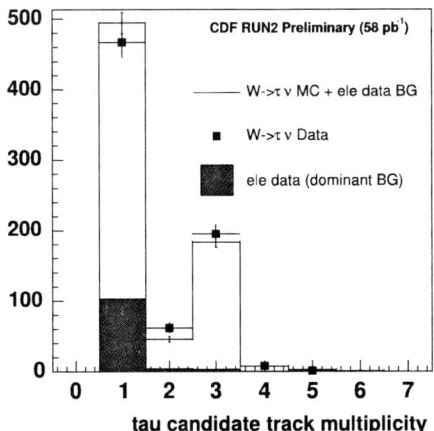

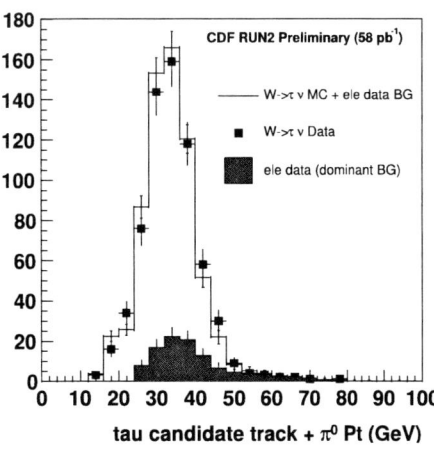

Fig. 53. *Left*: The distribution of the number of tracks associated with taus in the $W \to \tau \nu$ dominated data sample and Monte Carlo + electron background. *Right*: The distribution of the tau p_T in a $W \to \tau \nu$ dominated data sample and Monte Carlo + electron background

$I_{\mathrm{trk}}^{\Delta\Theta} = \sum_{\alpha_{\mathrm{trk}} < \Delta\Theta < 30°} p_T^{\mathrm{trk}}$. (ii) $N_{\mathrm{trk}}^{\Delta\Theta}$, is defined as the number of isolated tracks with $p_T > 1\,\mathrm{GeV}/c$. In a similar way, the tau candidate π^0 isolation is defined as $I_{\pi^0}^{\Delta\Theta} = \sum_{\alpha_{\pi^0} < \Delta\Theta < 30°} p_T^{\pi^0}$. Analogously to the track case, also N_{π^0} is defined as the number of π^0 candidates in the isolation cone.

Based on these variables, a $W \to \tau \nu + \mathrm{jets}$ data sample is selected in $58\,\mathrm{pb}^{-1}$. Figure 53 shows the distribution of the number of tracks associated with taus (left) and the distribution of the tau p_T (right) together with the signal simulation and electron background estimates, clearly demonstrating the strong enhancement in τ's.

Also DØ has worked out an equivalent τ lepton identification in Run-II, combining variables on track and calorimeter showers as well as isolation variables using an artificial neural network. As no results on top quark physics with τ leptons has been released yet, the τ identification in DØ is not described here. More details on the DØ τ identification can be found in [292].

3.3.6 Neutrinos

Neutrinos do not interact with any of the detector systems and can only be identified indirectly by the imbalance of the event in the transverse plane.

In CDF, the missing transverse energy, $\not{E}_T$, is defined as the magnitude of the vector

$$- \sum_i \left(E_{T,i} \cos \phi_i, E_{T,i} \sin \phi_i \right), \qquad (50)$$

where $E_{T,i}$ is the transverse energy, calculated with respect to the z vertex in the event, in calorimeter tower i with azimuthal angle ϕ_i. In the presence of any muon candidates, the $\not{E}_T$ vector is recalculated by subtracting the transverse momentum of the muon COT track and adding back in the small amounts of transverse energy in the calorimeter towers traversed by the muon. For all jets with $E_T \geq 8\,\mathrm{GeV}$ and $|\eta| < 2.5$, the $\not{E}_T$ vector is adjusted for the effect of the jet corrections as well as for electron and photon energy corrections.

In DØ, the transverse energy imbalance is reconstructed from the vector sum of all calorimeter cells which pass the T42 algorithm. Cells in the coarse hadronic system receive special treatment due to their high level of noise: they are only considered if clustered into a reconstructed jet. The momentum vector that balances this vector sum in the transverse plane is denoted the missing energy vector, and its magnitude is the raw missing transverse energy, $\not{E}_T^{\mathrm{raw}}$. The calorimeter response is different for electromagnetic particles and jets, and the respective corrections are propagated to the $\not{E}_T$ vector according to the presence of such objects resulting in $\not{E}_T^{\mathrm{CAL}}$. If a muon is present in the event, it will only deposit a small fraction of its energy in the calorimeter, and the $\not{E}_T$ vector is corrected accordingly. The expected muon energy deposition in the calorimeter is hereby taken from GEANT lookup tables, the muon p_T is measured in the tracking detectors. After all corrections, the magnitude of the missing transverse energy vector represents the quantity $\not{E}_T$ referred to throughout the rest of this review.

3.3.7 b-Tagging

Reconstructed and identified jets can be classified further according to the flavour, where light flavour jets originate from the hadronisation of a u-, d-, s-quark or a gluon, and heavy flavour jets originate from a c- or a b-quark. At least two techniques can be used to distinguish a heavy flavour jet from a light flavour jet:

- **soft lepton tagging (SLT)** the presence of a soft electron or muon within the jet cone indicates a semileptonic b or c hadron decay with a branching ratio of typically $\sim 10\%$ per lepton.
- **Lifetime tagging** identifying charged tracks which are significantly displaced from the primary vertex due to the finite lifetime of the b or c hadron decay.

CDF uses both, secondary vertex and soft-lepton tagging algorithms in their top quark physics analysis, which are briefly described in turn.

The *CDF SecVtx* algorithm relies on the displacement of secondary vertices relative to the primary vertex to identify b hadron decays. The Run-II algorithm is essentially

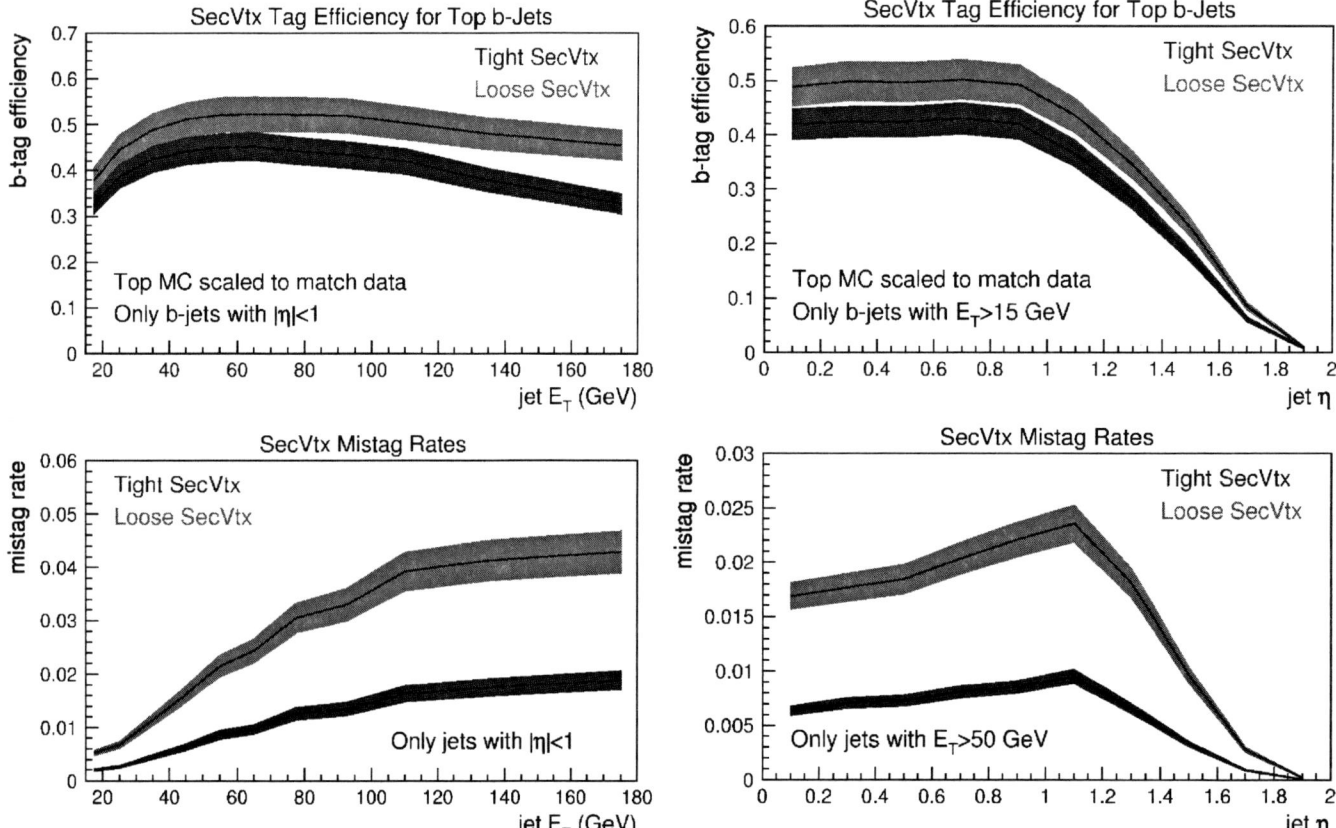

Fig. 54. *Top*: CDF efficiency to tag jets in $t\bar{t}$ Monte Carlo (scaled to the data efficiency) with matched b quarks for the 'tight' and the 'loose' tune of the SecVtx algorithm as a function of the jet E_T (*left*) and the jet η (*right*). *Bottom*: Mistag rate for jets resulting from light quark fragmentation as a function of the jet E_T (*left*) and the jet η (*right*)

unchanged from Run-I [293], but the track selection cuts are retuned for the CDF II detector. The primary vertex is determined on event-by-event basis either as the vertex nearest the high-momentum lepton, or in the control data samples it is refitted from high quality, low impact parameter tracks associated with the vertex of highest total scalar sum of transverse track momentum. The transverse beam profile of $30\,\mu$m at $z = 0$, rising to ≈ 50–$60\,\mu$m at $|z| = 40$ cm is used as a constraint in the fit. The resulting transverse vertex position uncertainty ranges from 10–$32\,\mu$m depending on the number of reconstructed tracks and the topology of the event.

The secondary vertex tag, evaluated on a per-jet basis, considers tracks within the jet cone and requires a minimum track p_T, number of silicon hits on the tracks, quality of those hits, and a maximum χ^2/ndf for the final track fit to reject poorly reconstructed tracks. A jet is defined "taggable" if it has two good tracks. Displaced tracks in the jet are selected based on the significance of the impact parameter with respect to the primary vertex and are used as input to the SecVtx algorithm. SecVtx uses at least three tracks with $p_T > 0.5$ GeV/c and impact parameter significance $|d_0/\sigma_{d_0}| > 2.5$ to reconstruct a secondary vertex. If unsuccessful, it performs a second pass with tighter track requirements ($p_T > 1$ GeV and $|d_0/\sigma_{d_0}| > 3$) to reconstruct a two-track vertex. To reduce the background from false secondary vertices (mistags), a good secondary vertex is re-

quired to have a decay length significance $L_{2D}/\sigma_{L_{2D}} > 3$ (positive tag) or $L_{2D}/\sigma_{L_{2D}} < -3$ (negative tag), where the decay length $\sigma_{L_{2D}}$, the total estimated uncertainty on L_{2D} including the error on the primary vertex, is estimated to be typically $190\,\mu$m. The negative tags are useful for calculating the false positive tag rate.

Figure 54 (top) shows the efficiency to tag jets in top quark Monte Carlo samples which have been matched to b-quarks, using both the "tight" and "loose" tunes of the SecVtx tagger. The efficiency is obtained by multiplying the tag rate for such jets in the Monte Carlo by the data/MC scale factors of 0.909 ± 0.06 for the tight tagger and 0.927 ± 0.066 for the loose tagger. The bands represent the systematic error on the data/MC scale factors. The decrease in efficiency at high jet E_T is due to the declining yield of good silicon tracks passing the quality cuts. Figure 54 (bottom) shows the mistag rate for jets resulting from light quark fragmentation. These have been measured from inclusive jet data. Some of the CDF top quark physics analyses presented in this review use an older version of the SecVtx algorithm with slightly lower tagging efficiency and slightly higher mistag rates.

The *Jet Probability – JP* is an alternative lifetime b-tagger, used to determine whether a jet has been produced from the hadronisation process of a light or a heavy quark. The algorithm makes use of the information of tracks that are associated to a jet to determine the probability for this ensemble of tracks to be consistent with coming from the primary vertex of the interaction. In particular, the track impact parameters and their uncertainties are used. The impact parameter of a track is assigned with a positive or negative sign depending on the position of the track's point of closest approach to the primary vertex with respect to the jet axis. The probability distribution of a set of tracks originating from the primary vertex is by construction uniformly distributed between 0 and 1. For a jet coming from heavy flavour hadronisation, the distribution peaks at 0, due to tracks from long lived particles that have a large impact parameter with respect to the primary vertex. Particles in a jet coming from a light quark should originate from the primary vertex. Due to the finite tracking resolution, these tracks are reconstructed with a non-zero impact parameter and have equal probability to be either positive or negative signed. The width of the impact parameter distribution from these tracks is solely due to the tracking detector resolution and multiple scattering effects. The tracking resolution can be extracted from the data by fitting the negative side of the signed impact parameter distribution. The efficiency of the algorithm is measured in inclusive electron data and matching Monte Carlo samples using a double-tag method (equivalent to the procedure in *SecVtx Vertex b-tag*). The efficiency, averaged over the jet E_T, to tag a heavy flavour jet with $E_T > 10$ GeV is found to be 0.197 ± 0.012 for a jet probability cut of 0.01, where the uncertainty includes statistical and systematic errors. The resulting relative difference in the jet tagging efficiency between data and Monte Carlo (scale factor) is 0.787 ± 0.105 for the same jet probability cut. The scale factor for charm tagging is not determined, but assumed to be identical to that for *b*-tagging with an additional uncertainty of 20%.

The *CDF soft lepton tagging – SLT* algorithm relies on the identification of muons within jets originating from semileptonic *b*-decay. Muon identification at CDF proceeds by extrapolating tracks found in the central tracker, through the calorimeter to the muon chambers, and matching them to track segments reconstructed in the muon chambers. In order to retain sensitivity for muons embedded in jets, the muon SLT algorithm makes full usage of the muon-track matching information without any requirement on the calorimeter information, while the standard muon identification requires a muon candidate to be consistent with minimum ionising energy deposits in the calorimeter. High-quality tracks with impact parameter less than 3 mm, z-vertex origin within 60 cm from the centre of the detector and extrapolation within $3\sigma(p_T)$ in x-direction outside the muon chambers are considered as a possible muon-tag candidate. Furthermore, candidate muons are selected with the SLT algorithm by constructing a quantity L, based on a comparison of the measured track-muon matching variables (in x, z, and ϕ) with their expectations. To construct L, the sum Q, of the individual χ^2 variables

$$Q = \sum_{i=1}^{n} \frac{(x_i - \mu_i)^2}{\sigma_i^2}, \qquad (51)$$

is formed, where μ_i and σ_i are the expected mean and width of the distribution of the matching variable x_i. The sum is taken over n selected variables, described below. L is then constructed by normalising Q according to

$$L = \frac{(Q - n)}{\sqrt{\text{var}(Q)}}, \qquad (52)$$

where the variance var(Q) is calculated using the full covariance matrix for the selected variables. The normalisation is chosen to make L independent of the number of variables n. For sufficiently large n, the distribution of L tends to a Gaussian centred at zero with unitary width. The correlation coefficients between each pair of variables are measured from $J/\psi \to \mu\mu$ data. The selected variables are the full set of matching variables, x, z, ϕ. Depending on the detector region of the muon candidate, between two and five variables might be used as measured by the available muon chambers. All available matching variables are used in the calculation of L for a given muon candidate. By placing an appropriate cut on L, background is preferentially rejected because hadrons have broader matching distributions than muons since the track segments in the muon chambers from hadrons are generally a result of leakage of the hadronic shower. The width of the matching distributions depend on p_T due to multiple scattering and are measured from muon in J/ψ decays at low p_T and W and Z-boson decays at high p_T.

The SLT tagging algorithm is applied to jets with at least one "taggable" track. A taggable track is defined as any track, distinct from the primary lepton, passing the track quality requirements, with $p_T > 3$ GeV, within $\Delta R < 0.6$ of a jet axis and pointing to the muon chambers to within a 3σ multiple scattering window. The z-coordinate of the track at the origin must be within 5 cm of the event vertex. Jets are considered "SLT tagged" if they contain a taggable track, which is also attached to a track segment in the muon chambers and the resulting muon candidate has $|L| < 3.5$. Events are rejected if the primary lepton is of the opposite charge to a SLT muon tag and the invariant mass of the pair is less than 5 GeV/c^2 (J/ψ veto), or if the primary lepton is a muon that together with an oppositely-charged SLT muon tag forms an invariant mass between 8 and 11 GeV/c^2 (Υ veto) or between 70 and 110 GeV/c^2 (Z veto). Those vetoes reduce the $t\bar{t}$ acceptance by less than 1%. The SLT tagging efficiency is measured in J/ψ and Z data as a combination of the track reconstruction efficiency, the efficiency to reconstruct segments in the muon chambers and the muon identification efficiency. The resulting SLT tagging efficiency is parameterised as a function of track p_T and η, as shown in Fig. 55, and applied to muons in the $t\bar{t}$ Monte Carlo.

At present, DØ only employs lifetime tagging algorithms in top quark physics analyses. The soft lepton tagging algorithm has been worked out and is being studied in the context of top quark physics.

DØ uses a *secondary vertex tagging – SVT* algorithm to identify *b*-quark jets. Secondary vertices are reconstructed from two or more tracks satisfying the following requirements: $p_T > 1$ GeV, ≥ 1 hits in the SMT layers and impact

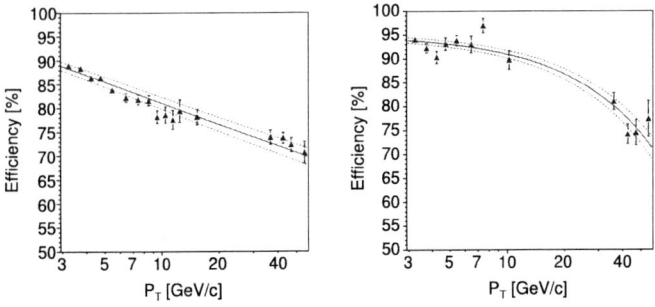

Fig. 55. The SLT efficiency for $|\eta_\mu| < 0.6$ (*left*) and $|\eta_\mu| \geq 0.6$ (*right*) as a function of p_T^μ, measured from J/ψ and Z data for $|L| < 3.5$. The *dotted lines* are the $\pm 1\sigma$ statistical uncertainty on the fit which is used in the evaluation of the systematic uncertainty

parameter significance $dca/\sigma_{dca} > 3.5$. Only those jets are considered taggable. The b-tagging efficiency is given with respect to taggable jets. Tracks identified as arising from

K_S^0 or Λ decays or from γ conversions are not considered. If the secondary vertex reconstructed within a jet has a decay length significance $L_{xy}/\sigma_{L_{xy}} > 7$, the jet is tagged as a b-quark jet. Events with exactly 1 (≥ 2) tagged jets are referred to as single-tag (double-tag) events.

Secondary vertices with $L_{xy}/\sigma_{L_{xy}} < -7$ appear due to finite resolution of their characteristics after reconstruction, and define the "negative tagging rate". The negative tagging rate is used to estimate the probability for misidentifying a light flavour (u, d, s quark or gluon) jet as a b-quark jets (the "mis-tagging rate").

Both the b-tagging efficiency and the mis-tagging rate are estimated using jets with ≥ 2 tracks satisfying less stringent requirements than those for SVT. In particular, the p_T cut is reduced from 1 GeV to 0.5 GeV for all but the highest p_T track, and no cut on dca/σ_{dca} of the track is made. These requirements have an efficiency per jet $> 80\%$ for $p_T > 30$ GeV and integrated over the rapidity y. The b-tagging efficiency is measured in a data sample of dijet events with enhanced heavy flavour content by re-

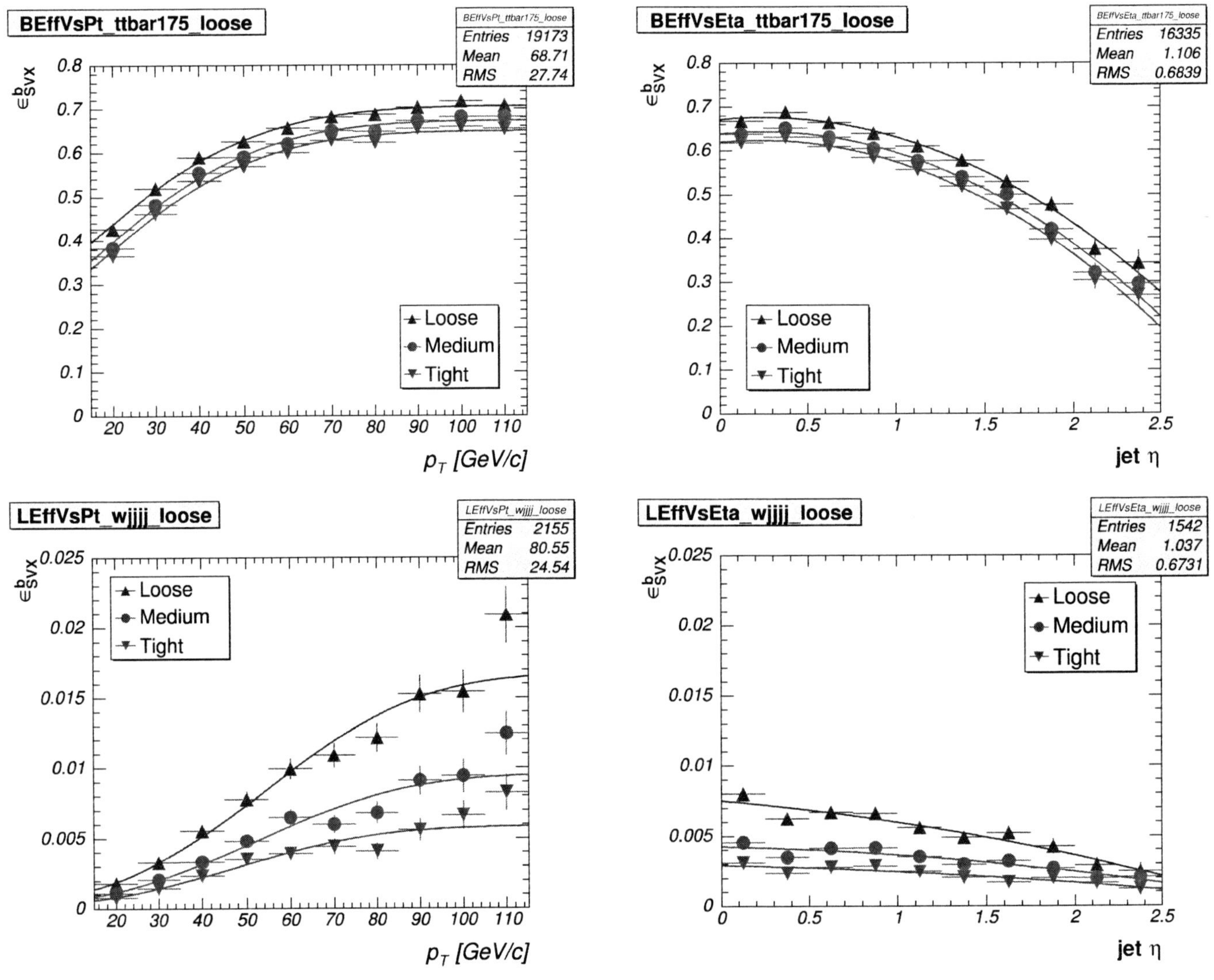

Fig. 56. DØ b-tagging efficiency in $t\bar{t}$ events (*top*) and the mis-tagging rate for $Wjjjj$ events (*bottom*) for taggable jets as a function of p_T (*left*) and η (*right*) for three different operating points of the SVT algorithm (loose, medium, and tight)

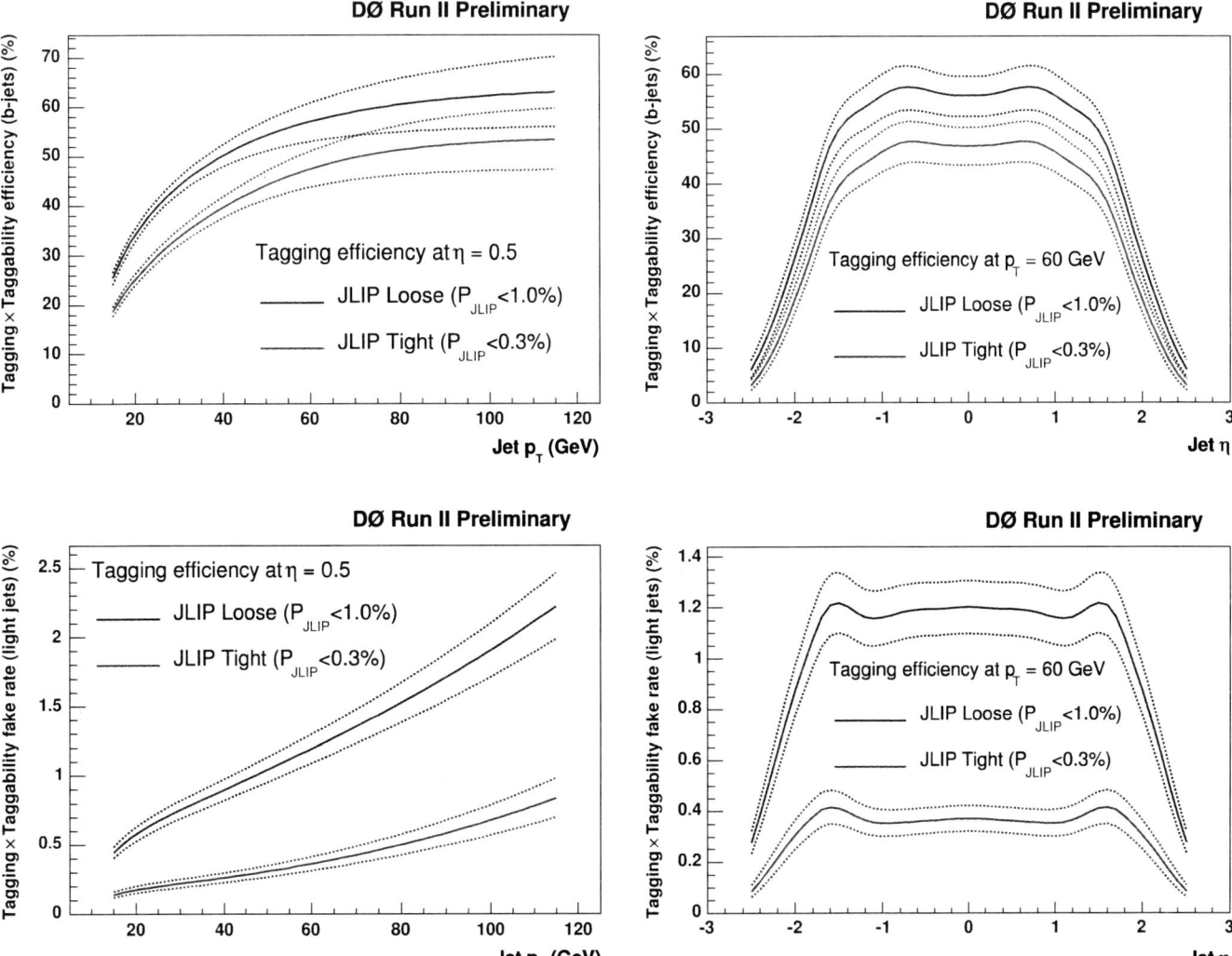

Fig. 57. DØ JLIP b-tagging efficiency in electron+jets data as a function of jet E_T (at $\eta = 0.5$) and jet η (at $E_T = 60$ GeV) for the Loose and Tight probability cuts. These *curves* include the taggability efficiency. The *first row* corresponds to the b-jets tagging efficiency and the *second row* to the light jet mistag rate. The *dashed curves* correspond to the total $\pm 1\sigma$ systematic *error band*

quiring a jet with an associated muon at high transverse momentum relative to the jet axis. By comparing the SVT and muon-tagged jet samples, the tagging efficiency for semileptonic b-quark decays ("semileptonic b-tagging efficiency") can be inferred. Using Monte Carlo (MC) simulation, the measured efficiency is corrected further to the tagging efficiency for inclusive b-quark decays. A similar procedure is used to estimate the c-tagging efficiency.

Figure 56 shows the b-tagging efficiency in $t\bar{t}$ events (top) and the mis-tagging rate for $Wjjjj$ events (bottom) for taggable jets as a function of p_T (left) and η (right) for three different operating points of the SVT algorithm (loose, medium, and tight). The DØ analyses on top quark physics use the tight SVT tagging algorithm.

The *DØ jet lifetime probability tagging algorithm – JLIP* [294] uses the signed impact parameter of tracks (representing the distance of closest approach of a track with respect to the primary vertex) within a jet to compute a probability for the jet to originate from the pri-

mary vertex; heavy quark jets are expected to have low values for the JLIP probability. Jets are tagged if their JLIP probability is smaller than a given cut. The probability distribution is expected to be flat for light quark jets and therefore the cut value gives approximately the mistag rate. In top quark analysis, two different cut values on the JLIP probability are used: Tight ($P_{\text{JLIP}} < 0.3\%$) and loose ($P_{\text{JLIP}} < 1.0\%$). Efficiencies (including taggability) for each probability cut are shown in Fig. 57.

4 Measurement of the $t\bar{t}$ production cross section at the tevatron

4.1 Introduction

The $p\bar{p} \to t\bar{t}$ production cross section is measured in all possible top quark decay modes, namely the di-lepton, the lepton + jets, and the all-jets channel (see Sect. 2). While generally 'lepton' here refers to electron and muon, also first analyses with identified τ-leptons in the final state are being performed. The cross section measurements can be categorised further according to their analysis technique, using topological or kinematic event characteristics for signal and background separation, exploiting multivariate techniques or being simple counting experiments, or using b-tagging on event- or on jet-basis. For the latter, several algorithms are available, looking for semileptonic decays of b-hadrons or reconstructing 3-dimensional secondary decay vertices or impact parameters. The counting experiments tend to be slightly less sensitive than the multivariate techniques. The latter, however, rely more on the assumption that $t\bar{t}$ production has only Standard Model contributions and on the modelling of these processes in Monte Carlo programs. The level of assumptions and systematic uncertainties varies between the different approaches, allowing to test the assumptions. In each channel, the $t\bar{t}$ production cross section is measured by

$$\sigma_{p\bar{p}\to t\bar{t}} = \frac{N - B}{\epsilon A \, B(t \to \dots) \int \mathcal{L}} , \qquad (53)$$

where N is the number of observed events, B is the estimated background contamination (based on data and Monte Carlo data), ϵ is the total selection efficiency for $t\bar{t}$ events assuming the Standard Model production mechanisms and assuming the measured world average of the top quark mass. A is the geometrical acceptance of the detector for $t\bar{t}$ events, $B(t \to \dots)$ the top decay branching ratio for the considered decay channel, and $\int \mathcal{L}$ the integrated luminosity of the data set.

CDF and DØ use the fixed order (LO) matrix element event generation PYTHIA or ALPGEN, and PYTHIA for the showering in the Monte Carlo event generation for the $t\bar{t}$ signal as well as for the vector boson+jets background processes. For systematic studies, CDF also uses HERWIG for the fragmentation step in the event simulation. In order to minimise the dependence on the Monte Carlo simulation, which can have rather large systematic uncertainties due to substantial normalisation and shape dependence of the LO factorisation scale, data is used wherever possible to quantify detector resolutions, reconstruction efficiencies or fake rates. Scale factors are determined from the comparison of data to Monte Carlo simulation for reference processes, of which the production of a Z-boson, possibly in association with jets, and a subsequent leptonic decay $Z \to \ell\ell$ is a very popular one as the lepton η and p_T spectra are similar to those in top quark events. In b-tagging analyses the kinematic distributions of the b-jets cannot be simulated in the LO Monte Carlo as precisely as for the inclusive jets due to the sensitivity to the factorisation scale (here the quark mass and the jet p_T set multiple scales). Nevertheless, the flavour decomposition (flavour fractions) of the jets in vector boson + jets events can be reliably calculated in ALPGEN and are therefore used in the b-tagging analyses. Background from QCD multijet production with fake identification of missing transverse energy, misidentified isolated high-p_T leptons or jets cannot be well

modelled in Monte Carlo and is more reliably estimated from data control samples.

Variations in m_{top} mainly change the lepton and jet p_T distributions and therefore the event selection efficiency. In multivariate techniques the m_{top} dependence can also change the overall cross section measurement through changes of kinematic reference distributions, or the level of correlations of kinematic variables. Those effects are studied and quantified separately, so that the cross section measurements can always be translated to the latest world average top quark mass value. In the case of multivariate analyses, N and B are determined from histogram shapes for signal and the background contributions, as obtained in Monte Carlo or control data samples, which are fit with floating normalisation to the respective distribution in data. In order to be able to combine $t\bar{t}$ cross section measurements easily, some measurements in different channels are designed to be orthogonal by vetoing events with the topology of the other channels. Those measurements can then be combined as truly independent measurements. Cross section combination of partially correlated measurements, for example in the lepton + jets channel using kinematic characteristics or using b-tagging, are more difficult due to the strong correlation of the results and are under study. The first combined results using this approach are available from CDF. The branching ratio calculations include all topologies and decays yielding the considered event final state. For example, the μ + jets channel also includes a $t\bar{t}$ contribution from $t\bar{t} \to \tau\nu_\tau + jets$ with $\tau \to \mu\nu\bar{\nu}$. In order to improve the sensitivity to the τ channel, for example in its one-prong hadronic decay mode, topologies with lepton + isolated track are also being performed.

The measurement of the $t\bar{t}$ production cross section is of high relevance for our understanding of top quark physics and serves several purposes:

– As described in Sect. 2.1, the Standard Model top quark pair production is a strong production mechanism, dominated at lowest order by the annihilation of gluons or quarks via an s- or t-channel gluon exchange. The comparison of the measured $t\bar{t}$ production cross section with the theory calculations is a test of multiscale QCD (m_{top} and jet p_T). The uncertainty of $\leq 15\%$ [113, 114, 116, 117] of those calculations, including higher order corrections and sophisticated resummation techniques, will soon be met or surpassed by the experimental measurements at the TEVATRON.

– The top quark is the most massive of nature's building blocks yet discovered. Because new physics associated with electroweak symmetry breaking will likely couple to an elementary particle in proportion to its mass, it is important to measure the top quark couplings as accurately as possible. In the strong interaction sector, the couplings are reflected in the $t\bar{t}$ production cross section in hadron collisions.

– A significant deviation from the Standard Model prediction would be an indication of new physics. Such an observation could be the result of additional production mechanisms, for example via an intermediate heavy resonance decaying into $t\bar{t}$ pairs [136–138], a Higgs boson decaying to $t\bar{t}$ [139], new top quark decay mechanisms, for example into supersymmetric particles [295, 296] or into a charged Higgs boson [297], a similar final state signature from a top-like particle [298–301], or non-Standard Model contributions in the background contribution. Those scenarios could lead to a measured $t\bar{t}$ production cross section apparently dependent on the $t\bar{t}$ final state. It is therefore necessary to precisely measure $\sigma_{t\bar{t}}$ in all decay channels and compare it with the Standard Model prediction.

– The $t\bar{t}$ production cross section measurements are the analyses, in which resolutions, detection efficiencies and background contamination from various sources are studied and quantified. They establish the top quark signal and provide the data samples and background estimates used in the top properties analyses. Experimentally, the $t\bar{t}$ cross section measurements are the basis of all top quark analyses.

In the following, the different $t\bar{t}$ cross section measurements by CDF and DØ are described and a summary is given at the end of this section.

4.2 CDF analyses

4.2.1 Dilepton channel analyses

Dilepton and lepton + track analysis. Using a data sample of $197 \pm 12\,\mathrm{pb}^{-1}$, CDF performs a measurement of the $t\bar{t}$ production cross section in Run-II using dilepton events with jets and missing transverse energy [302]. Two complementary analyses are carried out. In one analysis, both leptons are explicitly identified as either electron or muon (DIL analysis). In the other analysis, one lepton is explicitly identified as electron or muon, the other lepton is reconstructed as a high-p_T, isolated track (LTRK analysis). In the latter case, the lepton detection efficiency is significantly increased at the expense of a somewhat larger background expectation. Furthermore, the LTRK analysis also has an increased acceptance for single prong hadronic decays of the τ lepton from $W \to \tau\nu$, yielding 20% of its acceptance from taus, compared to 12% for the DIL analysis.

The $b\ell^+\nu_\ell\bar{b}\ell'^-\bar{\nu}_{\ell'}$ events are characterised by two high-p_T leptons, missing transverse energy ($\not{E}_T$) from the undetected neutrinos, and two jets from the hadronisation of b-quarks. CDF triggers on the dilepton events by requiring a central electron or muon with $E_T > 18\,\mathrm{GeV}$, or an end plug electron candidate with $E_T > 20\,\mathrm{GeV}$ in an event with $\not{E}_T > 15\,\mathrm{GeV}$. Offline two oppositely charged leptons with $E_T > 20\,\mathrm{GeV}$ are required. Both analyses require one lepton to satisfy tight selection criteria; the other lepton is identified as a 'loose' lepton. The DIL analysis requires the loose lepton to be an electron or muon as in the tight case but dropping the isolation requirement and the muon identification requirements are relaxed. The LTRK analysis defines a loose lepton as a well-measured, isolated track with $p_T > 20\,\mathrm{GeV/c}$ in the range $|\eta| < 1$ where the isolation requirement is now the tracking analogue of the calorimetric isolation employed for tight leptons. Candidate events must have $\not{E}_T > 25\,\mathrm{GeV}$. To reduce the occurrence of false

$\not{E}_T$ due to mismeasured jets or leptons, both analyses require in events with $\not{E}_T \leq 50\,\mathrm{GeV}$ that the $\not{E}_T$ vector points away from any jet and impose a minimum distance from the leptons. The LTRK analysis corrects the $\not{E}_T$ for all loose leptons whenever the associated calorimeter E_T is $< 70\%$ of the track p_T.

After removal of cosmic-ray muons and photon-conversion electrons, the dominant backgrounds to dilepton $t\bar{t}$ events are Drell–Yan ($q\bar{q} \to Z/\gamma^* \to \ell^+\ell^-$) production, "fake" leptons in $W \to \ell\nu + \mathrm{jets}$ events where a jet is falsely reconstructed as a charged lepton candidate, and diboson (WW, WZ, and ZZ) production. Drell–Yan events typically have small $\not{E}_T$. The LTRK analysis tightens the $\not{E}_T$ cut to $\not{E}_T > 40\,\mathrm{GeV}$ for events with dilepton invariant mass within $15\,\mathrm{GeV/c^2}$ of the Z-boson mass. The DIL analysis imposes a cut on the ratio of $\not{E}_T$ to the sum of the jet E_T's projected along the $\not{E}_T$ vector. The remaining Drell–Yan background is estimated from a comparison of a PYTHIA [223] simulation of that process and data. Selecting Drell–Yan candidates in the dilepton mass range 76–$106\,\mathrm{GeV/c^2}$ the number of events passing the nominal and Drell–Yan specific cuts is counted. After subtraction of expected non-Drell–Yan contributions, these two numbers provide the normalisation for the distribution of expected contributions inside and outside the Z-boson mass window. The fake lepton contribution is estimated by applying a fake lepton rate to a data sample of $W \to \ell\nu + jets$. This fake rate is determined from a multijet data sample, triggered by at least one jet with $E_T > 50\,\mathrm{GeV}$, where sources of real leptons such as W- or Z-decay are removed. The fake rate prediction is tested by applying the fake lepton rate to different samples of varying physics content sample. The fake rate is also tested on the like-sign dilepton sample, which is dominated by $W + \mathrm{jets}$ with one fake lepton. In all cases, good agreement is found, yielding confidence in this background estimate. The diboson backgrounds are modelled using PYTHIA [223] and ALPGEN + HERWIG Monte Carlo [224, 226, 227], normalised to the theoretical total cross section: $13.3\,\mathrm{pb}$ for WW, $4.0\,\mathrm{pb}$ for WZ, and $1.5\,\mathrm{pb}$ for ZZ [247]. The normalisation uncertainties are determined using different Monte Carlo calculations for the same process. The acceptance for the $t\bar{t}$ production is obtained from PYTHIA [223] Monte Carlo calculation assuming $m_{\mathrm{top}} = 175\,\mathrm{GeV/c^2}$. The CTEQ5L [94] parton distribution is used to model the momentum distribution of initial-state partons.

The predicted and observed numbers of oppositely charged dilepton events versus jet multiplicity are given in Table 9. The good agreement between data and background estimate in the background-dominated control region with $N_{\mathrm{jet}} = 0$ and $N_{\mathrm{jet}} = 1$ establishes confidence in the background estimate. The $t\bar{t}$ production cross section is measured in events with 2 or more jets. The DIL analysis additionally requires H_T, the scalar sum of the lepton p_T, jet E_T and $\not{E}_T$ to be $> 200\,\mathrm{GeV}$ in order to enhance the signal sensitivity further. Expected signal-to-background ratios of 3.1 for the DIL analysis and 1.7 for the LTRK analysis are found.

The systematic uncertainties, listed in Table 10, include uncertainties on the acceptance and efficiency for the sig-

Table 9. Expected background and $t\bar{t}$ contributions ($m_{\text{top}} = 175\,\text{GeV/c}^2$, $\sigma = 6.7\,\text{pb}$) compared to the observed number of events in data

	LTRK			DIL			
	$N_{\text{jet}} = 0$	$N_{\text{jet}} = 1$	$N_{\text{jet}} \geq 2$	$N_{\text{jet}} = 0$	$N_{\text{jet}} = 1$	$N_{\text{jet}} \geq 2$	$H_{\text{T}} > 200\,\text{GeV}$
Diboson	21.8 ± 5.2	6.3 ± 1.5	1.2 ± 0.3	11.4 ± 3.3	3.2 ± 0.9	1.1 ± 0.3	0.7 ± 0.2
Drell–Yan	26.5 ± 9.8	16.4 ± 6.0	4.2 ± 1.6	4.4 ± 1.9	2.9 ± 1.1	1.3 ± 0.5	0.9 ± 0.5
Fakes	16.5 ± 2.4	5.0 ± 1.0	1.5 ± 0.5	3.0 ± 1.2	2.4 ± 1.0	1.5 ± 0.6	1.1 ± 0.5
Total Bgd	64.8 ± 11.3	27.7 ± 6.3	6.9 ± 1.7	18.8 ± 4.0	8.5 ± 1.8	3.9 ± 0.9	2.7 ± 0.7
Expected $t\bar{t}$	0.3 ± 0.2	3.4 ± 0.6	11.5 ± 1.5	0.1 ± 0.0	1.3 ± 0.2	8.5 ± 1.2	8.2 ± 1.1
Total	65.1 ± 11.3	31.1 ± 6.3	18.4 ± 2.3	18.9 ± 4.0	9.8 ± 1.9	12.4 ± 1.6	10.9 ± 1.4
Observed	73	26	19	16	9	14	13

Table 10. Summary of systematic uncertainties

Signal and background uncertainties	LTRK	DIL
Lepton (track) ID	5% (6%)	5%
Jet energy scale – signal	6%	5%
Jet energy scale – background	10%	18%–29%
Initial/final state radiation	7%	2%
Parton distribution functions	6%	6%
Monte Carlo generators	5%	6%
WW, WZ, ZZ diboson estimate	20%	20%
Drell–Yan estimate	30%	51%
Fake estimate	12%	41%

nal and on the background normalisation. The dominant uncertainty for the signal is the jet energy scale. The background uncertainties are dominated by the limited statistics of high $\not{E}_{\text{T}}$ Drell–Yan events.

Using Table 9, $t\bar{t}$ production cross sections of $8.4^{+3.2}_{-2.7}{}^{+1.5}_{-1.1} \pm 0.5\,\text{pb}$ for the DIL analysis and $7.0^{+2.7}_{-2.3}{}^{+1.5}_{-1.3} \pm 0.4\,\text{pb}$ for the LTRK analysis are found, where the first two uncertainties are statistical and systematic and the third is due to the luminosity determination. These results are combined by dividing the analyses into three disjoint regions (DIL only, LTRK only, and the overlap). Taking the correlation of common systematics into account, the maximisation of a joint Poisson likelihood yields

$$\sigma_{t\bar{t}} = 7.0^{+2.4}_{-2.1}(\text{stat.})^{+1.6}_{-1.1}(\text{syst.}) \pm 0.4(\text{lum.})\,\text{pb}. \quad (54)$$

Various tests, requiring b-jet identification, changing the loose and tight lepton E_{T} and p_{T} cuts, or demand-ing two tight leptons all yield consistent results within uncertainties. Figure 58 shows the distribution of H_{T} for the LTRK analysis and its good agreement with the Standard Model, yielding a Kolmogorov–Smirnov probability of 75%. Also shown is the $\not{E}_{\text{T}}$ distribution of the thirteen events of the DIL analysis (ee: obs. 1, exp. 3.3 ± 0.5; $\mu\mu$: obs. 3, exp. 2.8 ± 0.5; $e\mu$: obs. 9, exp. 6.8 ± 0.8), yielding a Kolmogorov–Smirnov probability of 49%.

Global analysis of high-p_{T} dilepton sample. In a more global approach, using $184\,\text{pb}^{-1}$ of data, CDF measures simultaneously the cross section for $t\bar{t}$, WW and $Z \to \tau\tau$ [303] in a high-p_{T} dilepton sample. This method studies the shape of the 2-dimensional distributions in the $\not{E}_{\text{T}}$-N_{jet} plane for the main Standard Model contribution processes, namely $t\bar{t}$, WW, and $Z \to \tau\tau$, and fits them to the corresponding data distribution with floating normalisation. As all events are taken into account in the fit, this approach provides significant statistical gain over more traditional measurements, where strict selection criteria are imposed in order to reduce the background contamination in the sample.

The events are required to have two high-p_{T} oppositely charged leptons (electrons or muons) isolated from other activity in the event in terms of track and calorimeter isolation. Also cosmic-ray and conversion vetoes are applied. As Drell–Yan events, which present the dominant background in the ee and $\mu\mu$ channel, have no real missing transverse energy, in those two channels an additional cut on $\not{E}_{\text{T}}$ significance of $\not{E}_{\text{T}}^{\text{sig}} = \frac{\not{E}_{\text{T}}}{\sqrt{\sum E_{\text{T}}}} > 2.0\,\sqrt{\text{GeV}}$, where the scalar sum $\sum E_{\text{T}}$ runs over all (raw) calorimeter towers, and is corrected for the p_{T} of the muons. This cut is very effect-

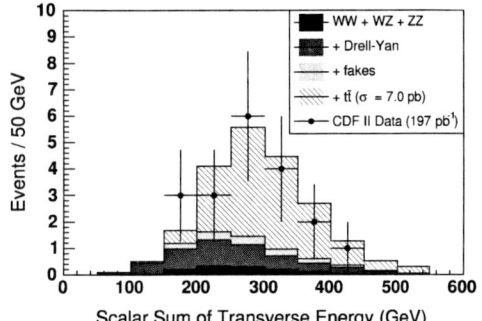

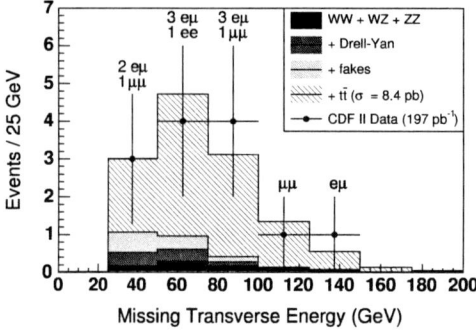

Fig. 58. *Left*: H_{T} distribution for events from the LTRK analysis with ≥ 2 jets. *Right*: $\not{E}_{\text{T}}$ for events from the DIL analysis with $H_{\text{T}} > 200\,\text{GeV}$ and ≥ 2 jets

Table 11. Summary of expected and observed numbers of events in $\sim 184\,\mathrm{pb}^{-1}$

	$e\mu$	ee	$\mu\mu$	$\ell\ell$
"Signal" processes				
$t\bar{t}$	5.4 ± 0.3	2.1 ± 0.1	2.3 ± 0.1	9.8 ± 0.5
WW	9.8 ± 0.6	3.8 ± 0.2	3.4 ± 0.2	16.9 ± 1.0
$Z\to\tau\tau$	32.1 ± 1.6	1.5 ± 0.1	1.3 ± 0.1	34.9 ± 1.7
"Background" processes				
$DY\to ee$	0.55 ± 0.26	10.4 ± 2.6	0.0	11.0 ± 2.7
$DY\to\mu\mu$	0.05 ± 0.01	0.0	5.3 ± 1.3	5.4 ± 1.3
WZ	0.38 ± 0.03	0.80 ± 0.04	0.66 ± 0.03	1.8 ± 0.1
ZZ	0.09 ± 0.01	0.65 ± 0.08	0.60 ± 0.07	1.3 ± 0.1
$W\gamma$	0.47 ± 0.10	0.27 ± 0.07	0.0	0.7 ± 0.1
$W+$ fake lepton	4.1 ± 2.8	0.40 ± 0.35	1.2 ± 1.3	5.7 ± 3.0
Total expected "Signal + Background" event count				
	53 ± 3	20 ± 3	15 ± 2	88 ± 5
Data				
	60	18	16	94

ive in reducing the Drell–Yan background while preserving most of the signal with real $\not{E}_{\mathrm{T}}$. It does, however, reduce $Z\to\tau\tau$ also. Therefore the $Z\to\tau\tau$ channel is only fit in the $e\mu$ channel, while its normalisation is fixed to the calculated Standard Model cross section for the fits in the ee and $\mu\mu$ channels. The standard CDF reconstruction and corrections are applied to $\not{E}_{\mathrm{T}}$ and jets.

The Standard Model processes are modelled in the following way: WW production is simulated using HERWIG [226, 227], $t\bar{t}$ is simulated using PYTHIA [223] and PYTHIA with TAUOLA [245] is used to simulate the $Z\to\tau\tau$ signal. Using those Monte Carlo simulations, the acceptances of the selection criteria including trigger efficiency and lepton ID scale factors are determined. The backgrounds considered are Drell–Yan ($Z/\gamma\to ee,\mu\mu$), $WZ, ZZ, b\bar{b}, W\gamma$, and $W+$ fake lepton, where the fake lepton is a misidentified jet. Since those contributions are much smaller than the considered signal processes, they are normalised to the expected Standard Model production cross sections. The rates for a jet to fake an electron or a muon ('fake rates'), are determined in data and applied to the $W+$ jets data sample. All the backgrounds are determined from Monte Carlo. Table 11 shows the expected event yield for signal and background processes and the observed data. The Monte Carlo processes are normalised to $184\,\mathrm{pb}^{-1}$, with slight variations between the channels. The errors include statistical and systematic uncertainties on the acceptances.

Figure 59 shows the data and all the expected Standard Model contributions discussed previously, in the $\not{E}_{\mathrm{T}}$-N_{jet} plane in the example of the $e\mu$ channel. The $t\bar{t}$, WW, and $Z\to\tau\tau$ distributions are normalised to unity for the fit. All other distributions are added together and normalised to the theoretical Standard Model cross section. For the ee and $\mu\mu$ channels, also the normalisation of the $Z\to\tau\tau$ contribution is fixed to its Standard Model cross section and not included in the fit due to the reduced event rate resulting from the $\not{E}_{\mathrm{T}}^{\mathrm{sig}}$ cut. As can be seen from Fig. 59, the distributions of the three signal samples fall into distinc-

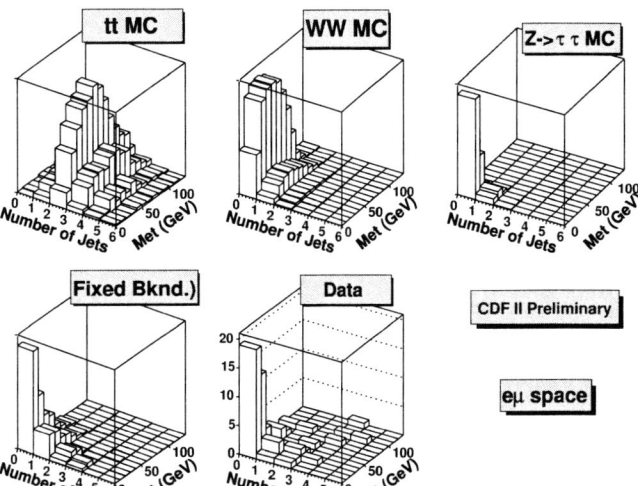

Fig. 59. The 2-dimensional distributions of the Standard Model "signal" sources, "background" sources (summed together) and from $184\,\mathrm{pb}^{-1}$ of data in the $\not{E}_{\mathrm{T}}$-N_{jet} plane for the example of the $e\mu$ channel

tively different regions of the $\not{E}_{\mathrm{T}}$-N_{jet} plane, allowing the effective extraction of the cross sections.

In each bin i in the $\not{E}_{\mathrm{T}}$-N_{jet} plane, the data is compared to the Standard Model contributions. The corresponding Poisson probability is formed in each bin $\rho_i = \frac{\mu_i^{n_i} e^{-\mu_i}}{n_i!}$, where n_i is the number of data events in bin i, and μ_i is the total expected number of events given by:

$$\mu_i = \alpha N_{t\bar{t},i} + \beta N_{WW,i} + \gamma N_{Z\to\tau\tau,i} + n_{\mathrm{other},i}\,. \quad (55)$$

The overall likelihood, which is maximised with respect to event numbers α, β, and γ for the signal processes with the normalised distributions $N_{t\bar{t},i}$, $N_{WW,i}$, and $N_{Z\to\tau\tau,i}$, is

$$\mathcal{L} = \prod_i \rho_i\,. \quad (56)$$

Table 12. Summary of systematic uncertainties on the acceptance for each 'signal' process. The term 'acceptance' here also includes the selection efficiencies

	$t\bar{t}$	WW	$Z \to \tau\tau$
Trigger efficiency	1%	1%	1%
Lepton ID	2%	2%	2%
Track isolation	4%	4%	4%
$\not{E}_{\mathrm{T}}^{\mathrm{sig}}$ (ee and $\mu\mu$ only)	3%	3%	–
Generator syst.	3%	4%	2%
Total	6.2%	2.8%	5.0%
Luminosity		6%	

Table 13. Summary of systematic uncertainties on the fitted cross sections from the $\not{E}_{\mathrm{T}}$-N_{jet} shapes

		$t\bar{t}$	WW	$Z \to \tau\tau$
Jet Energy Scale & $\not{E}_{\mathrm{T}}$	$e\mu$	13%	7.5%	3.5%
	$ee + \mu\mu$	12%	13%	–
Jet Multiplicity	$e\mu$	8%	2%	3%
	$ee + \mu\mu$	9%	8%	–
Generator	$e\mu$	5%	2%	4%
	$ee + \mu\mu$	5%	3%	–
PDF's	$e\mu$	1%	1%	1%
	$ee + \mu\mu$	1%	1%	–
Total	$e\mu$	16%	8%	6%
	$ee + \mu\mu$	16%	15%	

where i runs over all bins in the 2-dimensional $\not{E}_{\mathrm{T}}$-N_{jet} plane. Systematic uncertainties in all the acceptances and the luminosity are taken into account by multiplying the likelihood function in (56) by Gaussian constraint terms of the form

$$G_f = e^{-\frac{\left(A_f - \hat{A}_f\right)^2}{2\sigma_{A_f}^2}} \,, \qquad (57)$$

where f refers to a given acceptance or luminosity for each source, $\hat{A}_f$ is its expected value, σ_{A_f} is its uncertainty, and A_f is its value in the fit which is allowed to float. Consequently, in the fit α, β, and γ are actually of the form

$$\alpha_f = \sigma A_f \mathcal{L}_f \,, \qquad (58)$$

with the acceptances and luminosity now 'free', but with Gaussian constrained parameters in the fit. For the case of the $Z \to \tau\tau$ process, a correction factor is calculated in Monte Carlo, which relates the measured number of candidate events after cuts to the generated number of events with the true di-tau mass in the range $66 < M_{\tau\tau} < 116 \,\mathrm{GeV/c}^2$. The latter constraint yields approximately the $p\bar{p} \to Z \to \tau\tau$ cross section as opposed to the $Z/\gamma^* \to \tau\tau$ cross section. Table 12 summarises the systematic uncertainties on the acceptance for each signal process.

Another important effect of systematic uncertainties is the possible change in shape of the 2-dimensional Monte Carlo distributions. These effects include jet energy scale, jet multiplicities (ISR/FSR), modelling of $\not{E}_{\mathrm{T}}$ and $\not{E}_{\mathrm{T}}^{\mathrm{sig}}$, Monte Carlo generators, and parton distribution functions. Using pseudo-experiments, the expected effect of the modified shapes on the fitted cross section are determined. Table 13 summarises the resulting systematic uncertainties on the fitted cross sections.

Fitting the data in the $\not{E}_{\mathrm{T}}$-N_{jet} plane to linear combinations of the signal and background processes according to the procedure described above, where all but the process to be measured are normalised to their Standard Model cross section within Gaussian constraints, yields a $t\bar{t}$ production cross section of

$$\sigma_{t\bar{t}} = 8.6^{+2.5}_{-2.4} \,(\text{stat.} + \text{accept.}) \pm 1.1 \,(\text{shape syst.}) \,\mathrm{pb} \,. \qquad (59)$$

Similarly, the other cross section results are: $\sigma(WW) = 12.6^{+2.5}_{-3.0} \pm 1.3 \,\mathrm{pb}$ and $\sigma(Z \to \tau\tau) = 233^{+45}_{-42} \pm 17 \,\mathrm{pb}$. Various

tests have been performed on the fit, for example only fitting the $e\mu$ channel, or leaving the normalisation of all three signal processes floating in the fit. The results are all consistent with each other, while the chosen procedure for the extraction of the central cross section values yields the best precision on the statistical and acceptance errors.

4.2.2 Lepton + jets channel in kinematic analysis

In the lepton + jets channel, using $194 \,\mathrm{pb}^{-1}$ of data, CDF measures the $t\bar{t}$ production cross section using an artificial neural network technique to discriminate between top pair production and background processes [304]. This technique exploits the difference in event kinematics and topology between signal and background, using seven kinematical variables. As a cross check the analysis is also performed only using H_{T}, the scalar sum of the lepton p_{T}, $\not{E}_{\mathrm{T}}$, and the sum of the jet E_{T}'s. No b-tagging information is used. Therefore this analysis is complementary to the b-tagging analyses and exhibits different systematic uncertainties. The combination of the result from the neural network technique and the b-tagging analyses, described later in this section, significantly reduces the experimental uncertainty.

The events in the lepton + jets channel $p\bar{p} \to t\bar{t} \to W^+ W^- b\bar{b} \to \ell\bar{\nu}_\ell q\bar{q}' b\bar{b}$ are characterised by the presence of an isolated, high-p_{T} lepton (here only referring to electron or muon), large $\not{E}_{\mathrm{T}}$ from the neutrino, and four or more jets, out of which two are b-jets. Due to a minimum E_{T} requirement of 15 GeV on the jets within $|\eta| < 2.0$ and detector resolutions or jet identification inefficiencies towards the lower jet E_{T}, this analysis requires at least three jets, recovering some acceptance.

The event selection efficiency is determined using $t\bar{t}$ Monte Carlo events, generated with the PYTHIA [223] program, using the CTEQ 5L [94] parton distribution functions. These raw efficiencies are corrected for several effects, not sufficiently well-modelled in the simulation: the lepton trigger efficiency, measured from data; the fraction of the $p\bar{p}$ luminous region well-contained in the CDF detector, measured from data; the difference between the track reconstruction efficiency measured in data and simulation;

Table 14. The observed number of $W \to \ell\nu$ candidates in different jet multiplicity regions, compared to the expectation from PYTHIA$t\bar{t}$ Monte Carlo

Jet multiplicity	Electron	Muon	Total	Expected $t\bar{t}$
3	254	147	401	42.3
≥ 4	78	40	118	49.9

and the difference between lepton identification efficiencies measured in $Z \to \ell\ell$ data and PYTHIA Monte Carlo. The total acceptance of the event selection for $t\bar{t}$ is 7.11 $\pm$ 0.56%, mostly (relative 82%) coming from the lepton + jets channel (lepton = e, μ). This number also includes additional acceptance in the τ + lepton mode (7%), the τ + jets mode (6%), and the dilepton mode (5%) is included. The observed number of events in data and the expected number of $t\bar{t}$ events in different jet multiplicity regions is summarised in Table 14.

A variety of non-$t\bar{t}$ processes can also produce events that pass the $W+ \geq 3$ jets selection. These backgrounds can be grouped into three categories: production of a W-boson with associated jets, W + jets; other electroweak processes resulting in at least one high p_T lepton and jets; and generic QCD multijets processes. Since theoretical predictions for their total rate only exist at leading order and are associated with 50% uncertainty from the scale dependence, their contributions are estimated from the data itself. Only the shapes of the kinematic distributions from the Monte Carlo samples are used. The $W+ \geq n$ jets background is modelled using the $W + n$ parton ALPGEN + HERWIG Monte Carlo [224, 226, 227], where the larger jet multiplicities are modelled by the gluon radiation in the parton shower algorithm (HERWIG). A factorisation scale of $Q^2 = M_W^2 + \sum_i p_{\mathrm{T},i}^2$ is chosen for the parton distribution functions and for the evaluation of α_s, where $p_{\mathrm{T},i}$ is the transverse momentum of the i-th parton. ALPGEN + HERWIG is also used to model Z-boson and diboson (WW, WZ, ZZ) production with associated jets. PYTHIA [223] is used to simulate single top production. The sum of all electroweak background processes is labelled $W+$-jets. To estimate the rate of multi-jet background passing the selection requirements, assuming there is no correlation between the $\not{E}_T$ and the

isolation of the identified lepton, three control regions are compared:

- n_A: lepton isolation $I > 0.2$ and $\not{E}_T < 10$ GeV
- n_B: lepton isolation $I < 0.1$ and $\not{E}_T < 10$ GeV
- n_C: lepton isolation $I > 0.2$ and $\not{E}_T > 20$ GeV.

Correcting those numbers for their expected contamination from $W+$-jets and $t\bar{t}$ events, the multijet contamination of the signal region, defined by $\not{E}_T > 20$ GeV and lepton isolation $I < 0.1$, is estimated as $n_C \times n_B/n_A$, resulting in a multijet contamination of $\approx 3\%$ for the muons and 4%–8% for the electrons. The majority of the QCD multijet background in the electron sample comes from unidentified photon conversions, therefore increasing with jet multiplicity.

This analysis exploits the discrimination available from kinematic and topological properties to distinguish $t\bar{t}$ from background processes. Due to the large mass of the top quark, top pair production is associated with central, spherical events with large total E_T, unlike most of the background processes. Studying a large number of kinematic and topological variables, a combination of seven variables, summarized in Table 15, was found to give the best cross section precision. As a cross check, the analysis is also performed with only one single discriminant. For that purpose the total transverse energy in the event, H_T is chosen, since it is both one of the observables that provides good discrimination between events containing top decays and events from background processes, and since it has been commonly used in other top pair production cross section analyses. The sum of the jet transverse energies or the transverse energy of the third most energetic jet have similar statistical power. From a fit to the H_T distribution in the $W+ \geq 3$ jets sample, a statistical uncertainty in the range 19%–29% for 1σ is expected, in the $W+ \geq 4$ jets sample it is 25%–48%. The lower sensitivity is both due to lower statistics – 45% of the $t\bar{t}$ events that fail the 4th jet requirement – and reduced discrimination power – the increased jet activity means that the $W+ \geq 4$ jets events have larger H_T and are therefore more similar to top pair production.

The seven kinematic variables are combined using a feed-forward artificial neural network (ANN) with seven hidden nodes in a single hidden layer and one output node. The network is trained with 4000 PYTHIA $t\bar{t}$ and 4000

Table 15. The definition for the kinematical and topological properties considered in this analysis where Q_i are the eigenvalues of the normalised momentum tensor of the event, defined as $\frac{\sum_j p_j^a \, p_j^b}{\sum_j p_j^2}$ where the a, b indices run over the three spatial dimensions and the summation is taken over the five highest E_T jets, the lepton, and the missing transverse energy $\not{E}_T$

H_T	Scalar sum of transverse energies of jets, lepton and $\not{E}_T$
Aplanarity	$3/2 Q_1$ (Q_i are the eigenvalues of the normalised momentum tensor)
$\sum p_z / \sum E_T$	Ratio of total jet longitudinal momenta to total jet transverse energy
$\min(M_{jj})$	Minimum di-jet invariant mass of three highest E_T jets
$\eta_{\max}$	Maximum η of the three highest E_T jets
$\sum_{i=3}^n E_{T.i}$	Sum E_T of third highest E_T jet and any lower E_T jets
$\min(\Delta R_{jj})$	Minimum di-jet separation in η and ϕ for three highest E_T jets

$W + 3$ parton ALPGEN + HERWIG Monte Carlo events that pass the selection requirements, using back propagation as implemented in JETNET [305]. The comparisons of the data and Monte Carlo distributions in the seven kinematic input variables in the first, second, and third jet multiplicity bin yield good agreement. The resulting ANN output distributions for $t\bar{t}$, $W+$ jets and QCD multijet events are combined with floating normalisation and fit to the ANN distribution of the data using a binned maximum likelihood-fit:

$$L(\mu_{t\bar{t}}, \mu_w, \mu_q) = \prod_{i=1}^{N_{\text{bins}}} \frac{e^{-\mu_i} \mu_i^{d_i}}{d_i!}, \qquad (60)$$

where $\mu_{t\bar{t}}$, μ_w, μ_q are the parameters of the fit, representing Poisson means for the number of $t\bar{t}$, W-like, and multijet events in the data sample. The expected number of events in the i-th bin is $\mu_i = \mu_{t\bar{t}} P_{t\bar{t},i} + \mu_w P_{w,i} + \mu_q P_{q,i}$, where $P_{t\bar{t},i}$, $P_{w,i}$, $P_{q,i}$ is the probability for observing an event in the i-th bin from $t\bar{t}$, W-like and multijet processes respectively. The number of multijet background events, μ_q, is fixed to its expectation, while

its uncertainty is included in the systematic uncertainties. The fitted number of $t\bar{t}$ events is converted into the top pair production cross section, $\sigma_{t\bar{t}}$, using the acceptance estimate, $\epsilon_{t\bar{t}}$, including the branching ratio for $W \to \ell\nu$, and the luminosity measurement $\mathcal{L}$: $\sigma_{t\bar{t}} = \frac{\mu_{t\bar{t}}}{\epsilon_{t\bar{t}}\mathcal{L}}$. Figure 60 shows the H_T and the ANN distributions in the $W + \geq 3$ jets sample which serve as input to the cross section fit.

The top pair production cross section measurement is sensitive to systematic effects which have an impact on the signal acceptance, on the shape of various kinematic distributions, and the luminosity. Table 16 summarises the contributions from several sources of systematic uncertainties.

The jet energy scale and the choice of factorisation scale in the LO modelling of the $W+$jets background events are the dominant uncertainties, where the H_T fit is more sensitive to both sources while the ANN output fit is more stable.

Performing the likelihood fit on the ANN output distribution of the 519 selected events yields a $t\bar{t}$ production cross section of

$$\sigma_{t\bar{t}} = 6.6 \pm 1.1 \,(\text{stat.}) \pm 1.5 \,(\text{syst.}) \,\text{pb}, \qquad (61)$$

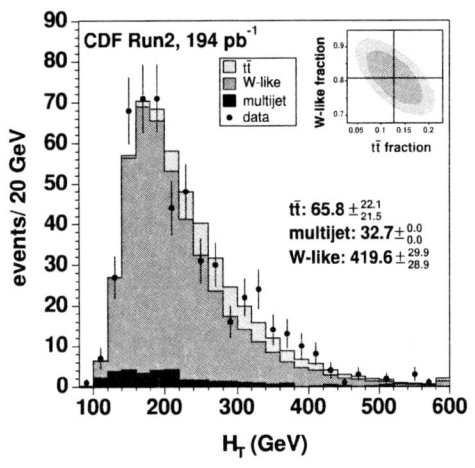
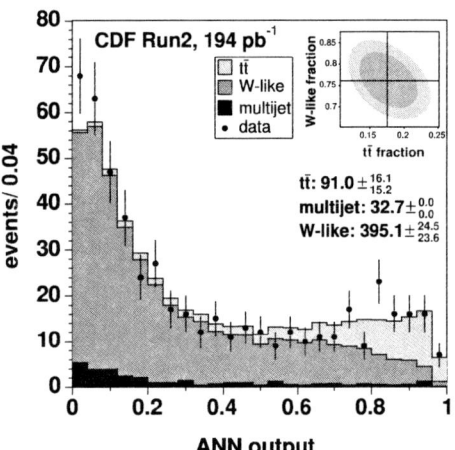

Fig. 60. Distribution of the observed H_T (*left*) and ANN output (*right*) in the $W + \geq 3$ jets sample compared with the result of the fit. The *insets* in both plots show the 1- and 2-standard deviation contours of the free parameters in the fit, normalised to the total number of events

Table 16. Systematic uncertainties of the acceptance and shape in % on the cross section, for fits to the ANN output (H_T) distribution in the $W+ \geq 3$ jets sample

Effect	Acc. (%)		Shape (%)		Total(%)	
Jet E_T scale	4.7	(4.7)	12.2	(21.4)	16.9	(26.1)
$W+$jets Q^2 scale	–	(–)	10.2	(24.6)	10.2	(24.6)
QCD fraction	–	(–)	0.6	(2.4)	0.6	(2.4)
QCD shape	–	(–)	1.1	(4.5)	1.1	(4.5)
Other EWK	–	(–)	2.0	(1.8)	2.0	(1.8)
$t\bar{t}$ PDF	1.5	(1.5)	2.9	(2.2)	4.4	(4.7)
$t\bar{t}$ ISR	2.1	(2.1)	1.9	(1.1)	3.0	(2.9)
$t\bar{t}$ FSR	1.7	(1.7)	1.0	(1.5)	2.7	(3.7)
$t\bar{t}$ generator	1.4	(1.4)	0.3	(1.0)	1.7	(2.4)
Lepton ID/trigger	2.0	(2.0)	–	(–)	2.0	(2.0)
Lepton isolation	5.0	(5.0)	–	(–)	5.0	(5.0)
Luminosity	–	(–)	–	(–)	5.9	(5.9)
Total					22.3	(37.8)

while the fit to the H_T distribution yields $\sigma_{t\bar{t}} = 4.8 \pm 1.6 \pm 1.8$ pb. The probability to find a difference equal to or larger than the observed difference between the two results is estimated to be 10%. Both results are consistent with the Standard Model expectation of $6.7^{+0.7}_{-0.9}$ pb. Their precision is similar to that of b-tagging analyses, where the artificial neural network technique reduces the expected statistical uncertainty by 30% and the estimated systematic uncertainty by 40% compared to only fitting the H_T distribution. Fitting only the $W + \geq 4$ jets sample yields consistent results. Also the comparison of the ANN output distribution in b-tagged or b-tag vetoed subsamples yields a good description of the data by the Monte Carlo distributions in shape, providing confidence in the measured $t\bar{t}$ production cross section.

A preliminary update of this analysis with 347 pb^{-1} of data yields an improved $t\bar{t}$ production cross section measurement of [306]:

$$\sigma_{t\bar{t}} = 6.0 \pm 0.8 \,(\text{stat.}) \pm 1.0 \,(\text{syst.}) \,\text{pb} . \qquad (62)$$

4.2.3 Lepton + jets channels in b-tag analyses

SecVtx vertex b-tag. Using 162 pb^{-1} of data, CDF performs a measurement of the $t\bar{t}$ production cross section in the lepton + jets channel, identifying heavy flavour quarks from top quark decays with a secondary vertex tagging algorithm [307]. Background contributions from fake W's, misidentified secondary vertices and heavy flavour production processes such as $Wb\bar{b}$ are estimated using a combination of Monte Carlo calculations and independent measurements in control data samples. An excess in the number of events which contain a lepton, missing transverse energy $\not{E}_T$, and three or more jets with at least one b-tag is the signal of $t\bar{t}$ production and is used to measure the production cross section $\sigma_{t\bar{t}}$, while the $W + 1$ jet and $W + 2$ jet bins, where the $t\bar{t}$ contribution is negligible, serve as checks of the background prediction.

This analysis has small acceptance for $t\bar{t}$ events with $W \to \tau\nu$ and subsequent leptonic τ decays, or with high-momentum semileptonic b-quark decays. These are included in the signal acceptance. Z-bosons and top to dilepton decays are removed by vetoing on the presence of a second lepton. Furthermore, Z-boson events are removed by eliminating events with one lepton and certain second objects which form an invariant mass between 76 and 106 GeV/c^2. Finally, jets are reconstructed as cone jets of size $\Delta R = 0.5$, with $E_T \geq 15$ GeV and $|\eta| \leq 2.0$. Cosmic ray and photon conversion vetoes are also applied. The overall acceptance $\times$ efficiency of this selection for $t\bar{t}$ events in the lepton + jets channel with three or more jets, including the leptonic branching ratios, is roughly 4% for the electron channel, and 1%–2% for the muon channel. As a final optimisation step, the selection incorporates an additional cut on the total transverse energy H_T of all objects in the event. Using the total sensitivity $(S/\sqrt{S + B + \sigma(B)^2})$ as figure of merit, where S is the signal expectation, B is the total background estimate, and $\sigma(B)$ is the absolute systematic error on the background estimate, a cut requiring $H_T > 200$ GeV is found to be optimal.

The understanding of acceptances, efficiencies, and backgrounds relies on detailed simulations of physics processes and the detector response. Most measurements of acceptances and efficiency rely on PYTHIA v6.2 [223] or HERWIG v6.4 [226, 227], both being leading order matrix element generators for the hard parton scattering, followed by parton shower simulation for the gluon radiation. For heavy flavour jets, the Monte Carlo program QQ v9.1 [244] is used to provide proper modelling of b and c hadron decays. The estimate of the b-tagging backgrounds due to higher-order QCD processes such as $Wb\bar{b}$ requires special care. This study of backgrounds in the b-tagged samples uses the ALPGEN program [224], which generates high multiplicity partonic final states using exact leading-order matrix elements. The parton level events are then passed to HERWIG and QQ for parton showering and b and c hadron decay.

Most of the non-$t\bar{t}$ processes found in the W + jets sample do not contain heavy quarks in the final state. Requiring that one or more of the jets in the event be tagged by the secondary vertex tagger ("SecVtx") keeps more than half of the $t\bar{t}$ events while removing approximately 95% of the background. Details on the SecVtx algorithm are given in Sect. 3.3.7.

This analysis requires a knowledge of the tagging efficiency for $t\bar{t}$ events, i.e. how often at least one of the jets in a $t\bar{t}$ event is positively tagged by the SecVtx. Because it is not possible to measure this directly in $t\bar{t}$ events a per-jet tagging efficiency is derived in a sample of jets whose heavy flavour fraction can be measured. Similarly, a matching sample of Monte Carlo jets is used to determine the tagging efficiency in the simulation for jets like those in the calibration sample. The ratio of efficiencies between data and simulation (scale factor) is then used to correct the tagging efficiency in $t\bar{t}$ Monte Carlo samples. So the geometrical acceptance and energy dependence of the tagger are taken from the simulation, with the overall normalisation determined from data. Studies of various control samples are performed. In one study, analysing low p_T inclusive electron data which is enriched in semileptonic decays of bottom and charm hadrons, with a double-tag technique, the away (electron) jet is required to be tagged by SecVtx, so that the other jets can be studied. In another study the single-tag rate of electron jets is used yielding consistent results. Combining all systematic and statistical errors a data to Monte Carlo tagging efficiency scale factor of 0.82 ± 0.06 is obtained.

A 'mistag' is defined to be a jet which does not result from the fragmentation of a heavy quark, yet has a SecVtx secondary vertex. Mistags are caused mostly by random overlap of tracks which are displaced from the primary vertex due to tracking errors, although there are contributions from K_S and Λ decays and nuclear interactions with the detector material as well. Contributions from these effects are measured directly from jet data samples without relying on the detector simulation. Because the SecVtx algorithm is symmetric in its treatment of the impact parameter d_0 and the decay length L_{2D} significance, the tracking-related mistags should occur at the same rate for $L_{2D} > 0$ and $L_{2D} < 0$. Therefore, a good estimate of the positive mistag rate due to resolution effects can be obtained from the negative tag rate. The sum of small corrections for a remaining asymmetry for positive and negative tags and the presence of particles with real lifetime and material interactions in the positive tag region are found to yield a correction factor of 1.2 ± 0.1. The rate of negative tags is measured in an inclusive sample of jet triggers and parameterised as a function of four jet variables – E_T, track multiplicity, η, and ϕ – and one event variable $\sum E_T$, the summed scalar E_T of all jets in the event with $E_T > 10$ GeV and $|\eta| < 2.4$. These parameterised rates are used to obtain the probability that a given jet will be negatively tagged. Assuming that the various contributions to systematic uncertainties are uncorrelated, they are added in quadrature to find a total systematic uncertainty of 8% on the negative tag rates, which combined with the uncertainty on the correction factor 1.2 ± 0.1 yields a total mistag rate relative uncertainty of 11%.

Heavy flavour production in association with a vector boson (e.g. $Wb\bar{b}$, $Wc\bar{c}$, Wc) contributes significantly to the non-$t\bar{t}$ background in the b-tagged lepton + jets sample. The relative fraction of W + heavy flavour production is well-defined and calculated in a matrix element Monte Carlo program (ALPGEN, [224]), while the normalisation of the W + jets production has large theoretical uncertainties and is therefore measured with collider data. The two results are combined to estimate the W + heavy flavour background. All heavy quark masses, spins and colour flows are treated properly inside ALPGEN [224]. Heavy flavour fractions calculated using ALPGEN are calibrated against fractions measured from jet data. The total W + heavy flavour contribution is estimated by multiplying the number of pretag W + jets events in data by the calculated W + heavy flavour fraction and the tagging efficiency in Monte Carlo (including the SecVtx efficiency scale factor between data and Monte Carlo). Because the event tagging efficiency depends on the number of heavy flavour jets in the fiducial region $|\eta| < 2.4$, the results are calculated separately for the case of 1 and 2 heavy flavour jets. The heavy flavour fractions for W + jets events, computed using an ALPGEN + HERWIG Monte Carlo sample, are defined to be the ratio of the observed W + heavy flavour and W + jets cross section. Inherent systematic uncertainties from the parton level selection and jet-parton matching prescription are found to be $\approx 21\%$. The heavy flavour fractions from ALPGEN are verified using an inclusive jet sample, without identified W-boson. This sample is a large related class of events whose production processes are described by Feynman diagrams similar to those in W + jets events. In particular, gluon splitting to heavy quark pairs accounts for part of the heavy flavour production in both samples. The contributions to the jet data sample from heavy and light partons are determined by fitting the pseudo-$c\tau$-distribution for tagged jets, thereby discriminating between jets from b, c, and light partons or gluons on a statistical basis. Pseudo-$c\tau$ is defined as $L_{2D} \times M_{vtx}/p_T^{vtx}$, where M_{vtx} is the invariant mass of all tracks in the secondary vertex and p_T^{vtx} is the transverse momentum of the secondary vertex four-vector. Measured heavy flavour fractions from the data are

consistently $50\% \pm 40\%$ higher than the ALPGEN prediction, for both b and c jets, where the uncertainty is dominated by the systematic uncertainties associated with the ALPGEN heavy flavour calculation. More detailed studies indicate that the gluon splitting contribution relative to other production mechanisms is well-modelled. Therefore the hypothesis of missing or underrepresented heavy flavour production diagrams is disfavoured. In fact, the measured ratio of 1.5 ± 0.4 between the heavy flavour fraction in the ALPGEN/HERWIG samples and the data is not inconsistent with other recent studies, which indicates that a K-factor might be necessary to account for higher-order effects [222]. Based on this calibration with the jet data sample, the expected $Wb\bar{b}$ and $Wc\bar{c}$ background contributions derived by ALPGEN are scaled by a factor 1.5 ± 0.4. Since the Wc background is produced through a different diagram, that contribution is not rescaled.

The non-$t\bar{t}$ events in the b-tagged W + jets sample are from direct QCD production of heavy flavour without an associated W-boson, mistags of light quarks in jets in the W + jets events, W + heavy flavour, and other low rate electroweak processes with heavy flavour such as diboson and single-top production. The non-W QCD multijet contamination is estimated using the same technique as described in Sect. 4.2.2, now applied to a b-tagged multijet sample, or applied to a pretagged sample, which is then scaled by the average tagging rate for QCD events. The background is estimated as the average of the two methods. Mistag background events are W + jets events where the tagged jet does not result from the decay of a heavy quark. To estimate the size of the mistag background, each jet is weighted with its mistag rate in the pretag sample, including the mistag correction factor of 1.2 ± 0.1 as described above. The production of W-bosons associated with heavy flavour in the processes $Wb\bar{b}$, $Wc\bar{c}$, and Wc is a significant part of the background for the tagged sample. The number of $Wb\bar{b}$, $Wc\bar{c}$ and Wc events is given by multiplying the heavy flavour fractions by the pretag event count, after subtracting the non-W backgrounds, and then multiplying by the tagging efficiencies. A number of backgrounds are too small to be measured directly, and are therefore predicted using Monte Carlo samples. This is done in particular with the diboson production (WW, WZ, and ZZ) in association with jets, the $Z \rightarrow \tau\tau$ production, and the single top production. As shown in Fig. 61, good agreement between background and data is found in the one and two jet bins, validating the background calculation, and good agreement when adding the $t\bar{t}$ contribution according to the measured cross section. Cross checks done on Z + jets samples confirm the reliable background description. The excess of tags in the three and four jet bins is attributed to $t\bar{t}$. With an efficiency for tagging at least one jet in a $t\bar{t}$ event (after all other cuts have been applied, including $H_T > 200$ GeV) of 53.4 ± 0.3 (stat.) ± 3.2 (syst.)%, the production cross section follows from the acceptance measurement and the background estimate: $\sigma_{t\bar{t}} = \frac{N_{obs} - N_{bkg}}{\epsilon_{t\bar{t}} \times \mathcal{L}}$, where N_{obs} and N_{bkg} are the number of total observed and background events, respectively, in the $W + \geq 3$ jet bins, $\epsilon_{t\bar{t}}$ is the signal acceptance, and $\mathcal{L}$ is the integrated luminosity.

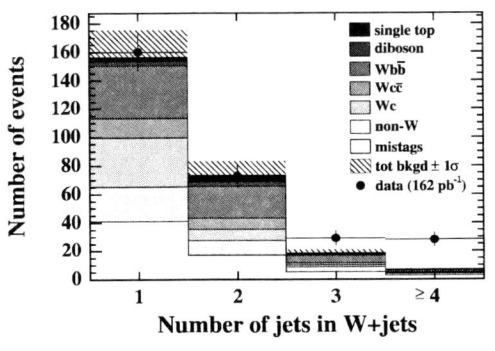 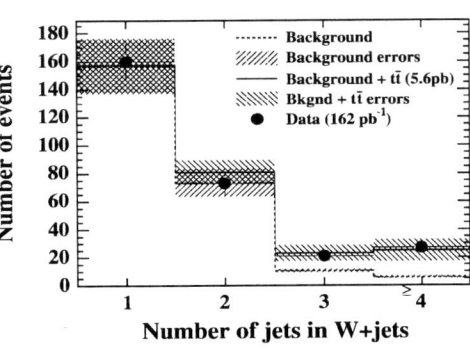

Fig. 61. *Left*: Number of events passing the selection criteria with at least one b-tagged jet before the H_T requirement is applied. *Right*: Background and $t\bar{t}$ expectation (based on measured $t\bar{t}$ cross section) as a function of jet multiplicity. Events with three or more jets are required to have $H_T > 200$ GeV

The resulting $t\bar{t}$ production cross section is measured to be:

$$\sigma_{t\bar{t}} = 5.6^{+1.2}_{-1.1} \,(\text{stat.})^{+0.9}_{-0.6} \,(\text{syst.})\,\text{pb}, \qquad (63)$$

where the systematic uncertainties are due to uncertainties on the signal acceptance (10% relative), luminosity measurement (6%), and background estimate (5%). Of the 57 tagged events in the three and four jet bins before the H_T cut, eight of these are double-tagged events with an expected background of 1.3. As a cross check, the $t\bar{t}$ production cross section is also extracted from this cleaner, but lower statistics sample and found to be $\sigma_{t\bar{t}} = 5.0^{+2.4}_{-1.9} \,(\text{stat.})^{+1.1}_{-0.8} \,(\text{syst.})\,\text{pb}$, in good agreement with the above result.

A preliminary update of this analysis with 318 pb^{-1} results in a $t\bar{t}$ cross section measurement of [308]:

$$\sigma_{t\bar{t}} = 8.9^{+0.9}_{-0.9} \,(\text{stat.})^{+1.2}_{-0.9} \,(\text{syst.})\,\text{pb for single-tags} \quad (64)$$

$$\sigma_{t\bar{t}} = 8.9^{+1.8}_{-1.6} \,(\text{stat.})^{+1.9}_{-1.3} \,(\text{syst.})\,\text{pb for double-tags}, \qquad (65)$$

using the tight SecVtx b-tag and assuming $m_{\text{top}} = 175$ GeV/c^2.

Loose vertex b-tag. Another CDF analysis measures the $t\bar{t}$ production cross section in 318 pb^{-1} using a loose vertex b-tag technique [308] and obtains for $m_{\text{top}} = 175$ GeV/c^2:

$$\sigma_{t\bar{t}} = 8.9^{+0.9}_{-0.9} \,(\text{stat.})^{+1.2}_{-0.9} \,(\text{syst.})\,\text{pb for single-tags}, \quad (66)$$

$$\sigma_{t\bar{t}} = 10.4^{+1.6}_{-1.4} \,(\text{stat.})^{+2.1}_{-1.4} \,(\text{syst.})\,\text{pb for double-tags}. \qquad (67)$$

The numerical similarity to the results of the tight b-tag analysis is a coincidence.

*Vertex **b**-tag and jet kinematic.* In an alternative approach, CDF measures the $t\bar{t}$ production cross section in 162 pb^{-1} of data, selecting lepton + jets events with at least one b-tagged jet and determining the background fraction from a fit of the transverse energy spectrum of the leading jet [309]. The analyses described so far rely on the ability of theoretical calculations to determine, for the background, the fraction of events that contain b-quark jets. This analysis instead measures the background fraction directly in the signal data sample. The transverse energy of the

highest or the second highest E_T jet is a good discriminator between signal and background. Typically in a $t\bar{t}$ event, these jets are the primary decay products (b-jets) of the very heavy top quarks and thus have a hard E_T spectrum. For most of the background sources, however, they are produced as QCD radiation, resulting in a much softer bremsstrahlung-like E_T distribution. In order to obtain the background spectrum, data that are kinematically similar to the final sample, but without a significant $t\bar{t}$ contamination, is needed. It is shown that the leading jet E_T spectra for the background processes are similar whether or not the events contain b-quark jets. Then the non-heavy flavour spectrum becomes the background template for measuring the $t\bar{t}$ fraction in the signal sample.

The event selection in terms of trigger, lepton identification and E_T requirement, $\not{E}_T$ reconstruction and cut, and jet reconstruction and E_T cut is identical to that in the CDF *SecVtx Vertex b-tag* analysis described above. Also in this analysis, at least one jet is required to be tagged by the SecVtx algorithm.

There are several background sources in the b-tagged $\ell + \not{E}_T + \ge 3$ jets sample. Their approximate contribution to the total background is determined from both data and theoretical calculations, although the precise composition is not needed for this analysis because the spectra are very similar to each other. The sources are the production of a W-boson accompanied by the QCD production of heavy flavour quarks ($W + \text{HF}: \sim 40\%$ of the total number of background events in the $W + \ge 3$ jets), misidentification in b-tagging mainly due to track and vertex resolution (mistag: $\sim 30\%$), a fake W-boson associated with a real or fake lepton (non-W: $\sim 20\%$), diboson production (WW, ZZ, WZ) and electroweak single-top production (*diboson + single*-top: $\sim 10\%$). The $W +$ heavy flavour and mistag shapes in the leading jet E_T are found to be the same in Monte Carlo studies with $Wb\bar{b}$, $Wc\bar{c}$, and Wc events and comparisons to data distributions. In this comparison, for the $W +$ heavy flavour events at least one jet is required to be b-tagged, whereas in the $W +$ light flavour events at least one jet is required to be taggable in order to avoid subtle kinematic differences based on jet rapidity. As found in the ALPGEN simulation of $Wb\bar{b}$ a small jet E_T dependence of the b-tagging efficiency is applied to the $W +$ light flavour shape as a correction to model the shape of the leading jet E_T distribution in $W +$ heavy flavour events. The shape of the leading jet E_T distribution from non-W events is determined from events with large $\rlap{/}{E}_T$, but no lepton isolation requirement, since the lepton isolation is not correlated with the E_T of other jets. This contribution is added to the other background with a relative normalisation taken from absolute estimates of the various background sources. The spectrum from single top production is taken from ALPGEN + HERWIG Monte Carlo, normalised to the theoretical cross section, whereas the diboson background is neglected because it is very small ($\sim 3\%$) and similar in shape to the other backgrounds. The performance of this background modelling is tested in W + jets events with 1 or 2 jets, where the $t\bar{t}$ contribution is very small. Very good agreement between the data sample and the background estimate is found, giving confidence in this method. For $W + \geq 3$ jets, i.e. the signal region, the background shape is determined as described above. A small $\sim 6\%$ $t\bar{t}$ signal contribution is subtracted iteratively from the leading jet E_T distribution. The shape of the resulting leading jet E_T distribution is fit for use in the final unbinned log-likelihood fit using a Landau distribution plus a Gaussian function (see the dot-dashed line in Fig. 62). The shape for the $t\bar{t}$ leading jet E_T is obtained from a HERWIG [226, 227] Monte Carlo (see the dashed line in Fig. 62). As expected, it is found to be significantly harder than that of the background. The obtained shape is fit to a Landau distribution plus two Gaussians.

To fit the data to a sum of the signal and background templates, an unbinned likelihood fit with the following form is used:

$$\mathcal{L} = \prod_{i=1}^{N} P(E_{T,i}; R)$$
$$= \prod_{i=1}^{N} [R \cdot P_{\text{signal}}(E_{T,i}) + (1 - R) \cdot P_{\text{background}}(E_{T,i})], \tag{68}$$

where the signal fraction $R = \frac{N_{\text{signal}}}{N_{\text{signal}} + N_{\text{background}}}$ is the one free parameter in the fit, $P_{\text{signal}}(E_{T,i})$ is the signal probability density as a function of E_T, and $P_{\text{background}}(E_{T,i})$ is that of the background. The validity of this procedure has been successfully demonstrated with Monte Carlo pseudo-experiments. Figure 62 shows the best fit to the 57 events in the $W + \geq 3$ jet data sample. The solid curve is the best fit, with the individual components shown as dashed ($t\bar{t}$) and dot-dashed (background) curves. The insert contains $-\ln(\mathcal{L}/\mathcal{L}_{\text{max}})$ as a function of the signal fraction. The signal fraction obtained is $R = 0.68^{+0.14}_{-0.16}$. Cross check fits of the second leading jet E_T (Fig. 62 right) or the fitting of the sum of the first and second leading jet E_T yield consistent results, demonstrating the robustness of the fit.

The $t\bar{t}$ production cross section is obtained from $\sigma_{t\bar{t}} = \frac{N_{\text{obs}} R_{\text{fit}}}{A_{t\bar{t}} \epsilon_{t\bar{t}} \int \mathcal{L} \, dt}$, where N_{obs} is the number of candidate $W + \geq 3$ jet events with at least one b-tag (57 events), R_{fit} is the signal fraction determined from the likelihood fit, and $A_{t\bar{t}}$ is the geometric acceptance for $t\bar{t}$ events in the CDF detector, including the branching fractions. The selection efficiency $\epsilon_{t\bar{t}}$ includes the trigger, event vertex position, event b-tagging, lepton identification, veto on cosmic ray photon conversion, dilepton and Z boson events. $\int \mathcal{L} \, dt$ is the integrated luminosity. $A_{t\bar{t}} \epsilon_{t\bar{t}}$ is determined from PYTHIA [223] Monte Carlo to be 4.02 ± 0.03 (stat.) ± 0.43 (syst.)%. Template shape uncertainties affect the signal fraction determination, while other effects mostly impact the acceptance. Table 17 summarises the systematic uncertainties on the $t\bar{t}$ production cross section. If the systematic affects both the template shape and the acceptance, the uncertainty is taken to be 100% correlated. The largest uncertainty originates from the effect of the jet energy scale on the $t\bar{t}$ simulation, while it does not contribute to the background template shape systematics, because that is determined from data.

The measured total $t\bar{t}$ production cross section is:

$$\sigma_{t\bar{t}} = 6.0^{+1.5}_{-1.6} \, (\text{stat.})^{+1.2}_{-1.3} \, (\text{syst.}) \, \text{pb}, \tag{69}$$

for an assumed $m_{\text{top}} = 175 \, \text{GeV}/\text{c}^2$.

Vertex tag with kinematic fit. In a data sample of 311 pb^{-1}, CDF has measured the $t\bar{t}$ production cross section using a missing E_T + jets selection and requiring at least one jet per event to be b-tagged [310]. The signal is mainly produced by lepton (e, μ, τ) + jets $t\bar{t}$ decays. This channel does not use explicit lepton identification requirements

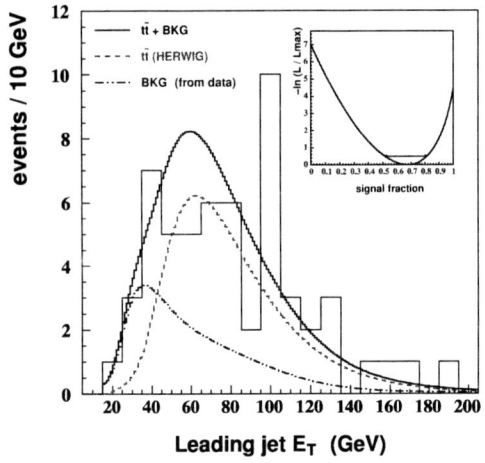
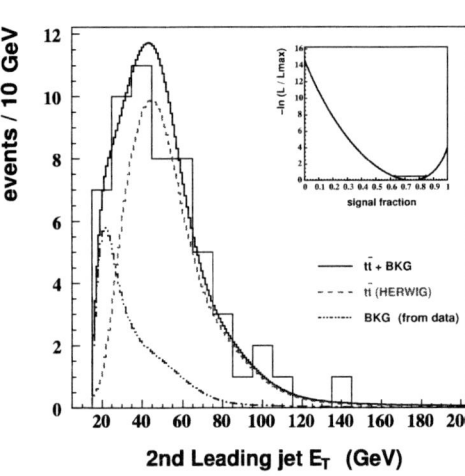

Fig. 62. The 57 seven candidate events (*histogram*) with the best fit curve (*solid*). The best fit composition, $t\bar{t}$ (*dashed*) and background (*dot-dashed*), is also shown. The *inset* shows the $-\ln(\mathcal{L}/\mathcal{L}_{\text{max}})$ as a function of the signal fraction. *Left*: Leading jet E_T spectrum; *right*: Second leading jet E_T spectrum

Table 17. Systematic uncertainties for the $t\bar{t}$ cross section from shape and acceptance affects in the $l + \text{jets}\,b\text{-tag} + \text{kinematics}$ analysis

Source	Shape (%)	Acceptance (%)	Total (%)
Jet energy scale	± 10.8	± 4.5	± 15.3
Absolute b-tag efficiency	–	± 7.4	± 7.4
Background statistics	$+2.6$ -6.9	–	$+2.6$ -6.9
Luminosity	–	± 5.9	± 5.9
Lepton ID	–	± 5.0	± 5.0
b-tag effic. (E_T dependence)	± 1.9	± 2.5	± 4.4
PDF	± 3.4	± 0.8	± 4.2
Gluon rad., non-W shape, other acceptance syst., non-W rate, $t\bar{t}$ shape, single top product.	$\ldots$	$\ldots$	$\ldots$
Total	$+12.4$ -13.9	± 12.3	$+20.6$ -21.5

and, therefore, has significant acceptance to the $\tau + \text{jet}$ decay mode. This channel uses a data sample, which is independent from those used in the other CDF cross section measurements. It therefore adds information to the overall combined $t\bar{t}$ cross section measurement in CDF.

The events are selected based on a multijet trigger with $\not{E}_T \geq 10\,\text{GeV}$, and at least one good quality vertex with $|z_{\text{vtx}}| < 60\,\text{cm}$ to increase the taggability of the jets. At least one jet in the event is required to be b-tagged by the SecVtx algorithm. Events with a good, high-p_T electron or muon are vetoed to avoid overlap with the other $t\bar{t}$ to lepton + jets analyses.

The $t\bar{t}$ Monte Carlo sample, generated with PYTHIA and HERWIG for $m_t = 178\,\text{GeV}/c^2$ is only used to optimise the kinematic event selection and to determine the event selection efficiency and systematic uncertainties. The overall amount of background-produced b-jets in the kinematically selected sample is predicted using b-identification probabilities, measured from data. This tagging probability is calculated using 3-jet events collected by the multijet trigger before any kinematical selection. It is used to predict the amount of background tags in higher jet multiplicity events and in different samples defined by some kinematical selection criteria. In order to track possible changes in the sample composition resulting from the kinematical selection, the tagging rate, defined by the ratio $\mathcal{P} = \frac{\text{number of tagged jets}}{\text{number of taggable jets}}$, is parameterised as a function of the jet E_T, the number of tracks per jet, N_{trk}, and missing E_T projected along the jet direction, $\not{E}_T^{\,prj} = \not{E}_T \cos \Delta\phi(\not{E}_T, \text{jet})$. The performance of this tagging rate matrix and in particular its capability to correctly predict the amount of background tags is checked in control samples depleted of signal contamination. The discrepancy between observed and predicted tags is found to be good within 10%. This agreement is used as estimated systematic uncertainty associated with the background prediction.

Using the tagging matrix prediction, the kinematical selection is optimised by minimising the expected relative statistical uncertainty on the cross section measurement. The best choice of cuts is found to be: $4 \leq N_{\text{jet}}(E_T \geq 15\,\text{GeV}, |\eta| < 2.0) \leq 8$, the missing E_T significance $\not{E}_T / \sqrt{\sum E_T} \geq 4.0 \sqrt{\text{GeV}}$, $\min\Delta\phi(\not{E}_T, \text{jet}) \geq 0.4\,\text{rad}$. With these requirements and a resulting signal selection ef-

ficiency of $(4.88 \pm 0.84)\%$, 597 data events are selected, out of which 106 have at least one b-tag, for a total of 127 tags. The expected number of tags provided by the positive tagging matrix parameterisation is $N_{\text{exp}}^{\text{tags}} = 67.4 \pm 2.7\,(\text{stat.}) \pm 6.7\,(\text{syst.})$. This estimate receives a small contribution from $t\bar{t}$ in the pre-tagging sample, which is iteratively estimated to be 10.0 events, yielding a top-free background determination of $N'_{\text{exp}} = 57.4 \pm 8.1$.

In order to further establish the $t\bar{t}$ presence in the selected data, a binned likelihood fit to kinematical distributions of the final data sample is performed. The $\not{E}_T$ and $\Delta\phi(\not{E}_T, \text{tagged jet})$ distributions, shown in Fig. 63, are fitted. The templates are constructed from the $t\bar{t}$ inclusive Monte Carlo sample for the signal, and from the positive tagging matrix application to data for background. The likelihood function in this fit is defined as:

$$\mathcal{L} = e^{-\frac{(L - \bar{L})^2}{2\sigma_L^2}} e^{-\frac{(\epsilon_{\text{kin}} - \bar{\epsilon}_{\text{kin}})^2}{2\sigma_{\epsilon_{\text{kin}}}^2}} e^{-\frac{(\epsilon_{\text{tag}} - \bar{\epsilon}_{\text{tag}})^2}{2\sigma_{\epsilon_{\text{tag}}}^2}}$$
$$e^{-\frac{(N'_{\text{exp}} - \bar{N}'_{\text{exp}})^2}{2\sigma_{N_{\text{exp}}}^2}} \frac{(\sigma_{t\bar{t}}\epsilon_{\text{kin}}\epsilon_{\text{tag}}L + N'_{\text{exp}})^{N_{\text{obs}}}}{N_{\text{obs}}!}$$
$$e^{-(\sigma_{t\bar{t}}\epsilon_{\text{kin}}\epsilon_{\text{tag}}L + N'_{\text{exp}})} . \tag{70}$$

In Fig. 64, good agreement between observed and predicted background tags is demonstrated in the 3-jet bin, where the matrix is computed before the kinematical selection. In the 4 to 6 jet bins, the $t\bar{t}$ Monte Carlo contribution needs to be added in order to explain the distribution observed in the data. Note the large contribution from $t\bar{t}$ events with $\tau + \text{jet}$ decay topology.

The systematic uncertainties, determined from ensemble testing in Monte Carlo and from the uncertainties on the background estimates, determined in the control samples, are summarised in Table 18. The resulting $t\bar{t}$ production cross section is in this analysis measured to be:

$$\sigma_{t\bar{t}} = 5.9 \pm 1.2\,(\text{stat.})^{+1.4}_{-1.0}\,(\text{syst.})\,\text{pb}, \tag{71}$$

for an assumed $m_{\text{top}} = 178\,\text{GeV}/c^2$. This result can be interpolated to an assumed $m_{\text{top}} = 175\,\text{GeV}/c^2$:

$$\sigma_{t\bar{t}} = 6.1 \pm 1.2\,(\text{stat.})^{+1.4}_{-1.0}\,(\text{syst.})\,\text{pb}. \tag{72}$$

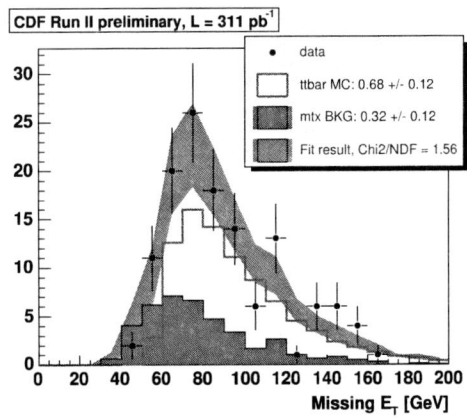

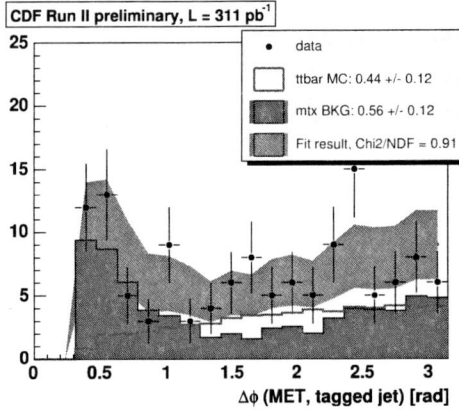

Fig. 63. $E\!\!\!/_T$ and $\Delta\phi(E\!\!\!/_T, \mathrm{jet})$ distributions for data after kinematical selection and with at least one positive SecVtx b-tag. The distributions are fitted to the sum of Monte Carlo and background templates, the latter being derived from the tagging matrix application to data

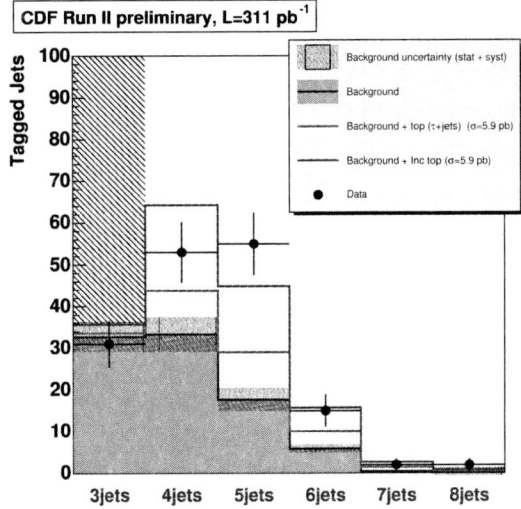

Fig. 64. Number of tagged jets versus jet multiplicity. Data (*points*), background iteratively corrected (*shaded histogram*) and $t\bar{t}$ expectation (*lines*) are shown after kinematical selection. The $t\bar{t}$ contribution is normalised to the fitted cross section

Table 18. Systematic uncertainties for the $t\bar{t}$ cross section $E\!\!\!/_T + \mathrm{jets}\ b$-tag $+$ kinematics analysis

Source	Method	Uncertainty
ϵ_{kin} systematics		
Trigger simulation	turn-on curves	14.8%
Generator	$\frac{\|\epsilon_{\mathrm{PYTHIA}} - \epsilon_{\mathrm{HERWIG}}\|}{\epsilon_{\mathrm{PYTHIA}}}$	8.2%
PDFs	MC reweighting	1.6%
ISR/FSR	sample comparison	2.0%
Jet Corrections	$\frac{\|\epsilon_{\mathrm{jetcorr},+1\sigma} - \epsilon_{\mathrm{jetcorr},-1\sigma}\|}{2\epsilon_{\mathrm{jetcorr}}}$	1.5%
ϵ_{tag} systematics		
SecVtx	$\frac{\|\epsilon_{\mathrm{tag},+1\sigma} - \epsilon_{\mathrm{tag},-1\sigma}\|}{2\epsilon_{\mathrm{tag}}}$	5.8%
Tagging matrix systematics		
Control samples	$N_{\mathrm{obs}}/N_{\mathrm{exp}}$	10.0%
Luminosity systematics		
Lumi measurement	–	6.0%

Soft muon b-tag. Based on $194\,\mathrm{pb}^{-1}$ of data, CDF measures the $t\bar{t}$ production cross section in the lepton $+$ jets channel, where b-jets are identified via their semileptonic decays to muons (SLT) [311]. The excess of events observed in the $W + \geq 3$ jets region beyond the expected background, dominated by $W +$ jets and QCD multijet production, both estimated from data, is attributed to $t\bar{t}$ production. The W plus one or two jet sample serves as a control sample with little signal contamination.

The lepton (electron or muon) $+$jets events are triggered and selected as described in Sect. 4.2.2: Electron E_T (muon p_T) greater than $20\,\mathrm{GeV}$, both isolated in calorimeter or tracker, respectively, and $E\!\!\!/_T > 20\,\mathrm{GeV}$. Jets with $E_T > 15\,\mathrm{GeV}$ and $|\eta| < 2.0$ are corrected for detector response variations in η, calorimeter gain instability, and multiple interactions in an event. In addition, the total transverse event energy H_T, which is the scalar sum of the electron E_T or muon p_T, the event $E\!\!\!/_T$ and jet E_T for jets with $E_T > 8\,\mathrm{GeV}$ and $|\eta| < 2.4$ is expected to be very large in $t\bar{t}$ events due to the large top quark mass and therefore required to be $H_T > 200\,\mathrm{GeV}$, rejecting 40% of the background while retaining more than 95% of the $t\bar{t}$ signal. This

leaves $337\ W$ plus three or more jet events in the data sample. To further improve the signal-to-background ratio, events with one or more b-jets are identified by searching inside jets for semileptonic decays of b hadrons into muons. Details of the CDF soft lepton tagging algorithm are described in Sect. 3.3.7.

There are 319 pretagged events with three or more jets, (211 in $e +$jets, 108 in $\mu +$jets), out of which 20 events have an SLT tag (15 in $e +$jets, 5 in $\mu +$jets).

W plus jets events enter the signal sample either when one of the jets is a b-jet or a c-jet with a semileptonic decay to a muon, or a light quark jet is misidentified as containing a semileptonic decay ("mistagged"). These background events are referred to as $W +$ heavy flavour and $W +$ "fakes", respectively. $W +$ heavy flavour events include $Wb\bar{b}$, $Wc\bar{c}$ and Wc production. Their contribution is measured from the pretagged sample by constructing a "tag matrix" that parameterises the probability that a taggable track with a given p_T, η and ϕ, in a jet with $E_T > 15\,\mathrm{GeV}$, satisfies the SLT tagging requirement. The tag matrix is constructed using jets in $\gamma +$jets events with one or more jets. The resulting tag probability is approximately 0.7% per taggable track. This technique relies on the assumption that the tagging rate in jets of the $\gamma +$jets sample is a good model for the tagging rate of the jets in $W+$jets events. The assumption is plausible because the SLT tagging rate in generic jet events is largely due to fakes. Only $\sim 20\%$ of the tags in the $\gamma+$jets sample are estimated to come from heavy flavour. Using Monte Carlo events, the heavy flavour contribution in $\gamma+$jets and $W+$jets was found to differ by approximately 30%, affecting the background prediction in $W+$jets events only at the few-percent level. The tag matrix is applied to all pre-tagged events in the signal region according to:

$$N_{\text{predicted}}^{\text{tag}} = \sum_{i}^{N_{\text{events}}} \left(1 - \prod_{j=1}^{N_{\text{trk}}} (1 - \mathcal{P}_j) \right), \quad (73)$$

where the sum runs over each event in the pretagged sample, and the subsequent product runs over each taggable track in the event. $\mathcal{P}_i$ is the probability from the tag matrix for tagging the i-th track with parameters $p_{\text{T},i}$, η_i and ϕ_i. The small $t\bar{t}$ contribution in the pretagged sample is corrected for. Also the fraction of events in the signal region, originating from QCD jet production with mistags, F_{QCD}, are subtracted from this number since their background contribution is estimated separately: $N_{\text{predicted}}^{Wj-\text{tag}} = (1 - F_{\text{QCD}})N_{\text{predicted}}^{\text{tag}}$. Various systematic studies on different data samples indicate that this procedure to estimate the tag rate is good to 10%.

Events with two or more jets in which the decay of a heavy flavour hadron produces a high-p_{T} isolated lepton, or in which a jet fakes such a lepton, are termed QCD events. Their contribution (absolute number of fraction F_{QCD} is estimated via a comparison of the event count in the pretagged sample, where the $\not{E}_{\text{T}}$ and the lepton isolation cuts are inverted (see Sect. 4.2.2). The number of tagged QCD events in the signal region is then estimated by multiplying F_{QCD} by the tagging probability for QCD events. They can be different from the ones designed for jets in $W+$jets: Mismeasurements in the jet energies and differences in kinematics between $W+$jets and QCD may affect the tagging probabilities. $W+$jets events have $\not{E}_{\text{T}}$ from the undetected neutrino, whereas QCD events have $\not{E}_{\text{T}}$ primarily from jet mismeasurement. This, however, is correlated with fake tags due to energy leakage from the calorimeter through calorimeter gaps or incomplete absorption of the hadronic shower, both of which can result in track segments in the CDF muon chambers. In QCD events the primary lepton is either a fake or a result of a semileptonic decay of heavy flavour, which typically enhances the tag rate. In QCD events, the ratio of predicted and observed number of tagged events is found to increase with $\not{E}_{\text{T}}$ and to be a function of the event H_{T}. Therefore the number of QCD background events is calculated as:

$$N_{\text{QCD}} = \langle F_{\text{QCD}}k \rangle N_{\text{predicted}}^{\text{tag}}, \quad (74)$$

where the brackets represent the product of F_{QCD} and the overall scale factor k, which is measured as a normalisation correction in QCD events, convoluted with the H_{T} distribution of QCD events from a control sample. The sys-

tematic uncertainties cover the difference between this convolution and the product of the average factors. The contribution of Drell–Yan events is estimated by scaling the observed number of tags in the Z-boson mass window, which is otherwise removed from the analysis, to the events outside this window: $N_{DY} = N_{\text{inside}}^{\text{tag}} R_{Z/\gamma^*}^{\text{out/in}}$. The ratio $R_{Z/\gamma^*}^{\text{out/in}}$ is measured in $Z+0$ jet data events without $\not{E}_{\text{T}}$ and H_{T} requirements for statistics reasons. It is assigned a 33% systematic uncertainty due to the variations observed in the corresponding ALPGEN samples for different jet multiplicities. Remaining background sources are due to WW, WZ, ZZ, $Z \to \tau\tau$ and single top production. They are small, and estimated from Monte Carlo normalised to the theoretical cross sections.

The acceptance and selection efficiency for $t\bar{t}$ events is obtained from PYTHIA [223] Monte Carlo, corrected for the ratio of lepton identification efficiencies in $Z \to \ell\ell$ decays from data and Monte Carlo. The SLT tagging efficiency is measured in data as a combination of the track reconstruction efficiency, the efficiency to reconstruct segments in the muon chambers and the muon identification efficiency. The resulting SLT tagging efficiency is parameterised as a function of track p_{T} and η and applied to muons in the $t\bar{t}$ Monte Carlo. Contributions to the $t\bar{t}$ event tagging efficiency from mistags are estimated by applying the mistag matrix described before, where the $\sim 20\%$ heavy flavour contribution in the initial $\gamma+$jets control sample is corrected for. The overall event SLT tagging efficiency in events with three or more jets is found to be 12%–15%. This procedure is tested in a di-jet Monte Carlo sample, which is found to describe the SLT tagging performance on a high-purity $b\bar{b}$ data sample very well. Figure 65 (left) shows the number of tagged events along with the expected background in $W+1$, 2, 3, and 4 or more jet events. Also shown (right) is the impact parameter significance distribution, defined as the impact parameter divided by its uncertainty, for SLT tracks and the expectation from signal plus background. The long-lived component from semi-leptonic b-hadron decays is readily apparent in the shape of the positive impact parameter distribution.

The systematic uncertainties on the signal acceptance and the background estimation are summarised in Table 19. The resulting $t\bar{t}$ production cross section is measured to be :

$$\sigma_{t\bar{t}} = 5.3 \pm 3.3\,(\text{stat.})^{+1.3}_{-1.0}\,(\text{syst.})\,\text{pb}, \quad (75)$$

for a $m_{\text{top}} = 175\,\text{GeV/c}^2$.

Jet probability b-tag. Using $162\,\text{pb}^{-1}$ of lepton+jets data, CDF has measured the $t\bar{t}$ production cross section using a jet probability tagging technique [312]. The lepton+jets selection is identical to the one described in Sect. 4.2.2. In addition, to improve the signal to background ratio, events are required to have one or more b-tagged jets, using the jet probability (JP) algorithm (Sect. 3.3.7).

The total acceptance of the selection cuts for $t\bar{t}$ events is determined in a PYTHIA [223] $t\bar{t}$ lepton+jets event sample, where the lepton identification efficiency is corrected

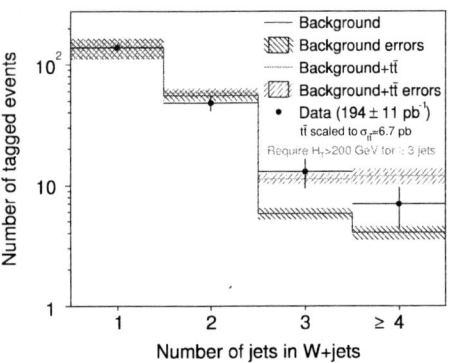
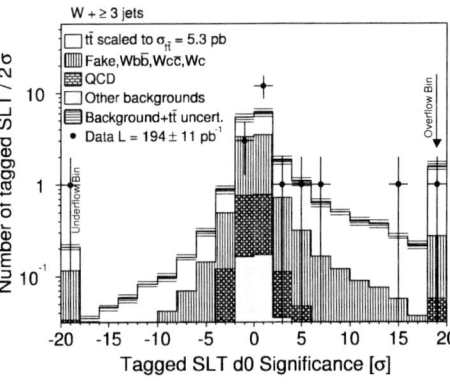

Fig. 65. *Left*: Number of tags in the $W + 1, 2, 3, \geq 4$ jet events together with the background estimate and the expected contribution from $t\bar{t}$ production (normalised to the theoretical cross section of 6.7 pb). *Right*: The impact parameter $(d0)$ significance for tagged events, compared with expectations from backgrounds plus $t\bar{t}$ for $W + \geq 3$ jets events

Table 19. Systematic uncertainties for the $t\bar{t}$ cross section signal acceptance, SLT b-tagging and the background modelling

Source	Syst. Uncertainty (%)	$\Delta\sigma_{t\bar{t}}$ (pb)
Acceptance modelling	± 8	$\left\{ \begin{array}{l} +1.10 \\ -0.70 \end{array} \right.$
SLT tagging efficiency	$+8, -11$	
Tag matrix prediction	± 10	± 0.68
QCD prediction	$\pm 19(e), \pm 67(\mu)$	± 0.14
Drell–Yan and other MC backgrounds	± 19	± 0.05
Luminosity	± 6	± 0.32
Total		$^{+1.3}_{-1.0}$

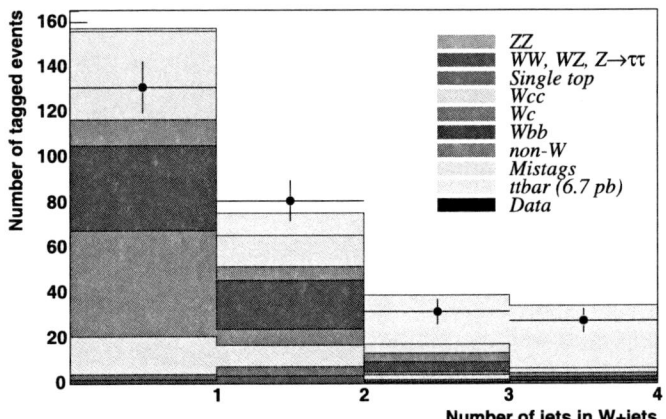

Fig. 66. Number of tags in the $W + 1, 2, 3, \geq 4$ jet events together with the background estimate and the expected contribution from $t\bar{t}$ production (normalised to the theoretical cross section of 6.7 pb)

Table 20. Systematic uncertainties for the $t\bar{t}$ cross section

Source	Syst. Uncert. (%)	$\Delta\sigma_{t\bar{t}}$ (%)
MC acceptance modelling	8.7	8.9
Tagging scale factor (b, c)	13, 20	16.6
Mistag rate prediction	$+20$	-3.1
Non-W fraction	50	0.8
Non-W prediction	50	7.4
$W +$ heavy flavour prediction	30	6.1
Other MC backgrounds	1.8	0.1
Luminosity	5.9	6
Total		± 22

for the ratio of efficiencies found in $Z \rightarrow \ell\ell$ data and Monte Carlo samples. The event tagging efficiency is evaluated by measuring the fraction of simulated $t\bar{t}$ events passing all kinematic cuts and having at least one positive tagged tight jet, where the data-to-Monte Carlo scale factor for the b-tagging efficiency, obtained in the electron-jet sample as described above, is included. The overall event tagging efficiency, averaged over the lepton flavours, is $57.24 \pm 0.21 \pm 3.85\%$, where the second, systematic error includes the uncertainty from the b-tagging data-to-Monte Carlo scale factor and the uncertainty from the scale factor for the b- and c-flavour tagging.

The dominant background comes from $W +$ jets events, where either one of the jets is a b-jet, or a light quark jet is mis-identified as a b-jet. Similar to the previously described analyses, a mistag rate, obtained in inclusive QCD multijet events and parameterised as a function of the jet E_T, N_{trk}, $\sum E_{T,i}$, η, Z_{vtx}, and ϕ is applied to the pre-tagged $W +$ jet data set to model the $W +$ heavy flavour background in the tagged sample. Also the non-W background of mis-identified isolated high-p_T leptons in QCD multijet events is determined from comparisons of event rates in a multijet sample at low $\not{E}_T$, with required or inverted lepton isolation. Other small backgrounds from several sources, mainly diboson production, $Z \rightarrow \tau\tau$, and single-top production, are estimated using the Monte Carlo prediction for acceptance and tagging efficiency, and are normalised to the theoretical cross section [148, 247]. Figure 66 shows the number of tags in the $W + 1, 2, 3, \geq 4$ jet events together with the components of the background estimate. Also shown is the $t\bar{t}$

contribution, normalised to the theoretical cross section of 6.7 pb.

Estimates of the systematic uncertainties on the signal acceptance, the b-tagging efficiency and the background modelling are summarised in Table 20. The resulting $t\bar{t}$ production cross section is measured to be :

$$\sigma_{t\bar{t}} = 5.7^{+1.3}_{-1.2} \, (\text{stat.}) \pm 1.3 \, (\text{syst.}) \, \text{pb} \,, \qquad (76)$$

for a $m_{\text{top}} = 175 \, \text{GeV c}^2$.

4.2.4 All-jets analyses

CDF has also measured the $t\bar{t}$ production cross section in the all jets channel [313], using $165 \pm 10\,\mathrm{pb}^{-1}$ of data. This channel is characterised by the hadronic decays of both W-bosons $W \to q\bar{q}'$, yielding an all-hadronic final state with a nominal 6-jet topology. This channel has a large branching ratio ($\approx 46\%$) and is fully reconstructed. However, the QCD multijet background dominates by three orders of magnitude. In order to improve the signal-to-background ratio, S/B, a two-step approach is taken. In the first step a set of kinematical requirements which favours $t\bar{t}$ candidates is applied, yielding S/B$\approx 1/24$. In the second step the presence of displaced secondary vertices (b-tags) in the event is required in order to increase the S/B and to extract the $t\bar{t}$ signal.

Based on a multijet trigger, which demands at least four calorimeter clusters with $E_\mathrm{T} \geq 15\,\mathrm{GeV}$ and a total transverse energy exceeding $125\,\mathrm{GeV}$, a number of pre-selection cuts are applied: (1) at least one good event vertex, (2) $E\!\!\!/_\mathrm{T}/\sqrt{\sum E_\mathrm{T}} < 6\sqrt{\mathrm{GeV}}$ to reject badly reconstructed events, (3) at least 4 jets (cone size 0.4) with raw $E_\mathrm{T} > 15\,\mathrm{GeV}$ and $|\eta| < 3.0$, (4) total transverse energy of these jets, $\sum E_\mathrm{T} \geq 125\,\mathrm{GeV}$, (5) removal of events with good, high-p_T central electrons or muons to avoid overlap with other channels. The following kinematic selection cuts are optimised on $t\bar{t}$ Monte Carlo events, generated with HERWIG 6.4 [226, 227] (CTEQ 5L as PDF), while additional samples for systematic studies are generated with PYTHIA 6.203 [223]. All background estimates are based on multijet data samples, where the $t\bar{t}$ contamination is very small.

The kinematic selection is optimised to achieve the best possible S/$\sqrt{\mathrm{S+B}}$, i.e. the minimum statistical error on the resulting cross section. The events are required to have four good jets with $E_\mathrm{T} \geq 20\,\mathrm{GeV}$ (corrected for detector and reconstruction effects), and $|\eta| \leq 2.0$. Further requirements are: (1) $6 \leq N_\mathrm{jets} \leq 8$, (2) $A + 0.0037 \sum_3 E_\mathrm{T} \geq 0.85$ (A is the aplanarity, $\sum_3 E_\mathrm{T}$ is obtained by removing the two jets with the highest E_T, E_T in GeV), (3) $C \geq 0.77$ (C is the centrality), (4) $\sum E_\mathrm{T} \geq 320\,\mathrm{GeV}$. This selection yields 1700 candidate events at an S/B $\approx 1/24$ with a signal selection efficiency of $\epsilon_k = 6.2 \pm 1.9\%$.

In order to further improve the S/B, the presence of secondary vertices reconstructed with at least two good quality tracks with hits in the Silicon vertex detector and a displacement significance larger than 3.0 is required. The cross section measurement is performed using the total number of tagged jets (not events) in order to avoid the explicit calculation of the background for double tagged events. The effects due to the presence of double tags are included in the systematic uncertainties. The average number of tags in a $t\bar{t}$ event passing the kinematical selection is $n_\mathrm{tag}^\mathrm{ave} = 0.763 \pm 0.005\,(\mathrm{stat.}) \pm 0.065\,(\mathrm{syst.})$. The background contamination is estimated using a multijet data sample, weighted with a tag rate function $P\frac{\#\,\mathrm{of\ tagged\ jets}}{\#\,\mathrm{of\ taggable\ jets}}$, which has been derived in a control sample with $N_\mathrm{jets} = 4$ and is parameterised in terms of jet-E_T, jet-η, number of tracks reconstructed in the vertex detector (N_trk), and aplanarity (A). A taggable jet is here defined as a jet in the vertex detector acceptance. The probability for jets to be taggable has been demonstrated to be independent of the jet multiplicity. Before kinematic selection cuts, when the multijet sample is still predominantly composed of background events only, the predicted and observed number of tagged jets in the jet multiplicity bins $N_\mathrm{jet} = 4\text{–}8$ are demonstrated to agree very well, giving confidence in the constructed tag rate function. The kinematic selection changes the event characteristics with respect to the events in the control sample. In comparison to a multijet sample with 'negated' kinematic selection, the overall systematic uncertainty in the background estimate is found to be 5%. The background estimate after the kinematical selection is summarised in Table 21. The presence of an excess in the signal region ($6 \leq N_\mathrm{jets} \leq 8$) is visible also in Fig. 67.

After the application of the kinematic selection, a total of $n_\mathrm{obs} = 326$ candidate tags are observed, whereas the total background expectation in the signal region, provided by the tag rate function, amounts to $n'_\mathrm{exp} = 264.7 \pm 17.2$ tags after correcting the background for the presence of $t\bar{t}$ events in the pretag sample. The resulting excess of 61 ± 25 tagged jets is attributed to $t\bar{t}$ production. The $t\bar{t}$ production cross section is determined by maximising the likelihood function:

$$\mathcal{L} = e^{-(L-\bar{L})^2/2\sigma_L^2} e^{-(\epsilon_k - \bar{\epsilon}_k)^2/2\sigma_{\epsilon_k}^2}$$
$$\times e^{-\left(n_\mathrm{tag}^\mathrm{ave} - n_\mathrm{tag}^\mathrm{\bar{a}ve}\right)^2/2\sigma_{n_\mathrm{tag}^\mathrm{ave}}} e^{-(b-\bar{b})^2/2\sigma_b}$$
$$\times \frac{(\sigma_{t\bar{t}}\epsilon_k n_\mathrm{tag}^\mathrm{ave} L + b)^n}{n!} e^{-\left(\sigma_{t\bar{t}}\epsilon_k n_\mathrm{tag}^\mathrm{ave} L + b\right)}, \quad (77)$$

where $\sigma_{t\bar{t}}\epsilon_k n_\mathrm{tag}^\mathrm{ave} L + b$ is the expected number of tagged jets and $n \equiv n_\mathrm{obs}$ is the observed number. The first four terms

Table 21. Observed and expected number of tagged jets after kinematical selection in the different jet multiplicity regions

N_jet	4	5	6	7	8
Nr. Events	60	420	773	630	253
Nr. tagged jets (bgd)	7.3 ± 0.7	61.3 ± 4.1	114 ± 7	113 ± 7	45.9 ± 2.4
Nr. tagged jets ($t\bar{t}$)	0.2 ± 0.1	7.1 ± 2.2	27.4 ± 8.5	19.3 ± 6.0	5.7 ± 1.8
Nr. tagged jets (bgd $+t\bar{t}$)	7.5 ± 0.7	68.4 ± 4.7	141.1 ± 11.0	122.3 ± 9.2	51.6 ± 3.0
Nr. tagged jets (data)	11	70	170	116	40
Nr. tagged jets (data–bgd)	3.7 ± 3.4	8.7 ± 9.3	56 ± 15	13 ± 13	-5.9 ± 6.8

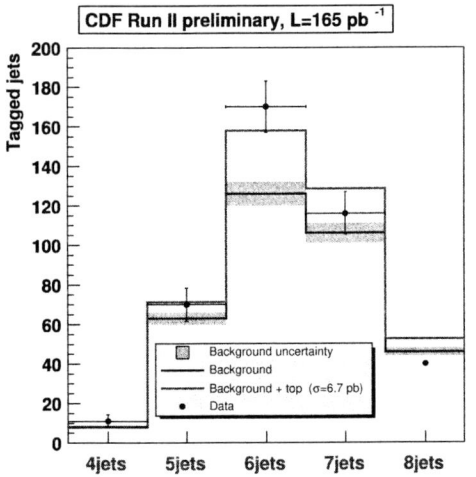 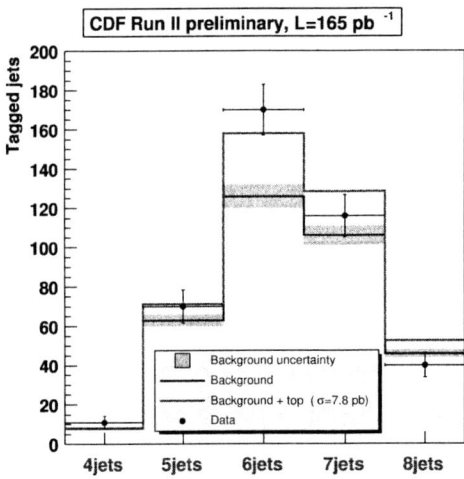

Fig. 67. Number of tagged jets versus jet multiplicity. Data (*points*), and background with uncertainty band. *Left*: $t\bar{t}$ expectation normalised to 6.7 pb; *Right*: $t\bar{t}$ expectation normalised to 7.8 pb, as obtained in the best fit

Table 22. Relative systematic uncertainties on the kinematical selection efficiency

Source	$\delta\epsilon/\epsilon$ (%)
Generator	2.9
ISR/FSR modelling	4.0
Parton distribution function	7.4
Jet energy scale	28.8
Total	30.1

represent Gaussian constraints on the luminosity L, the kinematic selection efficiency ϵ_k, the average number of tagged jets per $t\bar{t}$ event $n_{\text{tag}}^{\text{ave}}$, and the number of background tags $b \equiv n_{\text{exp}}'$. The central value of the cross section is given by the maximum of $\mathcal{L}$, that is $\sigma_{t\bar{t}} = \frac{n-b}{\epsilon_k n_{\text{tag}}^{\text{ave}} L}$. The systematic uncertainties on the kinematical selection efficiency ϵ_k is summarised in Table 22, where the jet energy scale is found to clearly dominate.

The result of the fit is then:

$$\sigma_{t\bar{t}} = 7.8 \pm 2.5\,(\text{stat.})^{+4.7}_{-2.3}\,(\text{syst.})\,\text{pb}\,. \qquad (78)$$

CDF has updated this analysis, using a larger data set of 311 pb^{-1} [314]. Due to the improved statistics and the more recent, better known jet energy scale, this analysis has reduced the total systematic uncertainties from 30% to 20%, yielding a final fit result for the $t\bar{t}$ production cross section of:

$$\sigma_{t\bar{t}} = 7.5 \pm 1.7\,(\text{stat.})^{+3.3}_{-2.2}\,(\text{syst.})\,\text{pb}\,. \qquad (79)$$

4.2.5 Combined $t\bar{t}$ cross section in CDF

CDF has worked out the combination of five different published measurements of the $t\bar{t}$ production cross section, using the 'best linear unbiased estimate – BLUE' method [315, 316]. The five individual measurements of the $t\bar{t}$ cross section are, as described above, the measurement in the dilepton channel [302], and the measurements in the lepton + jets channel using the secondary ver-

tex b-tagging [307], using the kinematic neural network approach [304], using the kinematic fitting of b-tagged events [309], and using the semileptonic b-tagging [311]. The general idea is to work out a global covariance matrix for the statistical and the systematic uncertainties, including their correlations. Inverting this matrix yields weights for each measurement in the combination. The statistical errors, the soft muon tag efficiency, the background statistics and other smaller uncertainties are treated as uncorrelated, while the signal acceptance, the luminosity, the SecVtx b-tagging efficiency, the shape of the jet E_T distributions, the shape of other variables in the signal sample, and the QCD multijet background are considered fully correlated.

In the systematic uncertainties, two types of uncertainties are distinguished: The *acceptance type* uncertainty is proportional to the combined cross section, i.e. $\Delta\sigma = \sigma\frac{\Delta A}{A}$, where A is the acceptance under consideration. This uncertainty needs to be determined in an iterative procedure, where usually 3–4 iterations are sufficient to yield a stable result. The second type of uncertainties is the *background type* uncertainty. This uncertainty does not depend on the combined cross section, so that uncertainties in a measurement which happen to fluctuate up are not artificially increased, i.e. $\Delta\sigma = \frac{\Delta b}{A L}$, where Δb is the uncertainty of the background count, A is the acceptance, and L is the integrated luminosity. The statistical correlations between the five measurements are determined via a sophisticated system of pseudo-experiments in the Monte Carlo, where each Monte Carlo event is subject to the event selection in each of the five analyses. Similarly, the overlap in the background of the five measurements is determined, but then treated as statistical correlations of the measurements. To test the stability of the method, all statistical correlations are varied by $\pm 10\%$, which is slightly larger than any single variation found in the pseudo-experiment studies. For ensembles, corresponding to 200 pb^{-1}, the largest observed change is 0.01 pb in the central cross section value and 0.02 pb in its uncertainty.

The resulting combined $t\bar{t}$ production cross section from CDF in Run-II is [317]:

$$\sigma_{t\bar{t}} = 6.0 \pm 0.9\,(\text{stat.}) \pm 0.7\,(\text{syst.})^{+0.4}_{-0.3}\,(\text{lumi.})\,\text{pb}\,, \qquad (80)$$

for $m_{\text{top}} = 175\,\text{GeV}/\text{c}^2$, yielding a 14% improvement in sensitivity with respect to the best single measurement.

Using the same procedure, CDF has also combined its preliminary Run-II $t\bar{t}$ cross section measurements, using the updated, preliminary measurements in the lepton + jets channel for the kinematic neural network ($347\,\text{pb}^{-1}$ [306]), the SecVtx b-tagging ($318\,\text{pb}^{-1}$ [308]), the vertex tag with kinematic fit ($311\,\text{pb}^{-1}$ [310]), and the updated measurement in the alljets channel ($311\,\text{pb}^{-1}$ [314]). The resulting combined $t\bar{t}$ production cross section from CDF in Run-II is [317]:

$$\sigma_{t\bar{t}} = 7.1 \pm 0.6\,(\text{stat.}) \pm 0.7\,(\text{syst.}) \pm 0.4\,(\text{lumi.})\,\text{pb}\,,\tag{81}$$

for $m_{\text{top}} = 175\,\text{GeV}/\text{c}^2$.

4.3 DØ analyses

4.3.1 Dilepton channel in topological analyses

DØ has measured the $t\bar{t}$ production cross section in the dilepton final state using $224\text{--}243\,\text{pb}^{-1}$ of data [318]. This analysis considers the e^+e^-, $e\mu$ and $\mu^+\mu^-$ final states. The electrons and muons may originate either directly from a W-boson or indirectly from a $W \to \tau\nu$ decay. The corresponding $t\bar{t}$ branching fractions (B) are 1.58%, 3.16%, and 1.57% [167] for the e^+e^-, $e\mu$ and $\mu^+\mu^-$ channels, respectively. The data used in this analysis are collected by requiring two leptons (e or μ) in the hardware trigger and one or two leptons in the software triggers. Two categories of backgrounds are distinguished: "physics" and "instrumental". Physics backgrounds are processes in which the charged leptons arise from electroweak boson decays and the $\not{E}_T$ originates from high p_T neutrinos. This signature arises in $Z/\gamma^* \to \tau^+\tau^-$ where the τ leptons decay leptonically and WW, WZ (diboson) production. Instrumental backgrounds are defined as events in which (a) a lepton within a jet fakes the isolated lepton signature, or (b) the $\not{E}_T$ originates from misreconstructed jet or lepton energies or from noise in the calorimeter.

Events are selected with at least two jets with $p_T^j > 20\,\text{GeV}$ and $|\eta| < 2.5$ and two leptons with $p_T^\ell > 15\,\text{GeV}$. Muons are accepted in the region $|\eta| < 2.0$, while electrons must be within $|\eta| < 1.1$ or $1.5 < |\eta| < 2.5$. The two leptons are required to be of opposite sign in the e^+e^- and $\mu^+\mu^-$ channels. In the $e\mu$ channel, $\not{E}_T > 25\,\text{GeV}$ is required, where the $\not{E}_T$ vector must not be in the direction of the muon. The overwhelming Drell–Yan background is reduced by cuts on the invariant dilepton mass. In the e^+e^- channel, events with dielectron invariant mass $80 \leq M_{ee} \leq 100\,\text{GeV}$ are vetoed, while $\not{E}_T > 35\,\text{GeV}$ ($\not{E}_T > 40\,\text{GeV}$) is required for $M_{ee} > 100\,\text{GeV}$ ($M_{ee} < 80\,\text{GeV}$). In the $\mu^+\mu^-$ channel, all events with $\not{E}_T > 35\,\text{GeV}$ are accepted. The final selection in the $e\mu$ channel requires the total transverse event energy, $H_T^\ell = p_T^{\ell_1} + \sum p_{T,j} > 140\,\text{GeV}$, where $p_T^{\ell_1}$ denotes the p_T of the leading lepton, rejecting the largest background from di-τ production and diboson production. The e^+e^- analysis uses a cut on sphericity $\mathcal{S} = 3(\epsilon_1 +$

$\epsilon_2)/2 > 0.15$, where ϵ_1 and ϵ_2 are the two leading eigenvalues of the normalised momentum tensor. This requirement rejects events in which jets are produced in a planar geometry through gluon radiation. The final selection applied in the $\mu^+\mu^-$ channel further rejects the $Z/\gamma^* \to \mu\mu$ background. For each $\mu^+\mu^-$ event the χ^2 of a fit to the $Z \to \mu\mu$ hypothesis given the measured muon momenta and known resolutions is computed. Selecting events with $\chi^2 > 2$ is more effective than selecting on the dimuon invariant mass for this channel.

Signal acceptances and efficiencies are derived from a combination of Monte Carlo simulation and data. $t\bar{t}$ production is simulated using ALPGEN [224] + PYTHIA [223]. B-hadron and τ lepton decays are modelled via EVTGEN [243] and TAUOLA [245], respectively. Lepton trigger and identification efficiencies as well as lepton momentum resolutions in the Monte Carlo are scaled to those measured in $Z \to \ell\ell$ data. Also the jet reconstruction efficiency, jet energy resolution and $\not{E}_T$ resolution in the Monte Carlo are adjusted to their measured value in data. Backgrounds from $Z/\gamma^* \to \tau\tau$ and diboson production are determined from simulations, using PYTHIA and ALPGEN, respectively. $Z/\gamma^* \to \tau\tau$ is normalised to the DØ measurement [319], while the diboson simulation is normalised to the theoretical cross section [247].

Instrumental backgrounds are determined from the data. Fake electrons can arise from jets comprised essentially of a leading π^0/η and an overlapping or conversion-produced track. This background is estimated by calculating the fraction f_e of loose electrons which appear as tight electrons in a control sample dominated by fake electrons. In the e^+e^- channel the control sample consists of events that satisfy the trigger and have two loose electrons. In the $e\mu$ channel, the events in the control sample must satisfy the trigger and have one tight muon and one loose electron. The two determinations of f_e agree. The predicted number of events with a fake electron in the final sample is obtained by multiplying the number of e^+e^- ($e\mu$) events with one loose electron and one tight electron (muon) by f_e. An isolated muon can be mimicked by a muon in a jet when the jet is not reconstructed. The fraction f_μ of loose muons that satisfy the tight muon criteria is measured in a control sample dominated by fake muons. In the $\mu^+\mu^-$ channel, the control sample is defined as events that have two loose muons, where the leading muon must fail the tight muon criteria to suppress physics processes with real isolated muons. The number of events with a fake muon is estimated by counting the number of events with one tight muon and a loose muon multiplied by f_μ. In the $e\mu$ channel, the contribution from events where both leptons are fake leptons is already accounted for by using f_e. The remaining contribution from events with a real electron and a fake muon, is determined by combining f_e and a fake rate f_μ obtained on a control sample that satisfies the $e\mu$ trigger.

The process $Z/\gamma^* \to \ell^+\ell^-$ ($\ell = e, \mu$), while lacking high p_T neutrinos, might have a significant amount of measured $\not{E}_T$ due to limited $\not{E}_T$ resolution. In the e^+e^- channel, this background is estimated by measuring a $\not{E}_T$ misreconstruction rate on a $\gamma + 2$ jets data sample and applying it to the simulation. In the $\mu^+\mu^-$ channel, the expected contri-

Table 23. Systematic uncertainties on $\sigma_{t\bar{t}}$

Source	$\Delta\sigma_{t\bar{t}}$ (pb)
Jet energy calibration	$+0.8 \quad -0.7$
Jet identification	$+0.3 \quad -0.6$
Muon identification	$+0.5 \quad -0.4$
Electron identification	± 0.3
Trigger	$+0.3 \quad -0.2$
Other	$+0.2 \quad -0.3$
Total	± 1.1

bution of $Z/\gamma^* \to \mu^+\mu^-$ background in the final sample is derived from events simulated with ALPGEN [224].

This analysis observes 5, 8, and 0 events in the e^+e^-, $e\mu$, and $\mu^+\mu^-$ channels, respectively. The probability to observe ≥ 5, ≥ 8, and exactly 0 events in the e^+e^-, $e\mu$, and $\mu^+\mu^-$ channels is estimated to be 22%, 43%, and 5%, respectively, using the measured $\sigma_{t\bar{t}}$ and taking into account systematic uncertainties. The significance of the observed $t\bar{t}$ signal over the background is 3.8 standard deviations. The systematic uncertainties are summarised in Table 23. In addition, a luminosity uncertainty of 6.5% is assigned to the luminosity measurement [283]. Figure 68 shows the observed jet multiplicity, $E\!\!\!/_T$, and the leading lepton p_T for the selected dilepton data events in comparison to the background estimate plus the expected $t\bar{t}$ contribution. Good agreement is found in all kinematic variables. The leading lepton p_T spectrum in the $t\bar{t}$ dilepton final state is studied by the CDF Collaboration (Sect. 8.2 [320]) and a mild excess is observed at low transverse momenta. This is not confirmed by this DØ measurement. In pseudo-experiments, starting with the predicted leading lepton p_T spectrum, normalised to the measured cross section, 31% of those experiments are less consistent with the parent distribution than the data. Therefore it is concluded that the data agree well with the prediction. To compute the cross section, in each channel the probability to observe the number of events seen in the data as a function of $\sigma_{t\bar{t}}$ given the number of background events and the signal efficiency is calculated. The combined cross section is the

value of $\sigma_{t\bar{t}}$ that maximises the product of the likelihoods in the three channels. Taking the systematic uncertainties and their correlations into account, the resulting top quark pair production cross section at $\sqrt{s} = 1.96$ TeV in dilepton final states is:

$$\sigma_{t\bar{t}} = 8.6^{+3.2}_{-2.7} \,(\text{stat.}) \pm 1.1 \,(\text{syst.}) \pm 0.6 \,(\text{lumi.}) \,\text{pb}, \tag{82}$$

for $m_{\text{top}} = 175 \,\text{GeV}/c^2$.

Using $370 \,\text{pb}^{-1}$ of data, DØ has updated this analysis with a preliminary result [321]. The main modifications with respect to the publication [318] are a slightly lower cut on the total transverse event energy in the $e\mu$ channel: $H_T^\ell = p_T^{\ell_1} + \sum p_{T,j} > 120 \,\text{GeV}$ instead of $> 140 \,\text{GeV}$, a result of a re-optimisation, and a different electron identification in the $e\mu$ channel. Previously, a "tight" electron had to have an electron likelihood value above some cut, where the likelihood compares several calorimeter and track based quantities with reference distributions for good electrons. Now, all electron candidate events are accepted, and the amount of fake electron background is fitted to the observed likelihood distribution in the data. First, the shape of the electron likelihood for real electrons is determined on a pure $Z/\gamma^* \to ee$ sample. The shape of the electron likelihood for fake electron background is determined in a sample dominated by fake electrons, which is selected in the following way: A muon is required to be anti-isolated instead of isolated both in terms of calorimeter and track activity, and $E\!\!\!/_T < 15 \,\text{GeV}$. The number of fake electrons in the selected sample is obtained by performing an extended unbinned likelihood fit to the observed distribution of electron likelihood in data. The likelihood in the fit is given by:

$$\mathcal{L} = \prod_{i=1}^{N} (n_e S(x_i) + n_{\text{fake}} B(x_i)) \frac{e^{-(n_e + n_{\text{fake}})}}{N!}, \tag{83}$$

where i is an index that runs over all selected events, x_i is the corresponding observed value of the electron likelihood, N is the total number of selected events, n_e is the number of events with an isolated electron, n_{fake} is

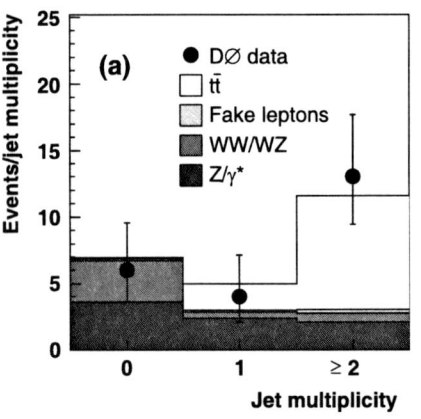 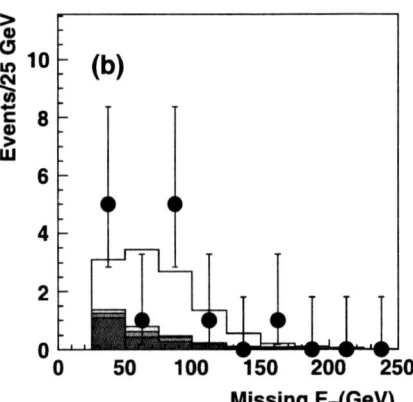 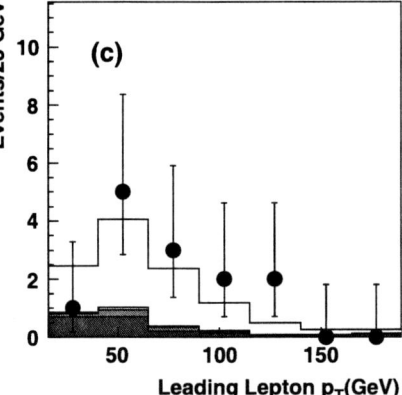

Fig. 68. Observed jet multiplicity (*left*-**a**), $E\!\!\!/_T$ (*middle*-**b**), and leading lepton p_T (*right*-**c**) in comparison to the background estimates plus the expected $t\bar{t}$ contribution, normalised to 7 pb

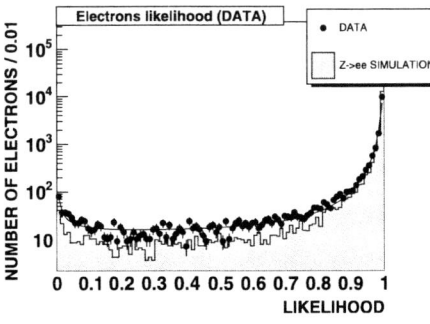 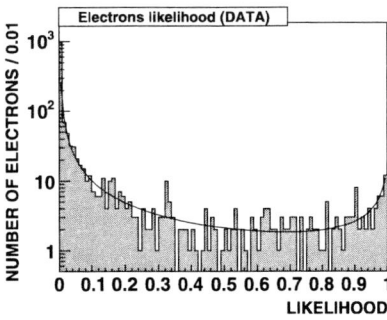 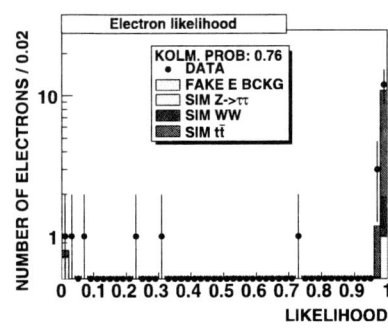

Fig. 69. Electron likelihood discriminant distribution for real electrons (*left*), fake electrons (*middle*), and the observed data (*right*)

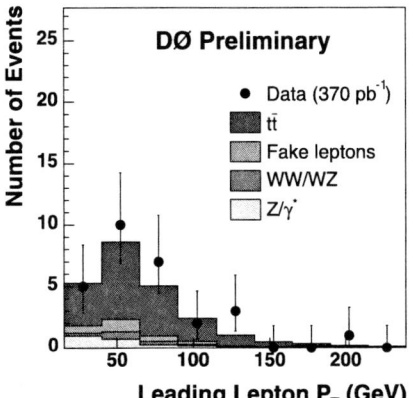 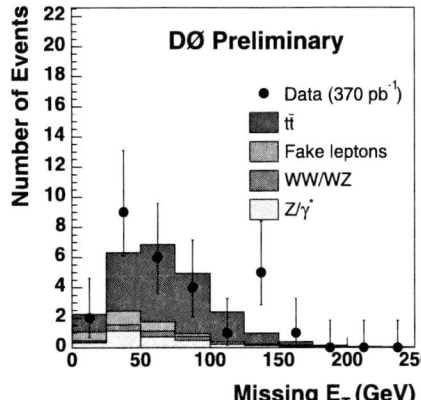 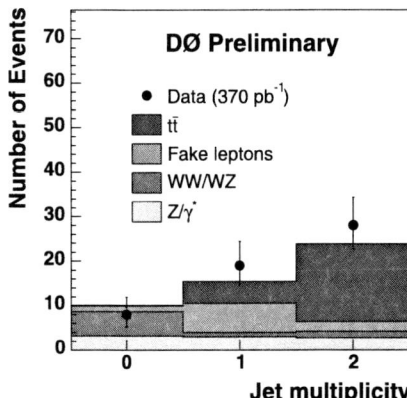

Fig. 70. Observed and predicted distributions for the various backgrounds and the signal. Leading lepton p_T (*left*), $\not{E}_T$ (*middle*), and jet multiplicity (*right*)

the number of events with a fake electron, S is the signal probability distribution function determined using real electrons, and B is the background probability distribution function derived from the sample dominated by fake electrons. Figure 69 shows the shapes of the electron likelihood discriminant distribution in the real and fake electron samples and the distribution observed in data.

Figure 70 shows some of the observed and predicted kinematic distribution for the various backgrounds and the $t\bar{t}$ signal for the leading lepton p_T (left), $\not{E}_T$ (middle), and the jet multiplicity (right). Good agreement between the data and the background plus top signal is found.

The resulting preliminary top quark pair production cross section is measured to be:

$$\sigma_{t\bar{t}} = 8.6^{+2.3}_{-2.0} \,(\text{stat.})^{+1.2}_{-1.0} \,(\text{syst.}) \pm 0.6 \,(\text{lumi.}) \,\text{pb}. \quad (84)$$

4.3.2 Lepton + jets channel in topological analysis

Using $230 \,\text{pb}^{-1}$ of lepton + jets data with four or more jets, DØ has measured the $t\bar{t}$ production cross section based on the kinematic characteristics of the events [322, 323]. Earlier versions of this analysis are described in [324–326]. This analysis exploits only the kinematic properties of the events to separate signal from background, with no assumptions about the multiplicity of final-state b-quarks,

thus providing a less model-dependent determination of the top quark production cross section.

The events in this channel are characterised by the presence of one high-p_T isolated electron (e + jets channel) or muon (μ + jets channel), large transverse energy imbalance due to the undetected neutrino ($\not{E}_T$), and at least four hadronic jets. "Loose" and "tight" electrons, "loose" and "tight" muons, jets and $\not{E}_T$ are reconstructed as described in the previous DØ analysis (see Sect. 4.3.1). At the trigger level, a single electron with transverse momentum (p_T) greater than $15 \,\text{GeV}$, and a jet with $p_T > 15 \,\text{GeV}$ ($20 \,\text{GeV}$) for the first (second) half of the data is required for the e + jets channel. For the μ + jets channel, a single muon detected outside the toroidal magnet (effective $p_T > 3 \,\text{GeV}$ requirement), and a jet with $p_T > 20 \,\text{GeV}$ ($25 \,\text{GeV}$) for the first (second) half of the data is required.

The "tight" selected sample consists of 87 (80) events that have only one tight electron (muon) with $p_T > 20 \,\text{GeV}$, $\not{E}_T > 20 \,\text{GeV}$ and not collinear with the lepton direction in the transverse plane, and at least four jets each with $p_T > 20 \,\text{GeV}$. Removing the tight requirements on the lepton identification results in 230 (140) events passing the selection for the "loose" sample. Monte Carlo simulation of $t\bar{t}$ and W + jets events are used to calculate selection efficiencies and to simulate kinematic characteristics of the events. Top quark signal and W + jets background processes are simulated using ALPGEN 1.2 [224] for the parton-level process, and PYTHIA 6.2 [223] for the subsequent hadronisation. Lepton and jet trigger and identification efficiencies derived from data are applied to the simulated events as correction factors. In the $e+$ jets and $\mu+$ jets channel, the fully corrected selection efficiency for top quark events is found to be $(11.6 \pm 1.7)\%$ and $(11.7 \pm 1.9)\%$ with respect to all $t\bar{t}$ final states that contain an electron or a muon originating either directly from a W-boson or indirectly from $W \to \tau\nu$ decay. The branching fractions of such final states are 17.106% and 17.036% [167] for the $e+$ jets and $\mu+$ jets channels, respectively.

The background within the selected samples is dominated by $W+$ jets events, which have the same signature as $t\bar{t}$ signal events. The samples also include contributions from multijet events in which a jet is misidentified as an electron ($e+$ jets channel) or in which a muon originating from the semileptonic decay of a heavy quark appears isolated ($\mu+$ jets channel). In addition, significant $\not{E}_T$ can arise from fluctuations and mismeasurements of the jet energies. This instrumental background is estimated using the matrix method [327] with the loose and tight samples described above. The loose sample consists of N_s signal events and N_b background events, where N_s is a combination of $W+$ jets and $t\bar{t}$ events. The tight sample consists of $\epsilon_s N_s$ signal events and $\epsilon_b N_b$ multijet background events, where ϵ_s and ϵ_b are the lepton selection efficiency for the tight sample relative to the loose sample, for signal and background, respectively. ϵ_s is measured from a combination of $t\bar{t}$ and $W+$ jets simulated events, corrected for the ratio of efficiencies in data and Monte Carlo for $Z \to \ell^+\ell^-$ events. ϵ_b is measured in a data sample with $\not{E}_T < 10$ GeV, which is dominated by multijet background. For the $e+$ jets channel, $\epsilon_s = 0.82 \pm 0.02$, and $\epsilon_b = 0.16 \pm 0.04$. For the $\mu+$ jets channel, $\epsilon_s = 0.81 \pm 0.02$, and $\epsilon_b = 0.09 \pm 0.03$.

To extract the fraction of $t\bar{t}$ events in the sample, a discriminant function is constructed, that makes use of the difference between the kinematic properties of the $t\bar{t}$ events and the $W+$ jets background. QCD multijet events have similar kinematics to the $W+$ jets events. A set of variables is chosen to provide the best separation between signal and background, whilst having the least sensitivity to the dominant systematic uncertainties coming from the jet energy calibration and the $W+$ jets model. To reduce the dependence on modelling of soft radiation and underlying event, only the four highest-p_T jets were used to determine these

variables. The optimal discriminant function was found to be built from six variables [323]: (i) H_T, the scalar sum of the p_T of the four leading jets; (ii) $\Delta\phi(\ell, \not{E}_T)$, the azimuthal opening angle between the lepton and the missing transverse energy; (iii) $K_{T,\,min} = \Delta R_{jj}^{min} p_T^{min} / E_T^W$, where ΔR_{jj}^{min} is the minimum separation in $\eta - \phi$ space between pairs of jets, p_T^{min} is the p_T of the lowest-p_T jet of that pair, and E_T^W is a scalar sum of the lepton transverse momentum and $\not{E}_T$; (iv) the event centrality $\mathcal{C}$, defined as the ratio of the scalar sum of the p_T of the jets to the scalar sum of the energy of the jets; (v) the event aplanarity $\mathcal{A}$ [328], constructed from the four-momenta of the lepton and the jets; and (vi) the event sphericity $\mathcal{S}$ [328], constructed from the four-momenta of the jets.

The discriminant has the following general form:

$$\mathcal{D} = \frac{S(x_1, x_2, \dots)}{S(x_1, x_2, \dots) + B(x_1, x_2, \dots)}, \quad (85)$$

where $x_1, x_2, \dots$ is a set of input variables and $S(x_1, x_2, \dots)$ and $B(x_1, x_2, \dots)$ are the probability density functions for the $t\bar{t}$ signal and background, respectively. Neglecting the correlation between the input variables, the discriminant function can be approximated by the expression:

$$\mathcal{D} = \frac{\prod_i s_i(x_i)/b_i(x_i)}{\prod_i s_i(x_i)/b_i(x_i) + 1}, \quad (86)$$

where $s_i(x_i)$ and $b_i(x_i)$ are the normalised distributions of variable i for signal and background, respectively. By construction, the discriminant peaks near zero for the background, and near unity for the signal. It is modelled using simulated $t\bar{t}$ and $W+$ jets events, and a data sample selected by requiring that the leptons fail the tight selection criterion, representative of the multijet background. Figure 71 shows the distribution of the discriminant function for data along with the fitted contributions from $t\bar{t}$ signal, $W+$ jets, and multijet background events.

The systematic uncertainties are summarised in Table 24.

A binned Poisson maximum-likelihood fit [323] of the modelled discriminant function distribution (in 10 bins) to that of the data yields the top quark cross section $\sigma_{t\bar{t}}$ and the numbers of $W+$ jets and multijet background events in the selected data sample. The multijet background is

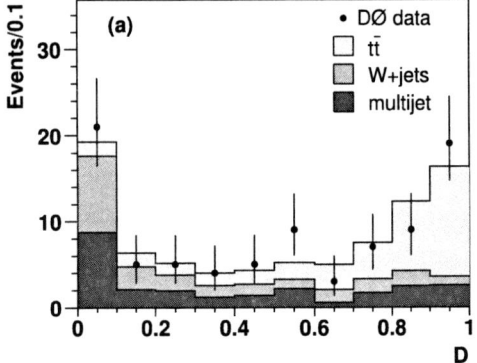
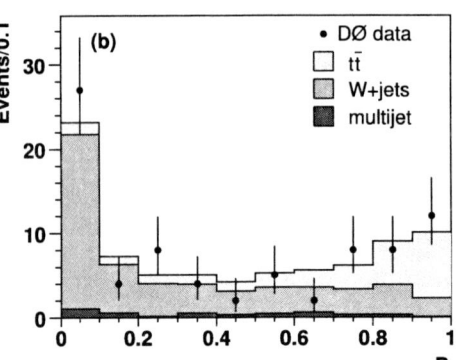

Fig. 71. Discriminant distribution for data overlaid with the result from a fit of $t\bar{t}$ signal, and $W+$ jets and multijet background (*left*) in the $e+$ jets channel and (*right*) in the $\mu+$ jets channel

Table 24. Systematic uncertainties on $\sigma_{t\bar{t}}$ (pb)

Source	$e+\text{jets}$	$\mu+\text{jets}$	$l+\text{jets}$
Lepton identification	± 0.3	± 0.2	± 0.2
Jet energy calibration	$+1.8 -1.2$	$+1.0 -0.7$	$+1.4 -1.0$
Jet identification	$+0.2 -0.2$	$+0.2 -0.1$	$+0.2 -0.1$
Trigger	$+0.1 -0.1$	$+0.4 -0.3$	$+0.3 -0.2$
Multijet background	± 0.3	± 0.03	± 0.2
W background model	± 0.2	± 0.4	± 0.3
MC statistics	± 0.5	± 0.3	± 0.3
Other	± 0.2	± 0.1	± 0.2
Total	$+1.9 -1.3$	$+1.2 -1.0$	$+1.6 -1.1$

constrained within errors to the level determined by the matrix method:

$$\mathcal{L}\left(N_t^{t\bar{t}}, N_t^W, N_t^{\text{QCD}}\right) = \left[\prod_{i=1}^{10} P\left(n_i^{\text{obs}}, \mu_i\right)\right] \times P\left(N_{l-t}^{\text{obs}}, N_{l-t}\right), \quad (87)$$

where $P(n, \mu) = \frac{\mu^n e^{-\mu}}{n!}$ generically denotes the Poisson probability density function for n observed events given an expectation of μ. The t and l indices refer to the "tight" or the "loose" sample selection. Since the logarithm is a monotone function, the solution that maximises the expression in (87), also minimises:

$$-\ln\mathcal{L}\left(N_t^{t\bar{t}}, N_t^W, N_t^{\text{QCD}}\right) = \left[\sum_{i=1}^{10} -n_i^{\text{obs}} \ln \mu_i + \mu_i\right] - N_{l-t}^{\text{obs}} \ln N_{l-t} + N_{l-t}, \quad (88)$$

where on the right hand side any terms independent of $\sigma_{t\bar{t}}$ have been dropped. The number of observed preselected data events that populate the i-th bin in the likelihood discriminant distribution is denoted by n_i^{obs}, the corresponding expected number of events is expressed as a function of $N_t^{t\bar{t}}$, N_t^W, and N_t^{QCD} and is given by:

$$\mu_i\left(N_t^{t\bar{t}}, N_t^W, N_t^{\text{QCD}}\right) = P_i^{t\bar{t}} N_t^{t\bar{t}} + P_i^W N_t^W + P_i^{\text{QCD}} N_t^{\text{QCD}}, \quad (89)$$

where $P_i^{t\bar{t}}$, P_i^W, and P_i^{QCD}, respectively, are the probability density functions for the $t\bar{t}$, W+jets and QCD-multijets likelihood discriminant distributions, evaluated in the i-th bin. The second term in (88) is a Poisson constraint on the observed number of events in the "loose-minus-tight" preselected sample and effectively completes the incorporation of the matrix method in the likelihood, and thus, implicitly, the Poisson constraint on the number of QCD-multijet events. By introducing $N_{l-t} = N_l - N_t$, it is ensured that N_{l-t} and N_t are indeed statistically uncorrelated. $N_{l-t}^{\text{obs}} = N_l^{\text{obs}} - N_t^{\text{obs}}$ is the number of observed data events after the loose preselection minus the number

of observed preselected data events (referred to as "tight" here). The expected number of loose minus tight events, N_{l-t}, can be expressed in terms of $N_t^{t\bar{t}}$, N_t^W, and N_t^{QCD} using the matrix method, yielding:

$$N_{l-t} = \frac{1-\epsilon_s}{\epsilon_s} N_t^{t\bar{t}} + \frac{1-\epsilon_s}{\epsilon_s} N_t^W + \frac{1-\epsilon_b}{\epsilon_b} N_t^{\text{QCD}}. \quad (90)$$

Plugging this term and (89) into (88) yields the likelihood expression to be minimised. Using this fit, the measurement of the $t\bar{t}$ production cross section at $\sqrt{s} = 1.96$ TeV in each lepton channel separately yields:

$$\sigma_{t\bar{t}}^{e+\text{jets}} = 8.2^{+2.1}_{-1.9} \text{ (stat.)}^{+1.9}_{-1.3} \text{ (syst.)} \pm 0.5 \text{ (lumi) pb} \quad (91)$$

$$\sigma_{t\bar{t}}^{\mu+\text{jets}} = 5.4^{+1.8}_{-1.6} \text{ (stat.)}^{+1.2}_{-1.0} \text{ (syst.)} \pm 0.4 \text{ (lumi) pb} \quad (92)$$

for $m_{\text{top}} = 175$ GeV/c^2. The combined cross section in the kinematic lepton + jets analyses is determined by minimising the sum of the negative log-likelihood functions for both channels, yielding:

$$\sigma_{t\bar{t}}^{l+\text{jets}} = 6.7^{+1.4}_{-1.3} \text{ (stat.)}^{+1.6}_{-1.1} \text{ (syst.)} \pm 0.4 \text{ (lumi.) pb}, \quad (93)$$

for $m_{\text{top}} = 175$ GeV/c^2.

Combined topological cross section in DØ. DØ has also combined the $t\bar{t}$ production cross section measurements in the topological dilepton analysis (Sect. 4.3.1) and of the kinematic lepton + jets analysis described above. All those measurements are based on ≈ 230 pb^{-1}. In this combination, the Poisson likelihoods from the counting experiments in the dilepton channels and the likelihoods on the 10 bins of the kinematic discriminant in the lepton + jets channel with the multijet background constraint are multiplied. The overall likelihood maximum is determined by taking the correlations of systematic uncertainties in the signal and the various background sources into account separately. Four background processes (WW, $Z \to \tau\tau$, $Z \to \mu\mu$, $W\gamma$) are considered separately and considered fully correlated across all measurements. For the lepton + jets channel, also the correlations between sources of systematic uncertainties in the event selection and possible distortions in the shape of the likelihood discriminant reference distributions are considered. For each systematic uncertainty, the corresponding efficiencies or the shapes of the likelihood discriminant are changed and a new cross section is determined. These variations in the central value of the cross section are then added quadratically to obtain the total systematic uncertainty. By construction, this method of the cross section computation does not allow the systematic errors to influence the results of the fit.

The combined $t\bar{t}$ production cross section in topological/kinematic analyses is then found to be [329]:

$$\sigma_{t\bar{t}} = 7.1^{+1.2}_{-1.2} \text{ (stat.)}^{+1.4}_{-1.1} \text{ (syst.)} \pm 0.5 \text{ (lumi.) pb}, \quad (94)$$

for $m_{\rm top} = 175\,{\rm GeV}/c^2$. In the region $170\,{\rm GeV}/c^2$ to $180\,{\rm GeV}/c^2$ the cross section changes as a function of $m_{\rm top}$ as:

$$\sigma_{t\bar{t}}(m_{\rm top}) = \sigma_{t\bar{t}} - 0.1 \frac{\rm pb}{{\rm GeV}/c^2} \times (m_{\rm top} - 175\,{\rm GeV}/c^2). \tag{95}$$

4.3.3 Lepton + jets channel in b-tag analyses

Based on $230\,{\rm pb}^{-1}$ of data, DØ has measured the $t\bar{t}$ production cross section in the lepton + jets channel using a lifetime-based b-jet identification technique [330]. Electrons, muons, jets and $\not{E}_{\rm T}$ are reconstructed as in the kinematic lepton + jets analysis, described before in Sect. 4.3.2. In both channels the events are selected by requiring $\not{E}_{\rm T}$ to exceed $20\,{\rm GeV}$ and not be collinear with the lepton direction in the transverse plane. In the electron and muon channel, an isolated electron with $p_{\rm T} > 20\,{\rm GeV}$ and $|\eta| < 1.1$, or isolated muon with $p_{\rm T} > 20\,{\rm GeV}$ and $|\eta| < 2.0$ are required. Events with 3 or ≥ 4 jets are expected to be enriched in $t\bar{t}$ signal and used for the $\sigma_{t\bar{t}}$ measurement, whereas events with only 1 or 2 jets are expected to be dominated by background and used to verify the background normalisation procedure.

The main background in this analysis is the production of W-bosons in association with jets (W + jets), with the W-boson decaying leptonically. In most cases, the jets accompanying the W-boson originate from light (u, d, s) quarks and gluons (W + light jets). Depending on the jet multiplicity, between 2% and 14% of W + jets events contain heavy flavour jets resulting from gluon splitting into $b\bar{b}$ or $c\bar{c}$ ($Wb\bar{b}$ or $Wc\bar{c}$, respectively), while in about 5% of events, a single c quark is present in the final state as a result of the W-boson radiated from an s quark from the proton's or antiproton's quark sea (Wc). A sizeable background arises from strong production of two or more jets ('multijets'), with one of the jets misidentified as a lepton and accompanied by large $\not{E}_{\rm T}$ resulting from mismeasurements of jet energies. Significantly smaller contributions to the background arise from single top, Z + jets, and weak diboson (WW, WZ and ZZ) production. Only a small fraction of the background events contain b or c-quark jets in the final state. As a consequence, the signal-to-background ratio is significantly enhanced when at least one jet is identified as a b-quark jet.

b-quark jets are identified using a secondary vertex tagging (SVT) algorithm, which is described in Sect. 3.3.7. Events with exactly $1 (\geq 2)$ tagged jets are referred to as single-tag (double-tag) events. They are treated separately because of their different signal-to-background ratios.

$t\bar{t}$ production, and all background processes except multijets are simulated, using ALPGEN [224] to generate the parton-level processes, and PYTHIA [223] to provide fragmentation and to decay unstable particles except B hadrons and τ leptons, which are modelled via EVTGEN [243] and TAUOLA [245], respectively. Lepton and jet resolutions and identification efficiencies are adjusted to that measured in data (for example in $Z \to \ell\ell$ events). For all processes except the multijets background,

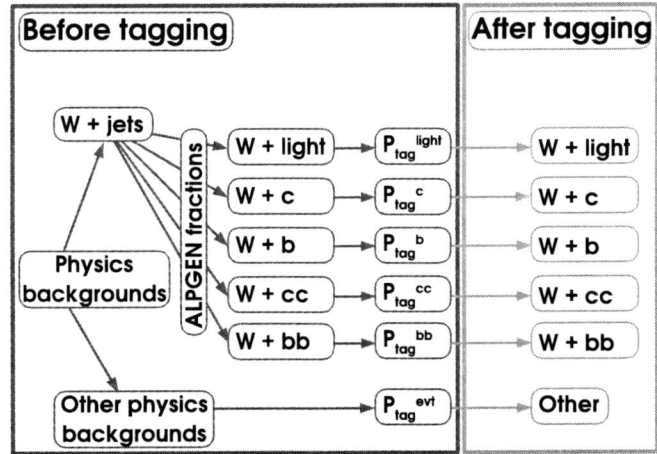

Fig. 72. Schematic of the b-tagging procedure: the background contamination is assessed by applying the tag rate functions, determined in data, to each jet in the simulation, taking the jet flavour, $p_{\rm T}$ and η into account

the total acceptance is computed using the MC simulation, where trigger, reconstruction and tagging efficiencies are taken from the data measurements. The tagging probability for a particular process depends on the flavour composition of the jets in the final state as well as on the overall event kinematics. It is estimated by applying the tagging rate measured in data to each jet in the simulation, taking into consideration its flavour, $p_{\rm T}$, and η. In the case of W + jets events, the simulation (ALPGEN [224]) is also used to estimate the fraction of the different W + heavy flavour subprocesses (Fig. 72).

The $t\bar{t}$ acceptance is computed for events with a true electron or muon arising from a $W \to \ell\nu$ ($\ell = e, \mu, \tau$) decay, corresponding to total branching fractions of 17.106% and 17.036% [167], respectively, in the electron and muon channels. In the electron channel, the total acceptance before tagging is estimated to be $(10.8 \pm 0.8)\%$ and $(14.2 \pm 1.7)\%$, for events with 3 and those with ≥ 4 jets, respectively. The corresponding numbers for the muon channel are $(9.9 \pm 1.0)\%$ and $(14.1 \pm 1.9)\%$. The estimated single-tag efficiencies are $(43.3 \pm 1.2)\%0$ and $(45.3 \pm 1.0)\%$ for events with 3 and those with ≥ 4 jets, respectively. The corresponding double-tag efficiencies are $(10.4 \pm 1.0)\%$ and $(14.2 \pm 1.3)\%$. Consequently, the average probability to tag a $t\bar{t}$ event with four or more jets is measured to be 60%. The number of multijet events is estimated from the data for each jet multiplicity using the matrix method [327], separately for the samples before and after tagging. Smaller contributions from single top, Z + jets, and diboson production (collectively referred to as "other bkg" in Fig. 73) are estimated from the simulation, normalised to the next-to-leading order theoretical cross section [162, 247, 331]. The number of tagged W + jets events is estimated as the product of the number of W + jets events in data before tagging and the average tagging probability for W + jets events (e.g. $\approx 4\%$ for single-tag and $\approx 0.4\%$ for double-tag events with ≥ 4 jets), correcting the number of W + jets events before tagging for the contribution of the other backgrounds.

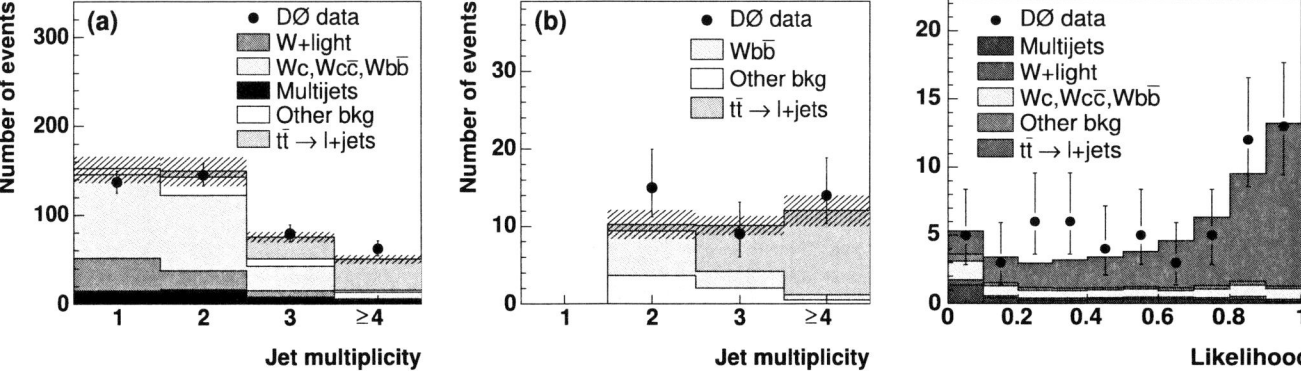

Fig. 73. Expected and observed number of single-tag (*left*) and double-tag events (*middle*). The hatched area represents the total uncertainty in the expectation. *Right*: Likelihood discriminant distribution (as defined in the kinematic analysis in Sect. 4.3.2) for the ℓ + jets events with four or more jets and one tag overlaid with the sum of the predicted background and $t\bar{t}$ signal

Table 25. Systematic uncertainties on $\sigma_{t\bar{t}}$ (pb)

Source	$\Delta\sigma_{t\bar{t}}$	
b-tagging efficiency	+0.6	−0.5
Jet energy calibration	+0.5	−0.4
Background modelling	±0.5	
Lepton selections	+0.5	−0.4
Jet identification	+0.3	−0.2
Multijet background	+0.3	−0.2
Mis-tagging rate	±0.1	
Total	+1.1	−1.0

Figure 73 shows the observed and expected number of events for each jet multiplicity. The excess over the background expectation in the third and fourth multiplicity bins is interpreted as the $t\bar{t}$ signal. The good agreement between observation and expectation in the first and second multiplicity bins validates the background estimation procedure.

The $t\bar{t}$ production cross section is calculated by maximising a likelihood function including a Poisson term for each of the eight independent channels considered: 3 and ≥ 4 jets, for single- and double-tag events in the electron and muon channels. At each step in the maximisation, the

multijet background in these eight tagged samples, and the corresponding samples before tagging, is constrained within errors to the amount determined by the matrix method. In addition, for each of the considered systematic uncertainties a Gaussian term is included, following the procedure described in [332]. In this approach, each source of systematic uncertainty is allowed to affect the central value of the cross section during the maximisation procedure. The systematic uncertainties are summarised in Table 25. In addition, a systematic uncertainty of 6.5% from the luminosity measured is assigned.

The $t\bar{t}$ production cross section measurement yields

$$\sigma_{t\bar{t}} = 8.6^{+1.6}_{-1.5} \,(\text{stat.} + \text{syst.}) \pm 0.6 \,(\text{lumi.}) \,\text{pb}$$
$$= 8.6^{+1.2}_{-1.1} \,(\text{stat.})^{+1.1}_{-1.0} \,(\text{syst.}) \pm 0.6 \,(\text{lumi.}) \,\text{pb}, \quad (96)$$

for $m_{\text{top}} = 175 \,\text{GeV}/c^2$.

Using $365 \,\text{pb}^{-1}$ of data, DØ has updated this analysis with a preliminary result [333]. Apart from the significantly increased statistics, there are no major modifications made to the published analysis described above. Figure 74 shows the observed and expected number of events for each jet multiplicity. The excess over the background expectation in the third and fourth multiplicity bins is interpreted as the $t\bar{t}$ signal. Again, the good agreement

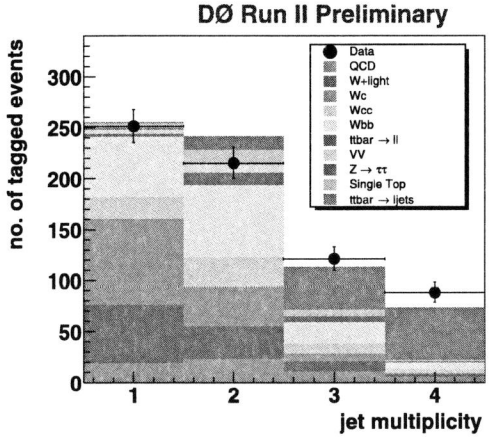

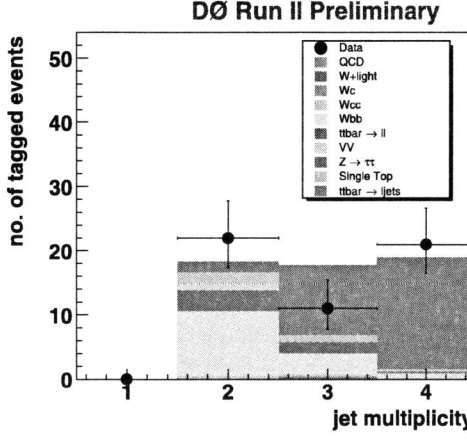

Fig. 74. Expected and observed number of single-tag (*left*) and double-tag events (*right*)

Table 26. Systematic uncertainties on $\sigma_{t\bar{t}}$ (pb)

Source	Offset (pb)	σ^+ (pb)	σ^- (pb)
Muon preselection	+0.02	+0.18	−0.15
Electron preselection	−0.02	+0.18	−0.15
Muon triggers	+0.07	+0.34	−0.28
Jet energy calibration	−0.07	+0.24	−0.21
Jet reco and jet ID	−0.09	+0.23	−0.18
SML b-tag efficiency (MC)	+0.03	+0.15	−0.14
SML b-tag efficiency (data)	+0.18	+0.40	−0.35
Heavy quark mass on W fractions	−0.00	+0.18	−0.19
W fractions matching + higher order effects	+0.01	+0.44	−0.44
Event statistics for matrix method	−0.02	+0.15	−0.15
Total		+0.9	−0.8

between observation and expectation in the first and second multiplicity bins validates the background estimation procedure.

The systematic uncertainties are summarised in Table 26. In addition, a systematic uncertainty of 6.5% from the luminosity measured is assigned.

This updated preliminary $t\bar{t}$ production cross section measurement yields

$$\sigma_{t\bar{t}} = 8.1^{+1.3}_{-1.2} \,(\text{stat.} + \text{syst.}) \pm 0.5 \,(\text{lumi.}) \,\text{pb}$$
$$= 8.1 \pm 0.9 \,(\text{stat.})^{+0.9}_{-0.8} \,(\text{syst.}) \pm 0.5 \,(\text{lumi.}) \,\text{pb},$$

$$(97)$$

for $m_{\text{top}} = 175 \,\text{GeV}/\text{c}^2$.

DØ has performed a simultaneous measurement of the $t\bar{t}$ production cross section and the ratio of top quark decay branching ratios $\mathcal{B}(t \to Wb)/\mathcal{B}(t \to Wq)$ in the lepton + jets channel, counting the number of ℓ + jets + $\not{E}_T$ events with 0, 1, and 2 b-jets in 230 pb^{-1} of data [334]. This analysis is essentially a combination of the $t\bar{t}$ cross section measurements in the lepton + jets channel using kinematic event characteristics (see Sect. 4.3.2) and using secondary vertex b-tagging (see Sect. 4.3.3). The analysis is described in more detail in Sect. 6.2.

The preliminary $t\bar{t}$ production cross section measurement in this analysis is

$$\sigma_{t\bar{t}} = 7.9^{+1.7}_{-1.5} \,(\text{stat.} + \text{syst.}) \pm 0.5 \,(\text{lumi.}) \,\text{pb}$$
$$= 7.9^{+1.4}_{-1.3} \,(\text{stat.})^{+0.9}_{-0.8} \,(\text{syst.}) \pm 0.5 \,(\text{lumi.}) \,\text{pb}, \quad (98)$$

for $m_{\text{top}} = 175 \,\text{GeV}/\text{c}^2$.

4.3.4 Dilepton channel in b-tag analysis

Based on 158 pb^{-1} of data, DØ has measured the $t\bar{t}$ production cross section in the $t\bar{t} \to e\mu$ + jets channel using secondary vertex b-tagging [335]. In the $e\mu$ channel, the final state is characterised by a high p_T isolated electron, a high p_T isolated muon, large missing transverse energy and two b-jets. The electrons, muons, jets, and $\not{E}_T$ are reconstructed in the same way as in the topological dilepton

analyses (see Sect. 4.3.1). Also the general event selection is identical to that analysis. The selection used is electron $p_T > 15 \,\text{GeV}$, muon $p_T > 15 \,\text{GeV}$, jet $p_T > 20 \,\text{GeV}$, and $\not{E}_T > 25 \,\text{GeV}$. The muon and electron tracks are further required to point to the primary vertex. Only events with one or more jets are considered.

$t\bar{t}$ events contain two b-jets while jets in the background processes originate predominantly from light quarks or gluons. Requiring at least one jet in the event to be b-tagged is therefore a powerful discriminant between signal and background. The b-tagging algorithm utilised in this measurement is the secondary vertex tagging algorithm (SVT; see Sect. 3.3.7). The measurements of tagging efficiency and tag rate function for mistags are described in Sect. 4.3.3.

The preselection efficiency for $t\bar{t}$ events is estimated using Monte Carlo events, which are generated with ALPGEN 1.2 [224] and PYTHIA 6.2 [223]. The W-bosons both decay to a lepton-neutrino pair, including all τ final states. The same Monte Carlo sample is used to estimate the tagging efficiency by taking the jet kinematics from the Monte Carlo and folding in the per jet tagging efficiency parameterisations determined in data.

Background processes that can produce the full $e\mu$ signature (one electron, one muon, jets and significant missing transverse energy) are rare. Decays of $Z/\gamma^* \to \tau\tau$ which subsequently decay to an electron and a muon is the largest physics background, but suffers from the low branching ratio of the two τ's to decay to leptons, as well as from soft lepton and neutrino p_T spectra. $(Z/\gamma^* \to \tau\tau)jj$ is generated using ALPGEN [224] followed by PYTHIA [223]. To evaluate the Z background for lower jet multiplicities, PYTHIA samples are used, normalised to the DØ measurement [336]. The $(Z/\gamma^* \to \tau\tau)jj$ is normalised to the ALPGEN cross section, corrected for the difference in the yield for Z events observed in data and Monte Carlo. Another physics background is the diboson production, that of WW being the most important since the leptons resemble very much the ones in $t\bar{t}$ events. This background has a very low cross section. The most powerful discriminant in this measurement is, however, the requirement of at least one b-tagged jet. $WW \to \ell\ell\nu\nu(jj)$ and $WW \to \ell\ell\nu\nu$ Monte Carlo samples are generated with ALPGEN followed by PYTHIA. The $WW \to \ell\ell\nu\nu$ sample is normalised to the next-to-leading-order cross section calculation [247], which provides a 35% larger cross section than that calculated with ALPGEN in LO. For consistency, also the $WW \to \ell\ell\nu\nu$ sample is normalised to the ALPGEN cross section, scaled up by 35%. The preselection efficiencies for the physics backgrounds are taken from the Monte Carlo. The b-tagging efficiency in the $Z/\gamma^* \to \tau\tau$ background is estimated in data, using $Z \to ee$ and $Z \to \mu\mu$ events. For the small WW background, the tagging efficiency is estimated in Monte Carlo, by taking the jet kinematics from the Monte Carlo, and folding in the per jet tagging efficiency parameterisation determined on data. Fake electron backgrounds are estimated by measuring the rate at which an electromagnetic jet is misidentified as an electron. This is done in a control data sample where the muon in the event is non-isolated. The electron fake rate is then applied

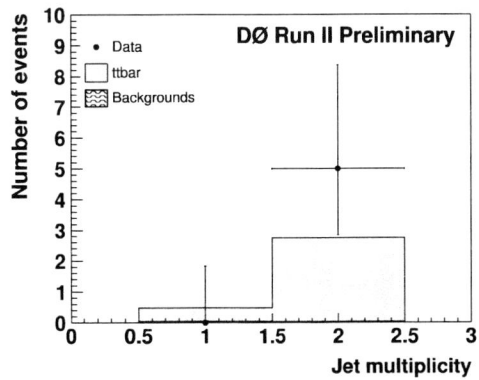
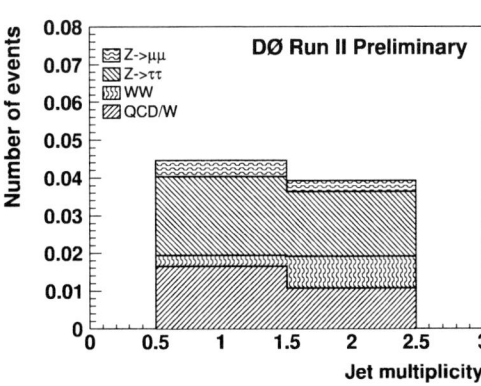

Fig. 75. The number of observed and predicted b-tagged events (*left*) and the predicted number of background events (*right*) as a function of the jet multiplicity. The $t\bar{t}$ signal is here normalised to 7 pb

to the data sample that fulfils all requirements except that the electron identification cuts have been omitted. The efficiency for the very small $Z \to \mu\mu$ background (where one of the muons emits a Bremsstrahlung photon) to pass the preselection cuts is estimated using Monte Carlo. The probability for a QCD or $W+$jets event to be b-tagged is estimated in data, from events containing an isolated muon passing all the muon identification cuts, and having $\not{E}_T < 10$ GeV. The b-tagging efficiency for $Z/\gamma^* \to \mu\mu$ background is determined in the same way as for the $Z/\gamma^* \to \tau\tau$ background.

Figure 75 shows the number of observed and predicted b-tagged events in the two jet multiplicity bins. Given the low statistics, the data is well described by the expected background (right) plus the $t\bar{t}$ signal, normalised to 7 pb.

The $t\bar{t}$ cross section is calculated by maximising the product of the Poisson likelihood functions constructed for each of the two multiplicity bins ($N_{\text{jet}} = 1$ and $N_{\text{jet}} \geq 2$), taking into account correlated sources of systematic uncertainties. Table 27 summarises the systematic uncertainties of this analysis. The preliminary $t\bar{t}$ production cross section measurement yields

$$\sigma_{t\bar{t}} = 11.1^{+5.8}_{-4.3}\,(\text{stat.}) \pm 1.4\,(\text{syst.}) \pm 0.7\,(\text{lumi.})\,\text{pb}\,, \quad (99)$$

for $m_{\text{top}} = 175$ GeV/c^2.

Table 27. Systematic uncertainties on $\sigma_{t\bar{t}}$ (pb) in the $e\mu$ channel with b-tagging

Source	$\Delta\sigma_{t\bar{t}}$ (pb)	
b-tagging efficiency in data	$+0.75$	-0.64
b-tagging efficiency in Monte Carlo	$+0.48$	-0.46
Decay model dependence of tagging efficiency		-0.32
Taggability	$+0.39$	-0.80
Jet energy calibration	$+0.51$	-0.38
Jet energy resolution		-0.023
Jet identification	$+0.26$	
Trigger	$+0.34$	-0.27
Top quark mass	$+0.43$	-0.37
Monte Carlo to data correction factors	± 0.46	
Monte Carlo and data statistics	± 0.34	
Total	± 1.4	

4.3.5 All-jets analyses

DØ has measured the $t\bar{t}$ production cross section in the all-jets (or all hadronic) channel, based on 162 pb^{-1} of data, using secondary vertex b-tagging and several kinematic and topological quantities, combined in artificial neutral networks, to extract the top quark signal [337]. This channel constitutes 46% of the total $t\bar{t}$ production cross section, larger than any other channel with one or both W-bosons decaying to leptons. It has the advantage that all partons from the $t\bar{t}$ process decay to particles that are visible in the detector. There are no energetic neutrinos produced. The all hadronic final state is characterised by the presence of at least six jets, two of which result from the hadronisation of b-quarks. Since the cross section of multijet production via the strong interaction is several orders of magnitude larger than the $t\bar{t}$ cross section, multijet background is overwhelming in this channel. The bulk of this background is rejected by the requirement of one identified b-jet in the event. Subsequently kinematic and topological variables are combined into a chain of artificial neural networks (ANNs) to extract the signal.

ALPGEN 1.2 [224] and PYTHIA [223] are used to simulate $t\bar{t}$ signal events, both in the all jets and in the lepton+jets channel. Note that this $t\bar{t}$ Monte Carlo is *only* used for neural network training. Events are triggered with a dedicated multi-jet trigger. The corresponding single-jet turn-on curve is applied to the $t\bar{t}$ Monte Carlo. In the preselection, six or more jets with $p_T > 15$ GeV and $|\eta| < 2.5$ are required. In order to obtain a data set orthogonal to the lepton+jets selection, events are rejected that contain isolated leptons. The algorithm used for b-quark jet identification is the secondary vertex tagger (SVT) (see Sect. 3.3.7). In this analysis exactly one tagged jet per event is required, because the probability to tag a second jet is larger, resulting from the change of flavour content in the sample after the first tag. The efficiencies to tag b, c and light quarks and gluon jets are measured using data. They are parameterised as a function of the jet p_T for different η bins and yield an overall $t\bar{t}$ event tagging efficiency of 46%.

The background is determined by applying mis-tagging tag rate functions (TRFs). Since both the tagged and the untagged sample are dominated by background, they are used to determine the probability that a 'background' jet will be tagged:

$$P_{\text{jet}}(p_T, \eta) = N f(p_T) g(\eta) \,. \tag{100}$$

As $f(p_T)$ and $g(\eta)$ are observed to be uncorrelated, they are factorised. $P_{\text{jet}}(p_T, \eta)$ is parameterised in four different bins of H_T. N is the normalisation for this probability density function, measured in data to be $N = 1.04 \pm 0.07$. Its uncertainty is used to estimate the uncertainty on the background prediction. Assuming that the tagging probabilities are uncorrelated between jets, the probability for an event to have one single tag is:

$$P_{\text{event}}(\text{tags} = 1) = \sum_{i=1}^{N_{\text{jet}}} P_{\text{jet}}(i) \prod_{j}^{i \neq j} (1 - P_{\text{jet}}(j)) \,. \tag{101}$$

The variables used to discriminate hadronic top signal from QCD multijet background can be distinguished into five categories, taken from the equivalent DØ Run-I analysis [338]:

(i) **Energy scale**: QCD background tends to have an overall lower transverse energy distribution, jets are less energetic and the total invariant mass of the events is smaller than in $t\bar{t}$ events. Even though the average jet energy is smaller in QCD events, the leading jets tend to be more energetic in QCD than in $t\bar{t}$ events. The variables used here are H_T, the scalar sum of all jet P_T's and $\sqrt{s}$, the invariant mass of the event.

(ii) **Soft non-leading jets**: As the QCD background mainly consists of hard 2-jet processes with extra soft gluon jets, the additional jets are expected to be softer in QCD background than in $t\bar{t}$ signal. The variables used are H_T^{3j}, the scalar sum of all jet p_T's except the leading two, $E_{T_{4,5}}$, the geometric mean of the fifth and sixth jet, and $\langle N_{\text{jet}} \rangle$, the p_T weighted jet multiplicity.

(iii) **Event shape**: These quantities describe the behaviour of the angles and sizes of jets in the event as a whole. Top events have a different shape than QCD background. The jets are almost spherically distributed in top events, while QCD events usually have

a more back-to-back jet distribution. The difference between signal and background is quantified by using aplanarity and sphericity [328].

(iv) **Rapidity distribution**: These quantities are used to identify where the set of jets in the event was observed in the detector. Because of their typical hard scatter origin, the jets in QCD background events are expected to be more back-to-back than top signal, while QCD events are also more likely to be boosted in the direction of the beam-line. This has a consequence that not all the jets in the event are expected to be central. The centrality, the ratio of the total transverse energy over the total energy in the event, and $\langle \eta^2 \rangle$, the p_T weighted jet variance from $\eta = 0$ are used.

(v) **Typical top properties**: The second neural network (NN2) is trained on properties which are typical for the top event structure, like the presence of W-bosons and b-quarks. The variables used are the W and top mass likelihoods $\mathcal{M}_{WW}$ and $\mathcal{M}_{t\bar{t}}$, and the minimal di-jet masses $M_{\text{min}}^{1,2}$ and $M_{\text{min}}^{3,4}$.

As an example, the H_T^{3j} distribution is shown in Fig. 76 (left) for tagged data events, along with the predicted background in those tagged events, and the hadronic $t\bar{t}$ signal Monte Carlo. The other variables have similar separation power. The known correlations of all the above listed kinematic variables is exploited for optimal discrimination between signal and background by using artificial neural networks (ANN).

Here, feed-forward neural networks (NNs) are trained by back propagation as implemented in JETNET [305]. All NNs have one output node and one middle layer with a number of nodes twice the number of the input layers. The NNs are trained on a small, randomly chosen fraction (2500 events, $\approx 1\%$) of the total background sample and the same number of simulated $t\bar{t}$ Monte Carlo events. Tagged data events are not used in the training of the NNs. However, the probability that an event is tagged, is used to select events for the random sample. Events with high tagging probability are more likely to be used for neural network training, so that the background training sample

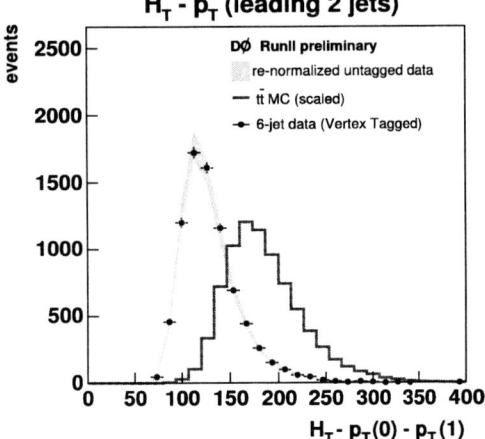

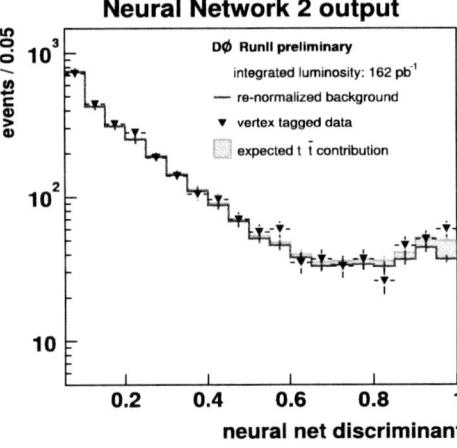

Fig. 76. *Left*: Distribution of H_T^{3j} for tagged data events (*points*), the predicted background in those tagged events (*grey shaded band*), and the hadronic $t\bar{t}$ signal Monte Carlo (*histogram*). *Right*: Neural network 2 output distribution for tagged data, expected background, and expected signal + background

contains taggable jets that are similar to the tagged events in the final data sample.

The kinematic variables described above are combined into a chain of neural networks:

– $NN0$: This neural network is used for rejection of obvious background. Events with an $NN0$ discriminant smaller than 0.05 are rejected. $NN0$ uses H_{T}, $\sqrt{s}$, H_{T}^{3j}, $\langle N_{\mathrm{jets}} \rangle$, sphericity, aplanarity, and centrality. $NN0$ does not have a major influence on the signal, but does reject a significant fraction of the background.

– $NN1$: The next neural network, $NN1$, includes the same variables as $NN0$, with additional information from $E_{\mathrm{T}_{4,5}}$ and $\langle \eta^2 \rangle$. As this neural network is trained only on events that pass the $NN0 > 0.05$ cut, it is a lot more sensitive to the difference between top-like background and hadronic $t\bar{t}$ signal. No cut is imposed on $NN1$, but it is used as input to the final neural network.

– $NN2$: The final neural network, $NN2$, combines the information from $NN1$ with variables which are sensitive to high-mass objects in the event, or to the $t\bar{t}$ hypothesis. The input variables for $NN2$ are $NN1$, $\mathcal{M}_{WW}$, $\mathcal{M}_{t\bar{t}}$, $M_{\mathrm{min}}^{1,2}$ and $M_{\mathrm{min}}^{3,4}$.

Figure 76 (right) shows the resulting $NN2$ output distribution for the expected background, signal + background and the actual observed tagged data distribution. The optimal $NN2$ cut, minimising the expected statistical error in $t\bar{t}$ Monte Carlo, has been found to be $NN2 > 0.75$. The resulting selection efficiency for hadronic $t\bar{t}$ events is $\epsilon_{\mathrm{all\text{-}jets}} = 0.058 \pm 0.001$ (stat.), with an additional marginal efficiency to top events in the lepton + jets channel of $\epsilon_{\mathrm{lepton+jets}} = 0.0018 \pm 0.0001$ (stat.). They are combined, according to the expected branching ratios (0.4619 for the all-jets and 0.4355 for the lepton + jets channels) and yield a final signal efficiency of

$$\epsilon \cdot \mathrm{BR} = 0.0275 \pm 0.0003 \, (\mathrm{stat.})^{+0.0086}_{-0.0079} \, (\mathrm{syst.}) \, . \tag{102}$$

The systematic uncertainties, dominated by jet energy calibration and top quark mass, are summarised in Table 28.

The number of observed events is $N_{\mathrm{observed}} = 220$ with an expected background of $N_{\mathrm{background}}^{\mathrm{expected}} = 186 \pm 5 (\mathrm{stat.})$. Note that the statistical error is so small because the background events are estimated from more than 186 events, weighted with TRFs. This preliminary $t\bar{t}$ production cross section measurement yields

$$\sigma_{t\bar{t}} = 7.7^{+3.4}_{-3.3} \, (\mathrm{stat.})^{+4.7}_{-3.8} \, (\mathrm{syst.}) \pm 0.5 \, (\mathrm{lumi.}) \, \mathrm{pb} \, , \tag{103}$$

for $m_{\mathrm{top}} = 175 \, \mathrm{GeV}/c^2$.

Using $350 \, \mathrm{pb}^{-1}$ of data, DØ has updated this preliminary analysis [339]. This analysis is very similar to the previous one, just described. One of the main modifications in this update is, that events are selected that have at least one b-tagged jet, so double-tagged events are now also included. The average probability to have at least one jet b-tagged by the SVT algorithm in a signal $t\bar{t}$ event is 61%. The background is estimated using the same technique of

Table 28. Systematic uncertainties on $\sigma_{t\bar{t}}$ (pb)

Signal source	$\Delta\epsilon$ (%)	
Vertex reconstruction	-1.0	$+1.0$
Jet identification	-9.8	–
Jet energy calibration	-28.3	$+28.1$
Jet energy resolution	-0.6	$+0.2$
Top quark mass $\pm 5 \, \mathrm{GeV}/c^2$	-7.6	$+5.9$
Trigger	-4.0	$+4.0$
SVT parameterisation	-4.1	$+3.6$
Total	-31.1	$+31.2$

Background source	$\Delta N_{\mathrm{background}}^{\mathrm{expected}}$ (%)
Statistical error on TRFs	± 3.6
Background modelling (TRFs)	± 6.6
Total	± 7.5

parameterised tag rate functions. In an additional selection requirement, multijet background originating from gluon splitting to heavy quark pairs is removed by demanding that the distance $\Delta R = \sqrt{\Delta\phi^2 + \Delta\eta^2}$ between two tagged jets is larger than 1.5. This requirement is necessary since the per-jet TRFs do not provide an adequate description of the correlations existing in $b\bar{b}$ production.

The other substantial modification in the analysis is the choice of 'only' six instead of the original twelve kinematic variables, which are now combined into 'only' one instead of the previous three artificial neural networks. These six kinematic variables are chosen:

– H_{T}, the scalar sum of the p_{T} of the four leading jets,
– Aplanarity $\mathcal{A}$, a linear combination of the eigenvalues of a normalised momentum tensor,
– $E_{\mathrm{T}_{5,6}}$, the geometric mean of the transverse energies of the fifth and sixth jet in the event,
– $\langle \eta^2 \rangle$, the weighted root-mean-square (RMS) of the η of the six leading jets in the event,
– $M_{\mathrm{min}}^{3,4}$, the second-smallest dijet mass in the event,
– the mass likelihood $\mathcal{M}$, a χ^2-like variable

$$\mathcal{M} = \frac{(M_{W_1} - M_W)^2}{\sigma_W^2} + \frac{(M_{W_2} - M_W)^2}{\sigma_W^2} + \frac{(M_{t_1} - M_{t_2})^2}{\sigma_{\mathrm{t}}^2} \, , \tag{104}$$

where $m_W = 83 \, \mathrm{GeV}$ and $\sigma_W = 13 \, \mathrm{GeV}$ are the expected central value and standard deviation of the W-boson mass peak, respectively, obtained from $t\bar{t}$ all hadronic Monte Carlo along with the resolution of the top quark mass, $\sigma_{\mathrm{t}} = 22 \, \mathrm{GeV}$. $\mathcal{M}$ is calculated for each possible assignment of jets to the W's and b-quarks, while only the permutation with the smallest $\mathcal{M}$ is used in this analysis.

The choice of variables is made such that variables that are known to be highly dependent on the jet energy calibration are avoided. These six variables are combined into one neural network called NN_{all}. Figure 77 shows the output distribution of NN_{all} for tagged data, the expected background and $t\bar{t}$ to all hadrons signal.

Table 29. Systematic uncertainties on $\sigma_{t\bar{t}}$ (pb)

Source	$\Delta\sigma_{t\bar{t}}$ (pb)	
Jet energy calibration	+1.12	−0.73
Jet identification	+0.68	−0.42
Trigger	+0.27	−0.05
Background prediction	+0.52	−0.50
$t\bar{t}$ tagging probability	+0.34	−0.29
Total	+1.48	−1.02

The output of the neural network is used to select the sample enriched in $t\bar{t}$ signal by applying the cut $NN_{\mathrm{all}} > 0.9$, which is optimised to minimise the fractional statistical error of the cross section measurement. After this cut, $N_{\mathrm{obs}} = 541$ events are observed with a predicted background of $N_{\mathrm{TRF}} = 494$. The $t\bar{t}$ production cross section is given by

$$\sigma_{t\bar{t}} = \frac{N_{\mathrm{obs}} - N_{\mathrm{TRF}}}{\epsilon_{t\bar{t}}\mathcal{L}\left(1 - \frac{\epsilon_{\mathrm{TRF}}}{\epsilon_{b\text{-}\mathrm{tag}}}\right)} , \qquad (105)$$

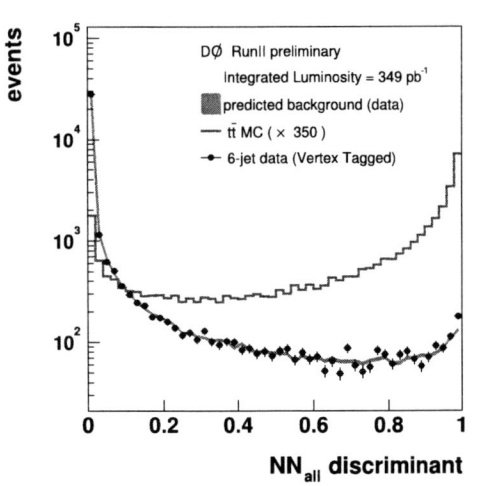
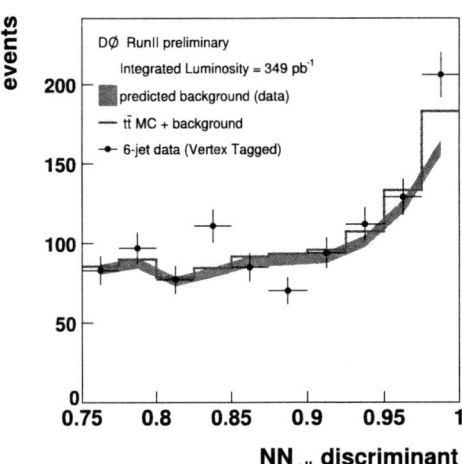

Fig. 77. *Left*: Output distribution of NN_{all} for tagged data (*points*), the expected background (*grey band*) and $t\bar{t}$ to all hadrons signal (*histogram*). *Right*: Same distribution, zoomed into the high discriminant region. The last four bins include the data with $NN_{\mathrm{all}} > 0.9$

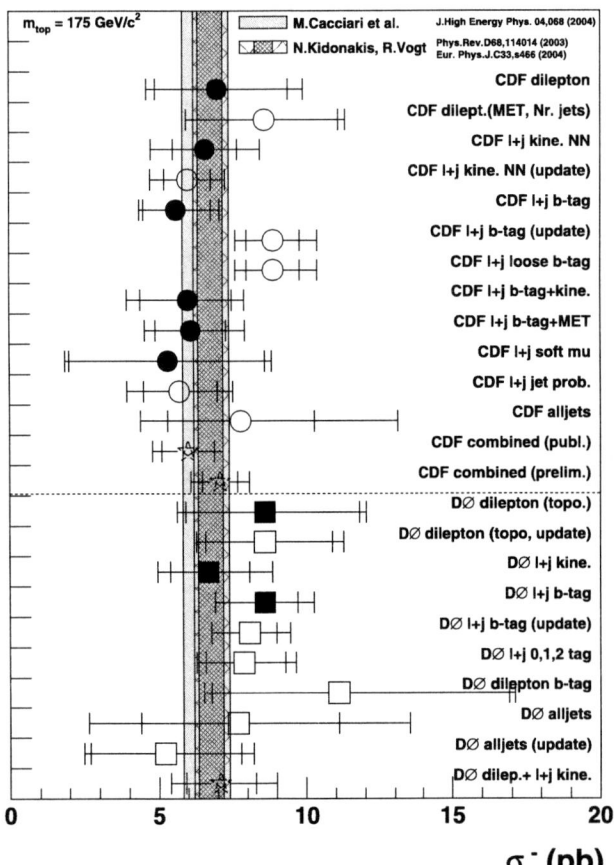
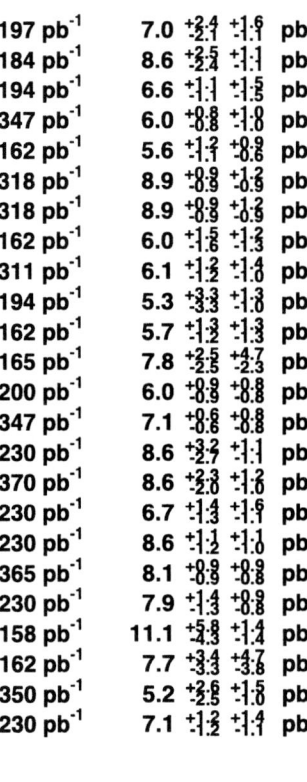

Fig. 78. Summary of all $t\bar{t}$ production cross section measurements from the TEVA-TRON in Run II at $\sqrt{s} = 1.96$ TeV. The *symbols* indicate published (•) and preliminary (○) CDF results, and published (■) and preliminary (□) DØ results. Preliminary combined cross section results are shown as *open stars*. The *inner error* bars represent statistical errors, the *outer error bars* the total errors including uncertainties from systematics and the luminosity measurement. The *vertical grey shaded areas* indicate the Standard Model expectation of $\sigma_{t\bar{t}} = 6.7^{+0.7}_{-0.9}$ pb [114] and $\sigma_{t\bar{t}} = 6.8 \pm 0.6$ pb [116, 118, 119]

where $\left(1 - \frac{\epsilon_{\mathrm{TRF}}}{\epsilon_{b\text{-tag}}}\right)$ is the correction factor to account for the bias introduced by using the entire selected sample, which includes $t\bar{t}$, to predict the number of background events. Table 29 summarises the systematic uncertainties in this analysis.

This updated preliminary $t\bar{t}$ production cross section measurement yields for $m_{\mathrm{top}} = 175\,\mathrm{GeV/c^2}$,

$$\sigma_{t\bar{t}} = 5.2^{+2.6}_{-2.5}\,(\text{stat.})^{+1.5}_{-1.0}\,(\text{syst.}) \pm 0.3\,(\text{lumi.})\,\mathrm{pb}\,. \quad (106)$$

4.4 Summary of cross section measurements

The results of the CDF and DØ measurements of the $t\bar{t}$ production cross section are summarised in Table 30 and shown in Figs. 78 and 79. All Run-II results are quoted for an assumed $m_{\mathrm{top}} = 175\,\mathrm{GeV/c^2}$, where the individual results also quote the top-quark mass dependence of the measured cross section (CDF) or include the uncertainty of the measured $t\bar{t}$ production cross section on the top quark mass, varied by $\pm 4.1\,\mathrm{GeV}$, in the quoted

Table 30. Cross section for $t\bar{t}$ production in $p\bar{p}$ collisions at $\sqrt{s} = 1.96\,\mathrm{TeV}$ from DØ and CDF (both for $m_t = 175\,\mathrm{GeV/c^2}$), and theory. Also shown are the final Run I measured values at $\sqrt{s} = 1.8\,\mathrm{TeV}$ from DØ ($m_t = 172.1\,\mathrm{GeV/c^2}$) and CDF ($m_t = 175\,\mathrm{GeV/c^2}$). The first set of errors are the statistical uncertainties, the second set are the systematic uncertainties including the luminosity uncertainty

$\sigma_{t\bar{t}}(pb)$	Source	Lumi (pb^{-1})	Ref.	Method
$7.0^{+2.4+1.6}_{-2.1-1.2}$	CDF Run II	197	[302]	$\ell\ell$
$8.6^{+2.5+1.1}_{-2.4-1.1}$	CDF Run II	184	[303]	$\ell\ell$ (MET, Nr. jets)
$6.6^{+1.1+1.5}_{-1.1-1.5}$	CDF Run II	194	[304]	ℓ + jets/kinematics
$6.0^{+0.8+1.0}_{-0.8-1.0}$	CDF Run II	347	[306]	ℓ + jets/kinematics (update)
$5.6^{+1.2+0.9}_{-1.1-0.6}$	CDF Run II	162	[307]	ℓ + jets/vtx b-tag
$8.9^{+0.9+1.2}_{-0.9-0.9}$	CDF Run II	318	[308]	ℓ + jets/vtx b-tag (update)
$8.9^{+0.9+1.2}_{-0.9-0.9}$	CDF Run II	318	[308]	ℓ + jets/loose vtx b-tag
$6.0^{+1.5+1.2}_{-1.6-1.3}$	CDF Run II	162	[309]	ℓ + jets/kin + vtx b-tag
$6.1^{+1.2+1.4}_{-1.2-1.0}$	CDF Run II	311	[310]	ℓ + jets/MET + vtx b-tag
$5.3^{+3.3+1.3}_{-3.3-1.0}$	CDF Run II	194	[311]	ℓ + jets/soft μ b-tag
$5.7^{+1.3+1.3}_{-1.2-1.3}$	CDF Run II	162	[312]	ℓ + jets/jet prob. b-tag
$7.8^{+2.5+4.7}_{-2.5-2.3}$	CDF Run II	165	[313]	all-jets/b-tag
$6.0^{+0.9+0.8}_{-0.9-0.8}$	CDF Run II	200	[317]	combined
$7.1^{+0.6+0.8}_{-0.6-0.8}$	CDF Run II	347	[317]	combined
$8.6^{+3.2+1.3}_{-2.7-1.3}$	DØ Run II	230	[318]	$\ell\ell$
$8.6^{+2.3+1.3}_{-2.0-1.2}$	DØ Run II	370	[321]	$\ell\ell$ (update)
$6.7^{+1.4+1.6}_{-1.3-1.2}$	DØ Run II	230	[322]	ℓ + jets/kinematics
$8.6^{+1.2+1.3}_{-1.1-1.2}$	DØ Run II	230	[330]	ℓ + jets/vtx b-tag
$8.1^{+0.9+1.0}_{-0.9-0.9}$	DØ Run II	365	[333]	ℓ + jets/vtx b-tag (update)
$7.9^{+1.4+1.0}_{-1.3-0.9}$	DØ Run II	230	[334]	ℓ + jets/0-2 b-tags
$11.1^{+5.8+1.6}_{-4.3-1.6}$	DØ Run II	230	[335]	$\ell\ell$/vtx b-tags
$7.7^{+3.4+4.7}_{-3.3-3.8}$	DØ Run II	162	[337]	all-jets/vtx b-tags
$5.2^{+2.6+1.5}_{-2.5-1.0}$	DØ Run II	350	[339]	all-jets/vtx b-tags (update)
$7.1^{+1.2+1.5}_{-1.2-1.2}$	DØ Run II	230	[329]	$\ell\ell + l$ + jets/(kine. comb.)
$6.7^{+0.7}_{-0.9}$	Theory ($\sqrt{s} = 1.96\,\mathrm{TeV}$)		[114]	$m_t = 175\,\mathrm{GeV/c^2}$
$6.8^{+0.6}_{-0.6}$	Theory ($\sqrt{s} = 1.96\,\mathrm{TeV}$)		[116, 118, 119]	$m_t = 175\,\mathrm{GeV/c^2}$
$6.5^{+1.7}_{-1.4}$	CDF Run I	105	[293]	all combined
5.7 ± 1.6	DØ Run I	110	[338, 340, 341]	all combined
$4.5 - 5.7$	Theory ($\sqrt{s} = 1.8\,\mathrm{TeV}$)		[113, 114, 116, 117]	$m_t = 175\,\mathrm{GeV/c^2}$
$5.2 - 6.2$	Theory ($\sqrt{s} = 1.8\,\mathrm{TeV}$)		[113, 114, 116, 117]	$m_t = 172.1\,\mathrm{GeV/c^2}$

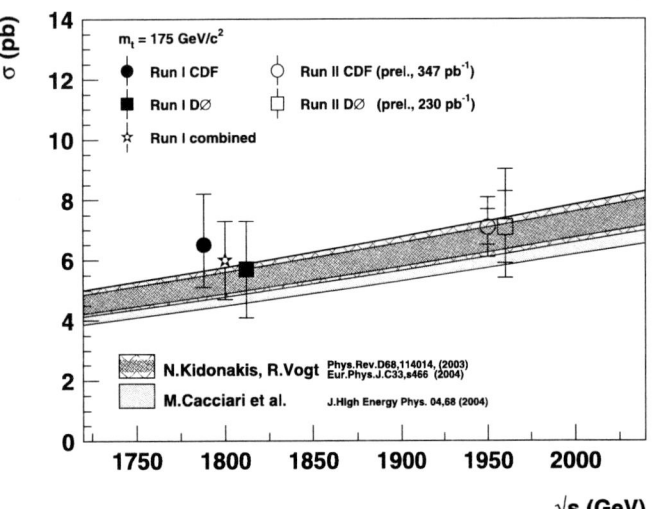

Fig. 79. Summary of $t\bar{t}$ production cross section measurements as a function of the centre-of-mass energy $\sqrt{s}$ from the TEVATRON in Run I (*full symbols*, $\sqrt{s} = 1.8$ TeV) and Run-II (open symbols, $\sqrt{s} = 1.96$ TeV). The Run-II point for CDF is a summary of all measurements with up to 347 pb^{-1}, for DØ it includes only the topological/kinematical measurement in the dilepton and lepton + jets channels with up to 230 pb^{-1}. The *inner error bars* represent statistical errors, the *outer error bars* the total uncertainties, adding in quadrature, the statistical, systematic and luminosity uncertainties. The *grey shaded areas* indicate the Standard Model expectation of $\sigma_{t\bar{t}}$ as a function of $\sqrt{s}$, as calculated in [114] and [116, 118, 119]

systematic uncertainties. For completeness, Table 30 also includes the combined $t\bar{t}$ production cross section measurements by CDF [293], DØ [338, 340, 341]. Cross section combinations in Run-II are very involved due to the large number of measurements in the various channels and the need for detailed understanding of their error correlations. First combinations of selected cross section measurements have been performed by CDF and DØ. Further studies by the Tevatron Electroweak Working Group towards a combined $t\bar{t}$ production cross section at Run II are ongoing and expected to be released in the coming months. Already now, individual measurements achieve a level of $\sim 17\%$ in total precision, while the uncertainty of the theoretical calculations is presently approximately 15%. The numerous cross section measurements are consistent between the different channels, between the two experiments, and between the experiments and the theory. Future analyses with the upcoming integrated luminosity of 1 fb^{-1} or more are expected to reach the level of $\approx 10\%$ or better in precision. Their combination will allow a detailed test of the $t\bar{t}$ production rate, the perturbative QCD calculations, and the comparison of the production rate times the top quark decay branching ratio in the different decay channels, implicitly testing for additional production mechanisms, or, in general, the presence of physics beyond the Standard Model.

5 Search for single top quark production

5.1 Introduction

In $p\bar{p}$ collisions at 1.96 TeV, top quarks are predominantly produced in pairs via strong interaction processes. Within the Standard Model (SM), top quarks are also expected to be produced singly by the electroweak interaction involving a Wtb vertex [144, 162, 170, 342]. Calculations of fully-differential NLO single-top quark cross sections have been performed in [148–151] and, including NLO top quark decay, in [152–156]. The measurement of the single-top cross section is particularly interesting as the production cross section is proportional to $|V_{tb}|^2$, where V_{tb} is the Cabibbo–Kobayashi–Maskawa (CKM) matrix element which relates top and bottom quarks. Assuming three quark generations, the unitarity of the CKM matrix implies that $|V_{tb}|$ is close to unity [167]. Furthermore, studying single top production provides information on the top quark polarisation, and will probe possible new physics in the top quark sector. In addition, the measurement of the single-top quark production cross section allows the determination of the density of (massless) b-quarks in the proton, which is important for some top and Higgs-boson production processes[18]. At the TEVATRON, the two relevant production modes are the t- [142] and the s-channel exchange [146] of a virtual W-boson. The most recent calculations at next-to-leading order (NLO) in the strong coupling [149, 152, 154, 155, 161], assuming $|V_{tb}| = 1$, predict cross sections of (1.98 ± 0.30) pb for the t-channel and (0.88 ± 0.14) pb for the s-channel mode for $m_{\text{top}} = 175$ GeV/c^2 at $\sqrt{s} = 1.96$ TeV [161]. Using these predictions, a measurement of the single-top cross section will allow for a direct determination of $|V_{tb}|$. Single-top searches test also exotic models which predict an anomalously altered single-top production rate [343–350]. Moreover, single-top quark processes result in the same final state as the Standard Model Higgs boson process $WH \rightarrow \ell\nu b\bar{b}$ and therefore impact on searches for the Higgs boson at the TEVATRON [351].

The experimental signature of single-top events consists of the W decay products plus two of three jets, including one b-quark jet from the decay of the top quark. In the s-channel events, a second b-quark jet from the Wtb vertex is expected. In the t-channel events, a second jet originates from the recoiling light-quark and a third jet is produced through the splitting of the initial-state gluon into a $b\bar{b}$ pair. Mostly, this third jet escapes detection, since it is produced in the high pseudorapidity (η) region and at low transverse energy E_{T}.

Results of searches for single-top quark production at $\sqrt{s} = 1.8$ TeV (Run-I) can be found in [352–355]. At the 95% CL, the DØ limit of the s-channel cross section is 17 pb, and the CDF limit is 18 pb. At 95% CL, the limit of the t-channel production cross section is 22 pb from DØ

---

[18] The measurement of the Zb/Zq ratio provides similar sensitivity to the density of massless b-quarks at high Q^2 scales in the proton, while the measurement of F_2^b at HERA tests the dynamics of massive b-quarks at low Q^2 in the proton.

and 13 pb from CDF. CDF places a 95% CL upper limit on the combined $s + t$ channel cross section at 24 pb. In this section, the published or preliminary Run-II analyses, utilising larger data sets, are discussed.

5.2 CDF analysis

CDF has performed a search for single-top quark production via the electroweak interaction in $162\,\mathrm{pb}^{-1}$ of Run-II data [356]. This publication includes two analyses: (1) a combined search for the t-channel plus s-channel single-top signal, (2) a separate search, where the rates for the two single-top processes are measured individually. The event selection of the two analyses resembles closely the one used in the CDF measurement of the $t\bar{t}$ cross section (see Sect. 4.2.3, [307]). In order to suppress QCD multi-jet background, only $W \to \mu\nu_\mu$ and $W \to e\nu_e$ candidates are selected. Jets are counted with transverse energy $E_\mathrm{T} \geq 15\,\mathrm{GeV}$ and $|\eta| \leq 2.8$. Only $W + 2$ jets events are accepted, where at least one of the jets must be identified as b-jets, using the SecVtx algorithm (Sect. 3.3.7). For optimised sensitivity, a cut on the invariant mass $M_{\ell\nu b}$ of the charged lepton, the neutrino and the b-tagged jet is applied: $140\,\mathrm{GeV}/\mathrm{c}^2 \leq M_{\ell\nu b} \leq 210\,\mathrm{GeV}/\mathrm{c}^2$. Here, the transverse momentum of the neutrino is set equal to the missing transverse energy vector $\mathbf{E}_\mathrm{T}$; $p_z(\nu)$ is obtained up to a 2-fold ambiguity from the constraint $M_{\ell\nu} = M_W$. From the two solutions, the one with the lower $|p_z(\nu)|$ is chosen. For the separate search, the sample is subdivided into events with exactly one b-tagged jet or exactly two b-tagged jets. For the 1-tag sample, at least one of the jets is required to have $E_\mathrm{T} \geq 30\,\mathrm{GeV}$. The total event detection efficiency ϵ_evt is determined from Monte Carlo events generated by MADEVENT [236, 237], followed by PYTHIA [223]. MADEVENT features the correct spin polarisation of the top quark and its decay products. For t-channel single-top production, two samples are generated, one $b + q \to t + q'$ and one $g + q \to t + \bar{b} + q'$ which are merged together to reproduce the p_T spectrum of the $\bar{b}$ as expected from NLO differential cross section calculations, yielding an improved model compared to PYTHIA.

There are two background components: $t\bar{t}$ and non-top background. The $t\bar{t}$ background is estimated based on Monte Carlo events generated with PYTHIA and normalised to the theoretical cross section $\sigma_{t\bar{t}} = 6.7^{+0.7}_{-0.9}\,\mathrm{pb}^{-1}$ [113, 114]. The primary source (62%) of the non-top background is the $W + $ heavy flavour process, where the flavour fractions are extracted from ALPGEN [224] and normalised to data. Additional sources are 'mistags' (25%), 'non-W' (10%), e.g. direct $b\bar{b}$ production are both estimated from data. The diboson (WW, WZ, ZZ) production (3%) is estimated from PYTHIA events normalised to the theory predictions [247]. Having applied all selection cuts, 42 events are observed in the combined search, 33 events in the 1-tag sample and six events in the 2-tag sample.

To extract the signal content in data, a maximum likelihood technique is used. t- and s-channel events are separated by using the $Q \cdot \eta$ distribution which exhibits a distinct asymmetry for t-channel events. Q is the charge of the lepton and η is the pseudorapidity of the untagged jet. The separate search defines a joint likelihood function for the $Q \cdot \eta$ distribution in the 1-tag sample and for the number of events in the 2-tag sample.

$$\mathcal{L}_\mathrm{sig}(\sigma_1, \ldots \sigma_4; \delta_1, \ldots, \delta_7) = \frac{e^{-\mu_\mathrm{d}} \mu_\mathrm{d}^{n_d}}{n_d!} \prod_{k=11}^{N_\mathrm{bin}} \frac{e^{-\mu_k} \mu_k^{n_k}}{n_k!}$$
$$\times \prod_{\substack{j=1 \\ j \neq \mathrm{sig}}}^{4} G(\sigma_j; \sigma_{\mathrm{SM},j}, \Delta_j)$$
$$\times \prod_{i=1}^{7} G(\delta; 0, 1)\,. \qquad (107)$$

Four processes are considered and labelled by the index j: t-channel ($j = 1$), s-channel ($j = 2$), $t\bar{t}$ ($j = 3$), and non-top ($j = 4$). The corresponding cross sections are denoted σ_j. The background cross sections are constrained to their Standard Model prediction $\sigma_{\mathrm{SM},j}$ by Gaussian priors of width σ_j. The index "sig" denotes the signal process, which is s- or t-channel, respectively. The μ_k are the mean number of events in bin k of the $Q \cdot \eta$ histogram ($N_\mathrm{bin} \equiv$ number of bins), while μ_d is the mean number of events observed in data, respectively.

Seven sources of systematic uncertainties are considered in the likelihood. The relative strength of a systematic effect due to source i is parameterised by the variable δ_i. Systematic effects change the acceptance and influence the shape of the $Q \cdot \eta$ distribution. When calculating $\mu_{k/d}$, the systematic shifts in the acceptance (Table 31) and in the shape of the template histograms, and their full correlations are taken into account. All variables except the signal cross section σ_sig are constrained to their expected values by Gaussian functions $G(x; x_0, \Delta_x)$ of mean x_0 and width Δ_x. The largest uncertainties are on the b-tagging efficiency (7%), luminosity (6%), top quark mass (4%), and the jet energy scale (JES) (4%).

To measure the combined t- plus s-channel signal in data, a kinematic variable is used whose distribution is very similar for the two single-top processes, but is different from background processes: H_T, which is the scalar sum of $\not{E}_\mathrm{T}$ and the transverse energies of the lepton and all jets in the event. A likelihood function similar to that in the separate search is

Table 31. Fractional changes in ϵ_evt of single-top processes in %. ϵ_trig is the trigger efficiency, ϵ_ID the lepton identification efficiency

i	Source	t-channel	s-channel	Combined
1	JES	$+2.4$ -6.7	$+0.4$ -3.1	$+0.1$ -4.3
2	ISR	± 1.0	± 0.6	± 1.0
3	FSR	± 2.2	± 5.3	± 2.6
4	PDF	± 4.4	± 2.5	± 3.8
5	MC (TopRex [241])	± 5.0	± 2.0	± 3.0
6	m_top	$+0.7$ -6.9	-2.3	-4.4
7	$\epsilon_\mathrm{trig}, \epsilon_\mathrm{ID}$, lumi	± 9.8	± 9.8	± 9.8

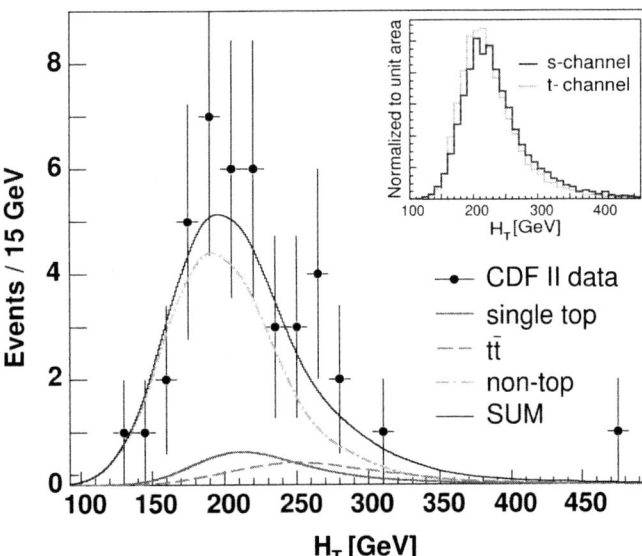

Fig. 80. H_T distribution for data (42 events) in the combined search compared to the smooth Monte Carlo predictions for signal and background

Table 32. Upper limits at the 95% CL and most probable value (MPV) of single-top cross sections in pb

	t-channel	s-channel	Combined
expected limit	11.2	12.1	13.6
observed limit	10.1	13.6	17.8
MPV ± HPD	$0.0^{+4.7}_{-0.0}$	$4.6^{+3.8}_{3.8}$	$7.7^{+5.1}_{-4.9}$

used. Figure 80 shows the H_T distribution observed in data compared to the Standard Model prediction.

Monte Carlo pseudo-experiments are performed to estimate the *a priori* sensitivity assuming Standard Model signal cross sections. Integrating out all variables except σ_{sig}, the marginalised likelihood $\mathcal{L}^*_{sig}(\sigma_{sig})$ is constructed. $\mathcal{L}^*_{sig}$ is interpreted as a posterior probability density function $p(\sigma_{sig})$. The 95% CL upper limit is calculated using a Bayesian method assuming a prior probability density, which is 0 if $\sigma_{sig} < 0$ and 1 if $\sigma_{sig} \geq 0$. The median and the expected upper limits define the sensitivity and are summarised in Table 32. The most probable value (MPV) and highest posterior density (HPD) intervals [167] in CDF II data are also shown in Table 32.

5.3 DØ Analyses

5.3.1 Analysis of 230 pb^{-1} of Run-II data

DØ has performed a search for single-top quark production in the s- and t-channel, using 230 pb^{-1} of Run-II data [357, 358]. Signal-like events are selected and the data are separated into independent analysis sets based on final state lepton flavour (electron or muon) and b-tag multiplicity (= 1 and ≥ 2 tags), where b-quark jets are tagged using re-

constructed displaced vertices in the jets. The independent analysis sets are combined in the final statistical analysis. Three methods are applied: a cut-based selection, and two multivariate analyses using neural networks and decision trees to separate the signals from the large backgrounds. Binned likelihood fits are performed on the neural network and decision tree outputs to obtain cross section limits.

In general the event selection follows that of the $t\bar{t}$ cross section measurement with b-tagging, but uses slightly softer selection criteria. Isolated electrons (muons) are required to have $E_T > 15$ GeV and $|\eta_{det}| < 1.1$ ($p_T > 15$ GeV and $|\eta_{det}| < 2.0$), missing E_T is required to be $\not{E}_T > 15$ GeV and the events must have between two or four jets with leading jet $E_T > 25$ GeV and $|\eta_{det}| < 2.5$, where the additional jets must have $E_T > 15$ GeV and $|\eta_{det}| < 3.4$. At least one jet is required to have a secondary vertex b-tag. s-channel and t-channel are separated by requiring at least one non b-tagged jet in the t-channel analysis.

The event selection acceptance for s- and t-channel single top production is estimated using events generated by the COMPHEP matrix element event generator [235]. The $W + $ jets background is estimated using ALPGEN [224] Monte Carlo events, normalised to the data sample before requiring a b-tagged jet. The fraction of $W + $ heavy flavour events is found using the MCFM program [221, 242]. Due to the normalisation procedure, also the $Z + $ jets background is included in the estimate. The diboson (WW and WZ) background is estimated using ALPGEN [224] Monte Carlo events normalised to the NLO cross section computed with MCFM. The $t\bar{t}$ background is estimated using Monte Carlo events generated with ALPGEN, normalised to the theoretical (N)NLO cross section $\sigma_{t\bar{t}} = 6.7 \pm 1.2$ pb [117] and PYTHIA [223]. The misidentified lepton background is estimated using multijet data samples as described in the $t\bar{t}$ cross section analysis. The acceptance for signal events with at least one b-tagged jet is $2.7 \pm 0.2\%$ and $1.9 \pm 0.2\%$ for the s-channel and t-channel, respectively.

All three analysis methods start from the same set of discriminating variables, which fall into three categories: individual object kinematics, global event kinematics, and variables based on angular correlations. These variables are selected based on an analysis of Feynman diagrams of signals [359] and on a study of single top quark production at NLO [154]. The full list of the 25 variables used can be found in [357, 358].

After event selection, neural networks are used to improve the signal-to-background separation. The networks are composed of three layers (input, hidden, output). For training and testing the MLPFit Package [360] is used. Testing and training event sets are created by dividing signal and background Monte Carlo samples. Based on studies to optimise the expected limits on the single top quark cross sections, two networks are found to be most effective in each channel. These networks correspond to the dominant backgrounds: $W + b\bar{b}$ and $t\bar{t} \to \ell + jets$. Therefore, eight separate neural networks are used corresponding to the combinations of signal-background pairs (s- and t-channels, $W + b\bar{b}$ and $t\bar{t}$, and the lepton flavours electron and muon). The input variables to each network are se-

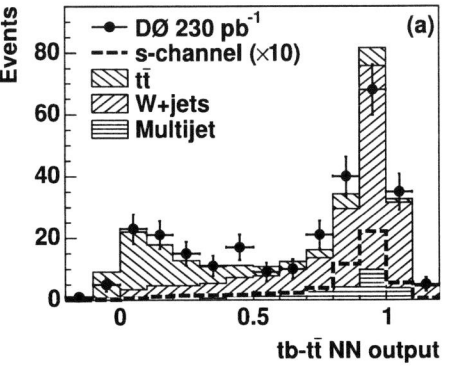
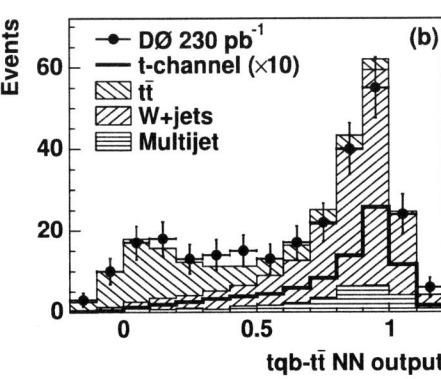
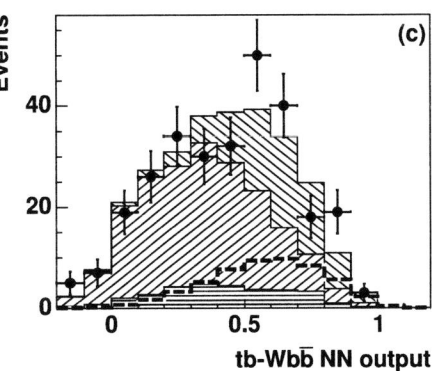
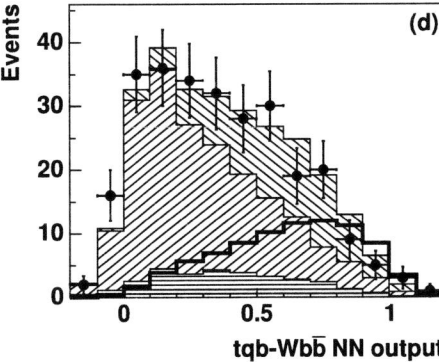

Fig. 81. Comparison of background, signal, and data for the neural network outputs, for the electron and muon channels combined, requiring at least one b-tag. The *upper row* shows the $t\bar{t}$ outputs, the *lower row* the $Wb\bar{b}$ outputs. The *left column* shows the s-channel outputs and the *right column* the t-channel outputs. Signals are multiplied by 10 for readability

lected by training with different combinations of variables and choosing the combination that produces the minimum testing errors.

The performance of the neural networks is illustrated in Fig. 81, where the output of the eight neural networks, combined for the electron and muon channel, is displayed for the corresponding signal-background pairs used in the training. Shown is also the comparison between data and the expected backgrounds and signals.

DØ has also used decision trees to evaluate the probability that a given event is a signal event. A decision tree is a binary tree with a simple selection cut implemented at each node [361]. Each event follows a unique path through the tree until it ends in one of the leaves. Each of these leaves is represented by a purity value, which is the ratio of signal and background events from the training samples that end up in this particular leaf. The distribution of purity values determines the decision tree output. The tree is trained using a procedure similar to the optimisation of a neural network. The same input variables and the same number of decision trees is used as in the neural network analysis. The decision trees separate signal and $t\bar{t}$ backgrounds efficiently, but give less separation for $W + $jets, especially in the s-channel.

In parallel to the multivariate techniques like neural networks and decision trees, a set of sequential cuts on the 25 considered variables is also performed. First, each variable is rated according to the best expected limit that an optimal cut on that single variable would achieve for each channel. The method used to choose the optimal cut point uses the signal events to seed the cut values and minimises the expected limit. Then, once each channel has the most

effective variables identified, they are combined (ANDed) in order, and for each subset the optimal cut values are recalculated. The set of variables and their optimised cuts that yields the lowest expected limit is then chosen for that particular channel.

The systematic uncertainties, summarised in Table 33, are evaluated for the Monte Carlo signal and background samples, separately for electrons and muons and for each b-tag multiplicity. The total uncertainty for the signal acceptance for single-tag events is 13% for the s-channel and 15% for the t-channel, and for double-tagged events it is 24% for the s-channel and 28% for the t-channel.

Table 33. Range of systematic uncertainty values for the various Monte Carlo signal and background samples in the different analysis channels

Source of systematic uncertainty	Uncertainty range (%)
Signal and background acceptance	
b-tag modelling	5–20
jet energy calibration	1–15
trigger modelling	2–7
jet fragmentation	5–7
jet identification	1–13
lepton identification	4
background normalisation	
theory cross sections	2–18
$W + $jets flavour composition	5–16
Luminosity	6.5

Table 34. Expected and observed 95% CL cross section limits for the single top production in the s- and t-channel for the three different analyses

Analysis	σ_s^{95} (pb)		σ_t^{95} (pb)	
	expect.	observed	expect.	observed
cut-based	9.8	10.6	12.4	11.3
decision tree	4.5	8.3	6.4	8.1
neural network	4.5	6.4	5.8	5.0

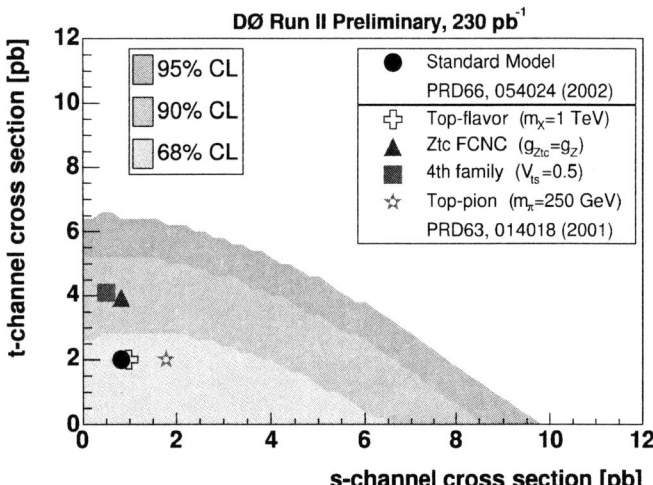

Fig. 82. Exclusion contours at 68%, 90% and 95% confidence level on the posterior density distribution as a function of both the s-channel and t-channel cross sections in the neural network analysis. The s-channel cross section is obtained from the tb muon data only and the t-channel cross section from tqb electron channel data only, such that the two likelihoods are independent. Several representative non-standard model contributions from [343] are also shown

The total uncertainty on the background is 10% for the single-tagged samples and 26% for the double-tagged samples.

The observed data are consistent with the background predictions for the three analysis methods and all eight analysis channels within uncertainties. Therefore, upper limits on the s-channel and t-channel production cross sections are set using a Bayesian approach. For the cut-based analysis, the inputs to the limit calculation are the integrated luminosity and the predicted and observed yields. For the neural network and decision tree analyses, the two-dimensional distributions of the $Wb\bar{b}$ versus $t\bar{t}$ filter outputs are used. A Poisson distribution for the observed counts, and a flat prior probability for the signal cross section are assumed. The priors for the signal acceptance and the background yields are multivariate Gaussians centred on their estimates and described by a covariance uncertainty matrix, taking into account correlations across different sources and bins.

The single-tagged and double-tagged, as well as the electron and muon channels, are combined in order to obtain better sensitivity to the single top cross sections. The

expected and observed cross section limits are summarised in Table 34. The improvement in limits from the cut-based analysis to the neural network and decision tree analyses comes from both, the use of multivariate techniques that take into account correlations in the data, and from the binned likelihood fits, which add shape information from the distributions.

The 95% CL upper limits on the single-top production cross section are also close to the region of sensitivity to models of physics beyond the Standard Model, such as a fourth quark generation with large $|V_{ts}|$, or a flavour-changing neutral current vertex [343], as shown in Fig. 82.

5.3.2 Analysis of 370 pb^{-1} of Run-II data

DØ has performed an updated, preliminary search for single top quark production in 370 pb^{-1} of Run-II data, using a likelihood discriminant method to separate signals and backgrounds [362].

The event selection follows very closely that of the 230 pb^{-1} analysis. The selection efficiency and background contamination is determined from Monte Carlo or data control samples which are essentially identical to the ones described in the 230 pb^{-1} analysis. The event selection requires at least one b-tagged jet, where the jet lifetime probability (JLIP) algorithm (Sect. 3.3.7) is used. The dataset is split into two orthogonal tagging schemes. The "single tag" sample corresponds to events containing exactly one tight and no extra loose b-tagged jets. The "double tag" sample is associated to events with at least one tight and another loose b-tagged jet.

After event selection, a final discriminating variable is constructed in order to efficiently characterise the signal type events and reject the background type ones. This likelihood variable is a robust statistical variable and is a more efficient way for separating signal from background than sequential cuts since a likelihood uses the entire shape of the signal and background distributions to distinguish between them. The use of a likelihood method is adequate as the sample size is reasonable and the distributions consist of essentially uncorrelated variables.

The final discriminating variable $\mathcal{L}(\mathbf{x})$ is constructed from a vector of measurements $\mathbf{x}$:

$$\mathcal{L}(\mathbf{x}) = \frac{\mathcal{P}_{\text{signal}}(\mathbf{x})}{\mathcal{P}_{\text{signal}}(\mathbf{x}) + \mathcal{P}_{\text{background}}(\mathbf{x})}, \qquad (108)$$

where $\mathcal{P}_{\text{signal}}(\mathbf{x})$ and $\mathcal{P}_{\text{background}}(\mathbf{x})$ are the probability density functions for the two categories of events. The optimal event-classification scheme selects events that have the largest values for the ratio of probabilities $\mathcal{P}_{\text{signal}}(\mathbf{x})/\mathcal{P}_{\text{background}}(\mathbf{x})$ to define a sample enriched in signal events. Signal events tend to have a value of $\mathcal{L}$ close to 1, background events have a value near 0.

The probability density functions $\mathcal{P}_{\text{signal}}(\mathbf{x})$ and $\mathcal{P}_{\text{background}}(\mathbf{x})$ are determined from the product of Monte Carlo one dimensional distributions of the input variables. Therefore, potential correlations between variables are not taken into account.

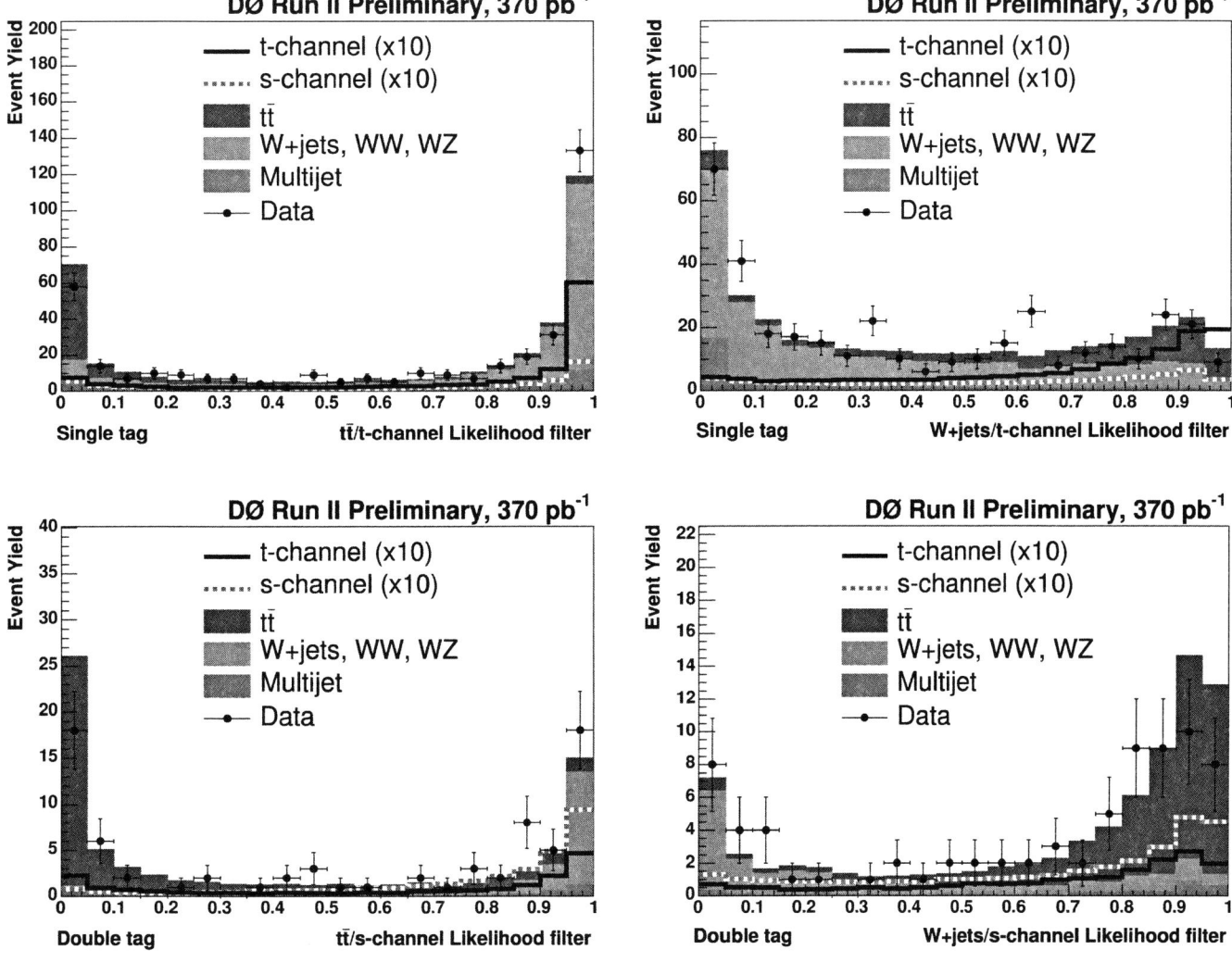

Fig. 83. Electron and muon channels combined. *Top*: Data to Monte Carlo comparison for the $tqb/t\bar{t}$ (*left*) and $tqb/W+$jets (*right*) filters for single tagged events. *Bottom*: Data to Monte Carlo comparison for the $tb/t\bar{t}$ (*left*) and $tb/W+$jets (*right*) filters for double tagged events

Many sets of variables were tested to build the likelihood discriminants, including transverse momenta, invariant masses and angular variables combining the different reconstructed objects (charged lepton, neutrino, jets). The final set of variables is selected to be the one with optimal discrimination of the likelihood variables. These variables are: (i) the transverse momentum of the leading, (ii) second leading and, (iii) if it exists, the third leading jet, (iv) the scalar sum of the missing transverse energy and the transverse energy of the lepton, (v) the invariant mass of the system of all jets, i.e. the four-vector sum of all jets in the event, (vi) the transverse mass of the W-boson, (vii) the invariant mass of the system of the W-boson and the leading tagged jet, i.e. the reconstructed top quark mass, (viii) the minimum angular separation between all jets, (ix) the cosine of the angle between the second leading jet and the charged lepton in the top rest frame, (x) the sphericity of the event, (xi) the centrality of the event, (xii) and finally, the $Q\eta$ variable. Examples

of likelihood filter outputs for signal and background are shown in Fig. 83.

Systematic uncertainties on the yields are evaluated separately for electron and muon channels for each b-tagged scheme. Sources of systematic uncertainties and their percentage range are summarised in Tables 35 and 36 for Monte Carlo yields ($t\bar{t}$, dibosons) and for yields normalised to data (multijet and $W+$jets).

The number of observed events is consistent with the background prediction for both muon and electron channels and for all b-tagging schemes, within the total uncertainties. Therefore, 95% CL upper limits are set using a Bayesian approach. The two-dimensional distributions of the signal/$W+$jets likelihood discriminant versus signal/$t\bar{t}$ likelihood discriminant is used. A Poisson distribution for the observed counts, and a flat prior probability for the signal cross section are assumed. The priors for the signal acceptance and the backgrounds are multivariate Gaussians centred on their estimates and described

Table 35. Averaged systematic uncertainties for Monte Carlo estimated yields

Systematic uncertainties (%)	
Luminosity	6.5
Cross section	$2(WW) - 18(t\bar{t})$
Branching fraction	2
Primary vertex reconstr.	2
Electron identification	4
Muon identification	5
Jet identification	1–4
Jet energy scale	1–5
Jet energy resolution	1
Jet fragmentation	5
Trigger modelling	2–7
Single (double) b-tag model	6 (17)
Sample statistics	1

Table 36. Right: Average systematic uncertainties for data normalised yields. The numbers for the double tagged sample are given in parentheses

Systematic uncertainties (%)	
Data normalisation	5–15
Single (double) b-tag modelling for W + jets events	9 (15)
Sample statistics	3 (2–17)

Table 37. Expected and observed 95% CL cross section limits for the single top production in the s- and t-channel for the likelihood discriminant analysis

Analysis	σ_s^{95} (pb)		σ_t^{95} (pb)	
	expected	observed	expected	observed
likelihood discriminant	3.3	5.0	4.3	4.4

by a covariance error matrix taking into account correlations across the different sources and bins. The four orthogonal analysis channels (electron and muon, single and double tag) are combined to enhance the sensitivity of the analysis. The expected and observed cross section limits are summarised in Table 37. Monte Carlo studies show that this likelihood discriminant method and the previously described neural network analysis, applied to identical data sets, have very similar sensitivity, even though the likelihood, in contrast to the neural network, does not take correlations between variables into account. The gain of the final limits comes mostly from the increased luminosity.

Given the presently studied analysis techniques, 2 fb^{-1} or more of highest quality Run-II data are estimated to be required for CDF and DØ to observe a first evidence for electroweak single-top quark production, assuming the production cross section of this process is consistent with the theoretical calculations in the Standard Model.

6 Top quark interactions with Gauge bosons

6.1 Top quark spin correlation

DØ has studied the spin correlation in $t\bar{t}$ production using 125 pb^{-1} of Run-I data at $\sqrt{s} = 1.8$ TeV [363]. At present there is no such study using Run-II data.

For a top quark mass of $m_t = 175$ GeV/c^2, the width of the top quark in the Standard Model is $\Gamma_t = 1.4$ GeV [163, 364], while the typical hadronisation scale is $\Lambda_{\rm QCD} \approx 0.22$ GeV [167]. The time scale needed for depolarisation of the top quark spin is of the order $m_t/\Lambda_{\rm QCD}^2 \gg 1/\Gamma_t$ [365], implying that the polarisation information should be transmitted fully to the decay products of the top quark. That is, the expected lifetime of the top quark is sufficiently short to prevent long distance effects (e.g. fragmentation) from affecting the $t\bar{t}$ spin configuration, which are determined by the short distance dynamics of QCD at production.

The observation of spin correlation in the decay products of $t\bar{t}$ systems is interesting for several reasons. First, it provides a probe of a quark that is almost free of confinement effects. Second, since the lifetime of the top quark is proportional to the Kobayashi–Maskawa matrix element $|V_{tb}|^2$, an observation of spin correlation would yield information about the lower limit on $|V_{tb}|^2$, without assuming that there are three generations of quark families [196]. Finally, many scenarios beyond the Standard Model [139, 164, 365–368] predict different production and decay dynamics of the top quark, any of which could affect the observed spin correlation.

In the decay of a polarised top quark, charged leptons or quarks of weak isospin $-\frac{1}{2}$ are most sensitive to the initial polarisation. Their angular distribution in the rest frame of the top quark is given by $(1 + \cos\theta)$, where θ is the angle between the polarisation direction and the line of flight of the charged lepton or down-type quark. Because of the experimental difficulties of identifying jets initiated by a down-type quark, only top quark events in dilepton channels are considered. The advantages associated with these channels are that: (1) objects sensitive to the polarisation of the top quark are clearly identified, (2) background is small compared to the lepton + jets channels, and (3) there are fewer ambiguities associated with assigning objects observed in the detector to their originating quarks. The disadvantages are that the number of events in the dilepton channels is small, and that it is necessary to reconstruct two neutrinos in an event whose combined transverse momenta give rise to the observed transverse momentum imbalance in the event.

The produced t and $\bar{t}$ quarks are expected to be unpolarised. However, their spins are expected to have strong correlation [196, 369] event by event and point along the same axis in the $t\bar{t}$ rest frame [193, 195]. In an optimised spin quantisation basis called the "off-diagonal" basis, contributions from opposite spin projections for top quark pairs arising from $q\bar{q}$ annihilations are suppressed at the tree level and only like-spin configurations survive. This spin quantisation basis can be specified using the velocity β^* and the scattering angle θ^* of the top quark with respect to the centre-of-mass frame of the incoming partons. The direction of the off-diagonal basis forms an angle ψ with respect to the $p\bar{p}$ beam axis that is given by [193, 195, 198]:

$$\tan\psi = \frac{\beta^{*2}\sin\theta^*\cos\theta^*}{1 - \beta^{*2}\sin^2\theta^*} . \quad (109)$$

This particular choice of basis is optimal in the sense that top quarks produced from $q\bar{q}$ will have their spins fully aligned along this basis. In the limit of top quark production at rest ($\beta^* = 0$), the t quark and $\bar{t}$ quark will have their spins pointing in the same direction along $\psi = 0$.

In $t\bar{t} \to$ dilepton events, defining θ_+ as the angle between one of the charged leptons and the axis of quantisation in the rest frame of its parent top quark, and similarly defining θ_- for the other charged lepton, the spin correlation can be expressed as [193, 195, 202]:

$$\frac{1}{\sigma}\frac{d^2\sigma}{d(\cos\theta_+)\,d(\cos\theta_-)} = \frac{1 + \kappa\cos\theta_+\cos\theta_-}{4} , \quad (110)$$

where the correlation coefficient κ[19] describes the degree of correlation present prior to imposition of selection criteria of effects of detector resolutions. For $t\bar{t}$ production at the TEVATRON, the Standard Model predicts $\kappa = 0.88$. In the off-diagonal basis, the correlation coefficient for $q\bar{q} \to t\bar{t}$ is $\kappa = 1$. When $gg \to t\bar{t}$ is included at $\sqrt{s} = 1.8$ TeV, the correlation is reduced to $\kappa = 0.88$ (at $\sqrt{s} = 1.96$ TeV $\kappa = 0.928$ in LO and $\kappa = 0.777$ in NLO [200]). The distribution is symmetric with respect to the exchange of θ_+ and θ_-, and it is therefore not necessary to identify the electric charge of the leptons. The physical meaning of κ in any spin quantisation basis corresponds to the fractional difference between the number of events in which the top quark spins are aligned and the number of events in which they have opposite directions.

The DØ event selection follows very closely that of the corresponding cross section analysis. The final sample consists of three $e\mu$ events, two ee events, and one $\mu\mu$ event, with expected backgrounds of 0.21 ± 0.16, 0.47 ± 0.09, and 0.73 ± 0.25 events, respectively. To study the distribution in $(\cos\theta_+, \cos\theta_-)$, the momenta of the two neutrinos must be deduced. The weighting scheme used is the neutrino weighting method (see the dilepton analyses in Sect. 7.1.1, [370]). For each kinematic solution, one can then boost the decay products into the rest frame of the original top quarks and calculate the relevant decay angles $(\cos\theta_+, \cos\theta_-)$. The event fitter returns many such solutions for an event, and the goal is to deduce the original value of $(\cos\theta_+, \cos\theta_-)$ from the reconstructed distributions.

To maximise the physical information present in the data, the full 2-dimensional phase space of $(\cos\theta_+, \cos\theta_-)$ is used in a 2-dimensional binned likelihood analysis. The phase space is split into a 3×3 grid, each side of which spans $1/3$ of the range of $\cos\theta_+$, and $\cos\theta_-$. The nine bins are populated for data with weights $(w_1, \ldots w_9)$ from the

---

[19] Note that the correlation coefficient κ in Sect. 2.4.4 is defined with reversed sign convention to be in agreement with the theoretical publications.

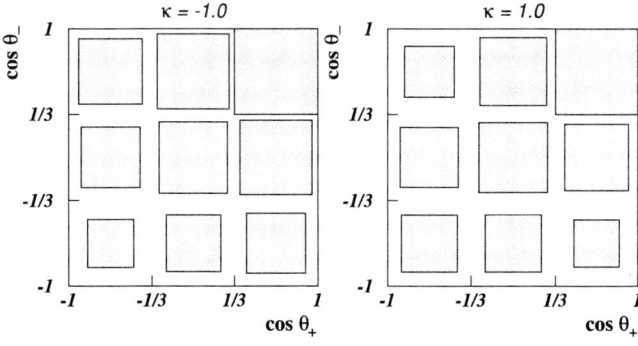

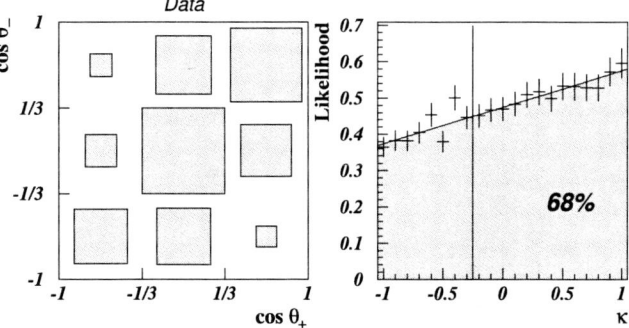

Fig. 84. Probability density for $t\bar{t}$ events in the dilepton channels in $(\cos\theta_+, \cos\theta_-)$ phase space. *Top left*: Monte Carlo events with $\kappa = -1$. *Top right*: Monte Carlo events with $\kappa = 1$. *Bottom left*: DØ data. *Bottom right*: the likelihood as a function of κ showing the 68% confidence limit of $\kappa > -0.25$. The *box area* is proportional to the summed weights in the bin

event fitter, with the distribution of weights for each event normalised to unity. Similar distributions are made for the generated Monte Carlo events using different values of κ for $t\bar{t}$ signal and an appropriate admixture of background. Comparisons of data with Monte Carlo, based on a likelihood, are used to extract κ.

Figure 84 shows the result. The probability densities for the Monte Carlo generator at $\kappa = -1$ and $\kappa = 1$ are shown for comparison. From the dependence of the likelihood on κ, a 68% confidence interval at $\kappa > -0.25$, based on the line fit in Fig. 84, in agreement with the Standard Model prediction of $\kappa = 0.88$.

6.2 Limits on BR($t \to Wb$)/BR($t \to Wq$) and $|V_{tb}|$

The Cabibbo–Kobayashi–Maskawa (CKM) matrix element $|V_{tb}|$ is indirectly constrained by the measurements of other CKM matrix elements ($|V_{ub}|$ and $|V_{cb}|$), to the interval $0.9990 < |V_{tb}| < 0.9992$ at 90% C.L. [167], based on the assumption of the unitarity of the CKM matrix and three quark generations. In this case the ratio of decay branching ratios $R = \mathcal{B}(t \to Wb)/\mathcal{B}(t \to Wq)$, where q can be a b, s or a d quark, can be expressed in terms of CKM matrix elements:

$$R = \frac{|V_{tb}|^2}{|V_{tb}|^2 + |V_{ts}|^2 + |V_{td}|^2} = |V_{tb}|^2 . \quad (111)$$

In the framework of the Standard Model, the ratio R is therefore constrained to be in the interval $0.9980 - 0.9984$ at the 90% CL [167]. The non-unitarity of the CKM matrix could for example arise from the existence of a fourth quark generation and lead to a deviation of R from unity, since the denominator in (111) would have to be modified to include a fourth entry $|V_{tX}|$.

R can be extracted from the relative rates of identified b-quarks in $t\bar{t}$ events. This is easily demonstrated under the assumption that all b-type quarks in a $t\bar{t}$ event are identifiable and uncorrelated, and no other quarks can be misidentified as b-quarks. Then, if each of the two top quarks in the event has a probability R to decay to a b-quark, and there is an efficiency ϵ_b to identify ("tag") the quark jet, the efficiencies to detect zero, one or two b-jets in the event are

$$\epsilon_0 = (1 - R\epsilon_b)^2 ,$$
$$\epsilon_1 = 2R\epsilon_b(1 - R\epsilon_b) ,$$
$$\epsilon_2 = (R\epsilon_b)^2 , \qquad (112)$$

and therefore

$$R\epsilon_b = \frac{2}{\epsilon_1/\epsilon_2 + 2} = \frac{1}{2\epsilon_0/\epsilon_1 + 1} = \frac{1}{\sqrt{\epsilon_0/\epsilon_2} + 1} , \qquad (113)$$

$$R\epsilon_b = \frac{2}{N_1/N_2 + 2} = \frac{1}{2N_0/N_1 + 1} = \frac{1}{\sqrt{N_0/N_2} + 1} , \qquad (114)$$

where N_i are the number of $t\bar{t}$ events observed with i tags. The naive assumptions are not true, but the equations illustrate the principle of the measurement – any two tagging rates determine $R\epsilon_b$, and three tagging rates overdetermine $R\epsilon_b$. The ratios of tag rates can only determine the product $R\epsilon_b$, as it is not possible to distinguish missing tagged jets due to tagging inefficiency ($\epsilon_b < 1$) from missing tagged jets due to the absence of b-quarks ($R < 1$). However, $R\epsilon_b$ can be determined independently of the total rate of $t\bar{t}$ production (i.e. the cross section) or be determined simultaneously with it, and it can be measured independently of the decay mode of the W-boson produced in top decay. ϵ_b can be estimated from $t\bar{t}$ simulation, calibrated with complimentary data samples, for use in extracting R. The measurement requires three basic steps. After identifying samples enriched in $t\bar{t}$ events, the background level in those events is estimated as a function of the number of b-tagged jets in the event. Then the expected tag rates (and, implicitly, their ratios) in $t\bar{t}$ events are predicted as a function of $R\epsilon_b$. Finally, the observed $t\bar{t}$ tag rates are compared to the expectations, yielding the most likely value of $R\epsilon_b$, and allowing to set a lower limit on R.

Based on $\sim 162\,\mathrm{pb}^{-1}$ of data, CDF follows the above detailed procedure and determines the ratio of branching ratios R from the relative numbers of $t\bar{t}$ events with different multiplicity of identified secondary vertices in the lepton + jets and the dilepton channels [371]. In both analyses, b-quark jets are identified ("tagged") by identifying displaced secondary vertices using the SECVTX algorithm (see Sect. 3.3.7). The event selection and background determination are essentially equivalent to the ones

Table 38. Summary of observed number of events with i-tags in the lepton + jets and dilepton samples, with estimates of nominal $t\bar{t}$ event-tagging efficiencies, background levels and expected event yields. The lepton + jets 0-tag background is measured with an ANN. The efficiency estimates and the 1-tag and 2-tag lepton + jets background estimates are given for $R = 1$

Lepton+Jets	0-tag	1-tag	2-tag
Eff. ($\epsilon_i(R+1)$)	0.45 ± 0.03	0.43 ± 0.02	0.12 ± 0.02
Bgd (N_i^{bgd})	62.4 ± 9.0	4.2 ± 0.7	0.2 ± 0.1
Total exp. (N_i^{exp})	80.4 ± 5.2	21.5 ± 4.1	5.0 ± 1.4
Obs. (N_i^{obs})	79	23	5

Dileptons	0-tag	1-tag	2-tag
Eff. ($\epsilon_i(R+1)$)	0.47 ± 0.03	0.43 ± 0.02	0.10 ± 0.02
Bgd (N_i^{bgd})	2.0 ± 0.6	0.2 ± 0.1	< 0.01
Total exp. (N_i^{exp})	6.1 ± 0.4	4.0 ± 0.2	0.9 ± 0.2
Obs. (N_i^{obs})	5	4	2

in the corresponding cross section analyses, described in Sects. 4.2.1–4.2.3. In particular, in the 1-tag and 2-tag subsamples of the lepton + jets sample, a priori estimates of the background are made with a collection of data-driven and simulation techniques. The background estimate requires a small correction for $R \neq 1$. The background estimate for $R = 1$ in these subsamples is given in Table 38. A novel feature of this measurement is the determination of the 0-tag lepton + jets event rate using event kinematics and an artificial neural net (ANN) technique, similar to the $t\bar{t}$ cross section measurement in the lepton + jets channel described in Sect. 4.2.2. An optimal signal to background discrimination is found with an ANN structure of nine input variables, one intermediate layer with ten nodes, and one output unit. The variables used are the transverse energies of the four leading jets, the minimum di-jet mass, the di-jet transverse mass with value closest to the mass of the W-boson, the scalar sum of the transverse energies of all leptons and jets, the total longitudinal momentum divided by the total transverse momentum, and the event aplanarity. A binned maximum likelihood fit of the ANN output distribution is performed for the fraction of $t\bar{t}$ events in the 0-tag subsample. Systematic uncertainties in the ANN-determined backgrounds are dominated by the understanding of the jet energy calibration, the renormalisation and factorisation scale, and the shape of the QCD template. They are strongly anti-correlated between the $t\bar{t}$ and W + jets measurements. The ANN-measured $t\bar{t}$ content in the lepton + jets sample without b-tagging requirement is found to be consistent with that in the earlier measurement of Sect. 4.2.2. Repeating the procedure in the 1-tag and 2-tag samples yields background estimates consistent with the a priori estimates, where the latter are statistically more precise and used in this analysis. In the dilepton channel, most of the jets in the background events arise from generic QCD radiation. To determine the background distribution across the i-tag subsamples, a parameterisation of the probability to tag a generic QCD jet, derived from jet-triggered data samples, is applied to the jets in the dilepton sample, correcting for the enriched $t\bar{t}$ content of the sample. The resulting estimates are given in Table 38.

The $t\bar{t}$ event tagging efficiency ϵ_i, defined as the probability to observe i-tags in a $t\bar{t}$ event, depends on the fiducial acceptances for jets that can potentially be tagged, and the efficiencies to tag those jets. Those efficiencies in turn depend on the species of the underlying quark in the jet. The efficiency ϵ_i depends strongly on R, as $R \neq 1$ implies fewer b-jets available for tagging, and more light-quark jets available instead. The jet acceptances and tagging efficiencies are used to parameterise $\epsilon_i(R)$. These quantities are estimated with a sample of simulated $t\bar{t}$ events from the PYTHIA [223] generator and the CDF detector simulation. The leading determiner of ϵ_i is the efficiency to tag a b-jet from the decay $t \to Wb$; $\epsilon_b = 0.44 \pm 0.04$ for b-jets falling within the fiducial acceptance and having at least two tracks with silicon information. The ϵ_i values also have small corrections from the efficiencies to tag jets from $W \to cs$ hadronic decays and from additional QCD radiation in $t\bar{t}$ events. The nominal values of ϵ_i for $R = 1$ are given in Table 38.

The expected event yield in each of the three tagged subsets of each of the lepton + jets and dilepton samples is

$$N_i^{\mathrm{exp}} = N_{\mathrm{inc}}^{t\bar{t}} \epsilon_i(R) + N_i^{\mathrm{bkg}}, \qquad (115)$$

where N_i^{bkg} is the number of background events in the i-tag subsample and $N_{\mathrm{inc}}^{t\bar{t}}$ is an estimate of the inclusive number of $t\bar{t}$ events in the sample, determined by

$$N_{\mathrm{inc}}^{t\bar{t}} = \sum_i (N_i^{\mathrm{obs}} - N_i^{\mathrm{bkg}}), \qquad (116)$$

where N_i^{obs} is the observed number of events in each subsample. In this construction, the measured value of R is independent of any assumption of the overall rate of $t\bar{t}$ production, and is thus sensitive only to the relative numbers of $t\bar{t}$ events with i tags. The full likelihood is a product of independent likelihoods for the lepton + jets and dilepton samples. Each likelihood is a product of Poisson functions comparing the N_i^{obs} to N_i^{exp} for each value of i, multiplied by Gaussian functions which incorporate systematic uncertainties in the event-tagging efficiencies and backgrounds, taking into account the correlations across the different subsamples.

The resulting likelihood as a function of R is shown in Fig. 85, along with the negative logarithm of the likelihood, yielding a central value of

$$R = 1.12^{+0.21}_{-0.19}\,(\mathrm{stat.})^{+0.17}_{-0.13}\,(\mathrm{syst.}). \qquad (117)$$

The dominant systematic uncertainties arise from the uncertainty on the background measurement in the 0-tag lepton + jets sample $\left(^{+0.14}_{-0.11}\right)$ and from the overall normalisation of the tagging efficiencies $\left(^{+0.09}_{-0.06}\right)$. Taken separately, the two final states of $t\bar{t}$ give consistent results for R; the lepton + jets sample alone yields $R = 1.02^{+0.23+0.21}_{-0.20-0.13}$, and the dilepton sample alone yields $R = 1.41^{+0.46+0.17}_{-0.40-0.13}$. These R results are consistent with the Standard Model expectation.

The ratio R can only take on physical values between zero and unity. The Feldman–Cousins prescription [372]

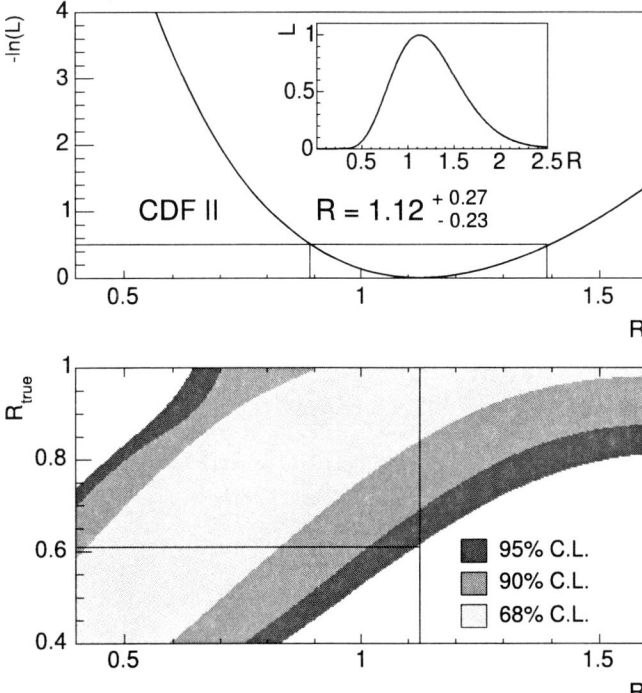

Fig. 85. *Top*: The likelihood as a function of R (*inset*) and its negative logarithm. The intersections of the *horizontal line* $\ln(L) = -0.5$ with the likelihood defines the statistical 1σ errors on R. *Bottom*: 95% (*outer*), 90% (*central*), and 68% (*inner*) CL bands for R_{true} as a function of R. The measurement of $R = 1.12$ (*vertical line*) implies $R > 0.61$ at the 95% CL (*horizontal line*)

is used to set a lower limit on R. Ensembles of pseudo-experiments are generated for different input values of $R(R_{\mathrm{true}})$, and varying the input quantities of the analysis, e.g. the background estimates, taking correlations into account. Using the likelihood-ratio ordering principle, the acceptance intervals as shown in Fig. 85 are found. With the measured value of R, this yields $R > 0.61$ at the 95% CL. Within the Standard Model, $R = \frac{|V_{tb}|^2}{|V_{tb}|^2 + |V_{ts}|^2 + |V_{td}|^2}$, up to phase-space factors. Assuming unitarity of the CKM matrix, the denominator is unity, yielding an estimate of $|V_{tb}| > 0.78$ at 95% CL. All of the measurements of R are consistent with the Standard Model expectations.

DØ measures simultaneously the ratio of branching ratios $R = \mathcal{B}(t \to Wb)/\mathcal{B}(t \to Wq)$ together with the top quark pair ($t\bar{t}$) production cross section ($\sigma_{t\bar{t}}$), using 230 pb^{-1} of data [334]. This analysis, which is an update of [373], determines R by selecting a sample enriched in $t\bar{t}$ events, with a high p_{T} isolated lepton (electron or muon), large missing transverse energy and three, four or more jets. The selected events are categorised into events with 0, 1 and 2 or more lifetime-tagged jets. From the number of observed events in the three categories and the kinematics of events with no lifetime-tagged jet, R and the $t\bar{t}$ pair production cross section are fit. The event selection, background determination and secondary vertex b-tagging algorithm are essentially equivalent to that of the corresponding cross section analyses, discussed in Sects. 4.3.2 and 4.3.3. Events with exactly 1(≥ 2) tagged jet are referred to as single-tag (double-tag) events. Events with exactly 0 tagged jet are referred to as zero-tag events.

In the Standard Model case with $R = \mathcal{B}(t \rightarrow Wb)/\mathcal{B}(t \rightarrow Wq) = 1$, the $t\bar{t}$ event tagging probabilities are computed assuming that each of the signal events contains two b-jets. In the present analysis the $t\bar{t}$ event tagging probability becomes a function of $\mathcal{B}(t \rightarrow Wb)/\mathcal{B}(t \rightarrow Wq)$. In general, for $R \neq 1$, a $t\bar{t}$ event might have 0, 1 or 2 b-jets from the two top quark decays, strongly affecting the event tagging probability and how $t\bar{t}$ events are distributed among the zero-, single- and double-tag samples. To derive the $t\bar{t}$ event tagging probability as a function of R, the event tagging probability in the three following scenarios is determined:

1. $t\bar{t} \rightarrow W^+ b\, W^- \bar{b}$ (further will be referred to as $tt \rightarrow bb$),
2. $t\bar{t} \rightarrow W^+ b\, W^- \bar{q}_l$ or its charge conjugate (referred to as $tt \rightarrow bq_l$),
3. $t\bar{t} \rightarrow W^+ q_l\, W^- \bar{q}_l$ (referred to as $tt \rightarrow q_l q_l$),

where q_l denotes either a d- or an s-quark. The probabilities $P_{n-\text{tag}}$ to observe $n - \text{tag} = 0, 1$ or ≥ 2 lifetime-tagged jets are computed separately for the three types of $t\bar{t}$ events, using the b-tagging efficiency for b-jets, c-tagging efficiency for c-jets and the light jet tagging probability for light flavour jets. The probabilities $P_{n-\text{tags}}$ in the three scenarios are then combined in the following way to obtain the $t\bar{t}$ tagging probability as a function of R:

$$
\begin{aligned}
P_{n\text{-tag}}(tt) = {} & R^2 P_{n\text{-tag}}(tt \rightarrow bb) \\
& + 2R(1-R)P_{n\text{-tag}}(tt \rightarrow bq_l) \\
& + (1-R)^2 P_{n\text{-tag}}(tt \rightarrow q_l q_l) , \quad (118)
\end{aligned}
$$

where the subscript $n - tag$ runs over 0, 1 and ≥ 2 tags.

The fraction of $t\bar{t}$ events in the $\ell + 4$ jets 0-tag sample is between 10% for $R = 1$ and 25% for $R = 0$. This is not significantly larger than the statistical error on the total number of events in the zero-tag sample. For this reason the information on the total number of observed events with zero-tag is a very poor constraint on $\mathcal{B}(t \rightarrow Wb)/\mathcal{B}(t \rightarrow Wq)$ and $\sigma_{t\bar{t}}$. Without a stronger constraint on the number of $t\bar{t}$ events in the 0-tag sample, a low value of $\mathcal{B}(t \rightarrow Wb)/\mathcal{B}(t \rightarrow Wq)$ and a rather large $t\bar{t}$ cross section (compared to a cross section measurement assuming $\mathcal{B}(t \rightarrow Wb)/\mathcal{B}(t \rightarrow Wq) = 1$) are still allowed by the data. To fully exploit the zero-tag sample further discrimination is needed. To this end a discriminant function is constructed that makes use of the differences between the kinematic properties of the $t\bar{t}$ events and the backgrounds. A set of four variables is selected. They are well-modelled by simulation in samples depleted in top events and provide good separation between signal and background. To reduce the dependence on modelling of soft radiation and underlying event, only the four highest p_T jets are used to determine these variables. In very close relation to the kinematic cross section measurement, discussed in Sect. 4.3.2, the discriminant function $\mathcal{D}$ is built from the following variables: (i) the event sphericity $\mathcal{S}$, constructed from the four-momenta of the jets; (ii) the event centrality $\mathcal{C}$, defined as the ratio of the scalar sum of the

p_T of the jets to the scalar sum of the energy of the jets; (iii) $K_{T_{\min}} = \Delta R_{jj}^{\min} p_T^{\min}/E_T^W$, where $\Delta R_{jj}^{\min}$ is the minimum separation in $\eta - \phi$ space between pairs of jets, $p_T^{\min}$ is the p_T of the lower-p_T jet of that pair, and E_T^W is a scalar sum of the lepton transverse momentum and $\not{E}_T$; (iv) H'_{T_2}, defined as the ratio H_{T_2}/H_z, where H_{T_2} is the scalar sum of the E_T for all jets excluding the leading jet and H_z is the scalar sum of the $|E_z|$ of all jets plus the absolute value of the energy of the lepton and the neutrino along the z-direction.

To measure $\mathcal{B}(t \rightarrow Wb)/\mathcal{B}(t \rightarrow Wq)$ and $\sigma_{t\bar{t}}$, a binned maximum likelihood fit is performed to the data. The data is binned in: (i) ten bins of the kinematic discriminant $\mathcal{D}$ in the $e + 4$ jets zero-tag events, (ii) ten bins of the discriminant $\mathcal{D}$ in $\mu + 4$ jets zero-tag events, (iii) two bins for the two zero-tag samples $e + 3$ jets, $\mu + 3$ jets, (iv) four bins for the four single-tag samples (electron and muon and 3 or 4 jets), (v) four bins for the four double-tag samples (electron or muon and 3 or 4 jets). In each bin the number of events is predicted as the sum of the expected background and the signal contribution. The signal contribution is a function of R and $\sigma_{t\bar{t}}$. To predict the number of events in each bin of the discriminant $\mathcal{D}$, its expected shape for the background and the $t\bar{t}$ signal is used. The contribution from QCD multijet events is constrained by an additional bin for each control sample. The result is a likelihood function which is the product of 30 Poisson terms in the signal bins (i) to (v) and 12 Poisson terms in the twelve control bins (zero-tag, single-tag and double-tag, with 3 or 4 jets and for $e + $jet and $\mu + $jet). Systematic uncertainties are incorporated into the likelihood by using nuisance parameters. The final likelihood function contains one Gaussian term for each nuisance parameter. The values of $\mathcal{B}(t \rightarrow Wb)/\mathcal{B}(t \rightarrow Wq)$ and $\sigma_{t\bar{t}}$ that maximise the total likelihood function are:

$$
\mathcal{B}(t \rightarrow Wb)/\mathcal{B}(t \rightarrow Wq) = 1.03^{+0.19}_{-0.17}\,(\text{stat.} + \text{syst.}),
$$
$$(119)$$

$$
\sigma_{t\bar{t}} = 7.9^{+1.7}_{-1.5}\,(\text{stat.} + \text{syst.}) \pm 0.5\,(\text{lumi})\,\text{pb}, \quad (120)
$$

and in good agreement with the Standard Model expectation. The result of the 2-dimensional fit is shown in Fig. 86 (left) in the plane $(\mathcal{B}(t \rightarrow Wb)/\mathcal{B}(t \rightarrow Wq), \sigma_{t\bar{t}})$, along with the 68% and 95% confidence level contours. In Fig. 86 (right) the observed number of events is compared to the prediction obtained with $\mathcal{B}(t \rightarrow Wb)/\mathcal{B}(t \rightarrow Wq) = 1$ and $\sigma_{t\bar{t}} = 7$ pb in the zero-, single- and double-tag samples and for events with exactly three jets.

Also lower limits on R and the CKM matrix element $|V_{tb}|$ are extracted, assuming $|V_{tb}| = \sqrt{R}$. Using a Bayesian approach with the following prior:

$$
\pi(R) = \begin{cases} 1 & \text{if } 0 \leq R \leq 1 \\ 0 & \text{if } R < 0 \text{ or } R > 1 \end{cases} \quad (121)
$$

yields $R > 0.81$ at 68% CL and $R > 0.64$ at 95% CL. For the CKM matrix element $|V_{tb}|$ limits of $|V_{tb}| > 0.90$ at 68% CL and $|V_{tb}| > 0.80$ at 95% CL are found.

Table 39 summarises all measurements of $R = \mathcal{B}(t \rightarrow Wb)/\mathcal{B}(t \rightarrow Wq)$ from Run-I or Run-II so far. At present,

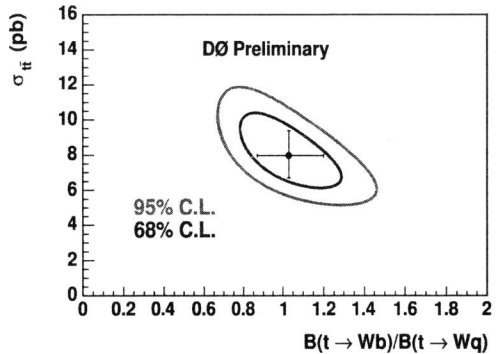 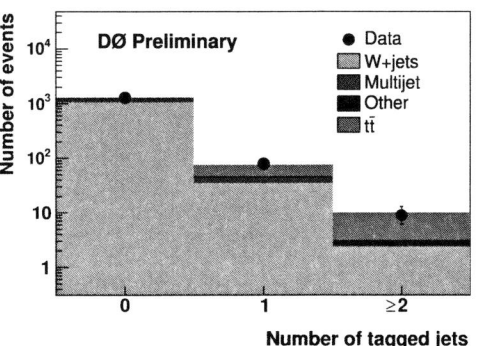

Fig. 86. *Left:* The 68% and 95% CL contours in the plane $(\mathcal{B}(t \to Wb)/\mathcal{B}(t \to Wq), \sigma_{t\bar{t}})$. The *point* indicates the best fit to the data. *Right:* Predicted and observed number of events in $l+3$ jets events in the zero-, single- and double-tag samples for $\mathcal{B}(t \to Wb)/\mathcal{B}(t \to Wq) = 1$ and $\sigma_{t\bar{t}} = 7$ pb

Table 39. Measurements of $R = \mathcal{B}(t \to Wb)/\mathcal{B}(t \to Wq)$ and $|V_{tb}|$ from CDF and DØ

R or $|V_{tb}|$	Source	Ref.
$R = 0.94^{+0.31}_{-0.24}$ (stat. + syst.)	CDF Run I	[374]
$R > 0.56$ (95% CL)	CDF Run I	[374]
$R = 1.12^{+0.27}_{-0.23}$ (stat. + syst.)	CDF Run II	[371]
$R > 0.61$ (95% CL)	CDF Run II	[371]
$R = 1.03^{+0.19}_{-0.17}$ (stat. + syst.)	DØ Run II	[334]
$R > 0.64$ (95% CL)	DØ Run II	[334]
$|V_{tb}| > 0.75$ (95% CL)	CDF Run I	[374]
$|V_{tb}| > 0.78$ (95% CL)	CDF Run II	[371]
$|V_{tb}| > 0.80$ (95% CL)	DØ Run II	[334]

all measurements are consistent with the Standard Model expectation. There is no indication for any deviation from it, and therefore no hint for a fourth quark generation or other physics beyond the Standard Model. Clearly more statistics is needed to make a more conclusive statement about the top quark decay branching ratio $R = \mathcal{B}(t \to Wb)/\mathcal{B}(t \to Wq)$ and $|V_{tb}|$. Ultimately, the most direct way to measure this CKM matrix element will be the precise measurement of the single-top production cross section, which is proportional to the square of $|V_{tb}|$ (see Sect. 5).

6.3 Top quark decays to tau leptons

The Standard Model's heavy third generation particles, the top and bottom quarks, the tau and the tau neutrino are intriguing. The high energies required to produce the third generation particles, particularly in the case of the top quark, have resulted in the particles being the least studied in the Standard Model. Current measurements leave room for new physics in the interactions and decays of these particles [375–377]. The high masses of the particles give rise to the hope that studying them could help shed light on the origin of fermion masses [378, 379].

CDF measures the rate of top-antitop events with a semi-leptonically decaying tau in $t\bar{t} \to e\tau bb\nu\nu$ and $t\bar{t} \to \mu\tau bb\nu\nu$ events in 193.5 pb^{-1} of Run-II data [380]. Semi-leptonic tau decays account for 64% of all tau decays. This analysis does not include taus decaying to electrons or

muons because these final states are difficult to differentiate from prompt leptons. CDF compares the observed with the predicted rate as a test of the Standard Model. Many extensions to the Standard Model predict identical final states which could lead to an anomalous rate. For example the charged Higgs decay from $t\bar{t}$, $t\bar{t} \to H^{\pm}Wb\bar{b}$, $H^{\pm} \to \tau^{\pm}\nu_{\tau}$ [381, 382]. This analysis is a search for any such anomalous processes that could show up in the final state as an enhanced (or suppressed) rate for tau leptons in top decays.

The events are selected by requiring a lepton with reconstructed, isolated electron E_T (muon p_T) greater than 20 GeV, a tau lepton with $p_T > 15$ GeV (see Sect. 3.3.5 for details on the tau identification), $\not{E}_T > 20$ GeV, two or more jets with $E_T > 25$ GeV and $|\eta| < 2$ and an additional jet with $|\eta| < 2$ and $E_T > 15$ GeV, $H_T > 205$ GeV, where H_T is the sum of the energy of the physics objects (electrons, muons, taus, jets and missing transverse energy) in the event, opposite charge of the primary lepton and the tau, and, finally, the event must survive the Z mass veto, with a window of 65 GeV to 115 GeV, which is made only on events with angular distribution between the jets and $\not{E}_T$ resembling the signature of $Z \to \tau\tau$.

The total acceptance is measured in a combination of data and Monte Carlo. The geometric times kinematic acceptance of the basic tau dilepton event selection is measured using the PYTHIA Monte Carlo program [223]. The efficiency for identifying the isolated, high-p_T lepton is scaled to the value measured in data using the unbiased leg in Z-boson decays. The tau identification is sensitive to some calorimeter quantities that are difficult to model in the Monte Carlo simulation. CDF uses PYTHIA Monte Carlo to generate $t\bar{t}$ events with the TAUOLA package [245] to correctly handle the tau polarisation. The tau ID efficiency for the acceptance is measured for signal and several backgrounds in the Monte Carlo, where the data to Monte Carlo normalisation is measured in $W \to \tau\nu$ events to within 6%. The geometric times kinematic acceptance with both the electron and muon channel included is $0.00080 \pm 0.0005(\text{stat.}) \pm 0.00014(\text{syst.})$. Systematic uncertainties considered include the jet energy calibration, electron and muon identification, the tau identification, Monte Carlo generators, the modelling of initial and final state radiation and the parton distribution functions.

The dominant background is from W-bosons produced in association with one or more jets where a jet passes all tau

Table 40. Summary of background and signal prediction in 193.5 pb^{-1} for the $tt \to \tau +$ lepton final state

Process	Expected number of events
$\gamma^*/Z \to \tau\tau +$jets	0.26 ± 0.06 (stat.) ± 0.05 (syst.)
$j \to \tau$ fakes	0.75 ± 0.12 (stat.) ± 0.20 (syst.)
$e \to \tau$ fakes	0.08 ± 0.03 (stat.) ± 0.02 (syst.)
$Z \to \mu\mu$	0.05 ± 0.03 (stat.)
WW	0.14 ± 0.02 (stat.) ± 0.03 (syst.)
WZ	0.02 ± 0.02 (stat.)
Total expected bgd	1.30 ± 0.14 (stat.) ± 0.21 (syst.)
Signal expectation	1.03 ± 0.06 (stat.) ± 0.17 (syst.)

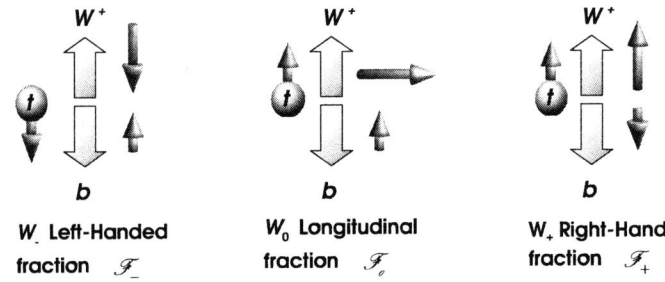

Fig. 87. Schematic of the helicity states of the W-boson in top quark decays

identification cuts to fake a tau. The rate of jets faking taus is calculated in four data samples as a function of jet E_T and calorimeter isolation and applied to loose tau candidates in a 50 GeV jet sample. Similarly the background from electrons faking taus is determined by measuring the fake rate in $Z \to ee$ events in data and applying this rate to events with tau candidates that fail the electron removal cut. Muons faking taus provide another non-tau background and are estimated from a $Z \to \mu\mu$ Monte Carlo. There is also background from processes that create real taus and which can mimic the signal final state, in particular $Z \to \tau\tau +$jets, and WW and WZ production. They are estimated from HERWIG Monte Carlo. The summary of all backgrounds and the expected signal is given in Table 40.

In total, 2.3 events are expected in 193.5 pb^{-1} and 2 events are observed, both in the $e + \tau$ channel. One of the events has a b-tagged jet, identified via the secondary vertex algorithm. Slightly over 50% of the $t\bar{t}$ events are expected to have a b-tagged jet. CDF measures the parameter r_τ where

$$r_\tau \equiv \frac{B(t \to b\tau\nu)}{B_{\mathrm{SM}}(t \to b\tau\nu)}. \tag{122}$$

An upper limit of

$$r_\tau < 5.0 \quad \text{at } 95\% \text{ CL} \tag{123}$$

is set. This measurement is consistent with the Standard Model prediction of $r_\tau = 1$. A larger data sample and continued improvement to the purity and efficiency of τ identification will allow r_τ to be further constrained.

6.4 Helicity of the W-boson in top quark decays

An important consequence of a heavy top quark is, to good approximation, that it decays as a free quark. Its expected lifetime is approximately 0.5×10^{-24} s, and it therefore decays about an order of magnitude faster than the time needed to form bound states with other quarks [164]. Consequently, the spin information carried by top quarks is expected to be passed directly on to their decay products, so that production and decay of top quarks provide a probe of the underlying dynamics, with minimal impact from gluon radiation and binding effects of QCD [368, 383–386].

Studies of decay angular distributions provide therefore a direct check of the $V - A$ nature of the Wtb coupling and information on the relative coupling of longitudinal and transverse W bosons to the top quark. The emitted b-quark can be considered essentially massless compared to the top quark ($m_\mathrm{b} \ll m_\mathrm{t}$). To conserve angular momentum, the spin of the b-quark, with its dominantly negative helicity (i.e., spin pointing opposite to its line of flight in the rest frame of the top quark) can therefore point either along or opposite to the spin of the top quark (Fig. 87). In the first case, the spin projection of the vector W-boson must vanish (i.e., the W is longitudinally polarised, or has zero helicity $\mathcal{F}_0$). If the spin of the b-quark points opposite to the top quark spin, the W-boson must then by left-handed polarised (have negative helicity $\mathcal{F}_-$). Hence, for massless b-quarks, a top quark can decay only to a left-handed or longitudinal W-boson. In the Standard Model, the fraction of decays to longitudinally polarised W-bosons is determined by the masses of the involved particles and expected to be [367, 383–385]

$$\mathcal{F}_0^{\mathrm{SM}} \approx \frac{m_\mathrm{t}^2}{2M_W^2 + m_\mathrm{t}^2 + m_\mathrm{b}^2} = 70.3 \pm 1.2\%, \tag{124}$$

for $m_\mathrm{t} = 175$ GeV/c^2, where M_W is the mass of the W-bosons and m_b is the mass of the bottom quark. The fraction $\mathcal{F}_0$ is enhanced due to the large coupling of the top quark to the Goldstone modes of the Higgs field. Fractions of left- or right-handed W-bosons are denoted as $\mathcal{F}_-$ and $\mathcal{F}_+$, respectively. In the Standard Model, the $\mathcal{F}_-$ is expected to be $\approx 30\%$ and $\mathcal{F}_+$ is suppressed by a factor $(m_\mathrm{b}/m_\mathrm{t})^2$, yielding at next-to-leading order a value of 3.6×10^{-4} [387]. Charge-conjugation symmetry implies that the $\bar{t}$ quark decays to either a longitudinally- or right-handed-polarised W^-. A measurement of $\mathcal{F}_+$ that differs significantly from this value would be an unambiguous indication of new physics. For example, an $\mathcal{F}_+$ value of 0.3 would indicate a purely $V + A$ charged current interaction. A possible theoretical model that includes a $V + A$ contribution at the tWb vertex is an $SU(2)_L \times SU(2)_R \times U_Y(1)$ extension of the Standard Model [368, 388, 389]. Direct measurements of the longitudinal fraction by CDF and DØ in Run I have found $\mathcal{F}_0 = 0.91 \pm 0.39$ [390] and $\mathcal{F}_0 = 0.56 \pm 0.31$ [391], respectively. In addition, measurements of the $b \to s\gamma$ decay rate have indirectly limited the $V + A$ contribution in top quark decays to less than a few percent [392–394]. However, direct measurements of the $V + A$ contribution are still necessary because the limit from $b \to s\gamma$ assumes that the electroweak penguin contribution is dominant.

The angular distribution ω of the W-boson decay products with weak isospin $I_3 = -1/2$ (charged lepton or d, s quark) in the rest frame of the W-boson can be described by introducing the angle θ^* with respect to the top quark direction [367, 383–385]:

$$\omega(\cos\theta^*) = \frac{3}{4}(1 - \cos^2\theta^*)\mathcal{F}_0 + \frac{3}{8}(1 - \cos\theta^*)^2\mathcal{F}_- + \frac{3}{8}(1 + \cos\theta^*)^2\mathcal{F}_+ \,. \quad (125)$$

Due to backgrounds and reconstruction effects, the distribution of $\cos\theta^*$ that is observed differs from $\omega(\cos\theta^*)$. However, the shape of the measured $\cos\theta^*$ distribution depends on $\mathcal{F}_+$ and this dependence can be used to measure $\mathcal{F}_+$.

CDF and DØ use various techniques to measure the helicity of the W-boson in top quark decays in lepton + jets events: Using a kinematic fit for the association of the final state particles with the quarks and leptons, the distribution of the helicity angle ($\cos\theta^*$) between the lepton and the b quark in the W rest frame, provides the most direct measure of the W helicity. The second method (lepton p_T) uses the different lepton p_T spectra from longitudinally or transversely polarised W-decays to determine the relative contributions. This method is also used by both experiments in the dilepton channel. A third method uses the invariant mass of the lepton and the b-quark in top decays ($M_{\ell b}^2$), which is directly related to $\cos\theta^*$, as a discriminant. Finally, the matrix element method (ME), described for the top quark mass measurement, has been applied, forming a 2-dimensional likelihood $\mathcal{L}(m_{\text{top}}, \mathcal{F}_0)$, where the mass-dependence is integrated out so that only the sensitivity to the W-helicity in the top quark decay is exploited.

In this section the recent Run-II analyses of the helicity of the W-boson in top quark decays are summarised. The results of all CDF and DØ analyses are in agreement with the Standard Model expectation, but within large statistical uncertainties.

6.4.1 CDF Analyses

In Run-II, CDF has measured the branching fraction of the top quark to longitudinally and right-handed polarised W-bosons, $\mathcal{F}_0$ and $\mathcal{F}_+$, using approximately $200\,\text{pb}^{-1}$ of data in the dilepton and the lepton + jets channel [395]. Two observables in the $t\bar{t}$ candidate events are used to measure the W helicity. Charged leptons from the decay of a longitudinally-polarised W-boson have a symmetric angular distribution $\propto (1 - \cos^2\theta^*)$ (see Fig. 88), where θ^* is the helicity angle between the charged lepton momentum in the W rest frame and the boost direction from the top quark rest frame into the W rest frame. Left-handed W's have an asymmetric distribution $\propto (1 - \cos\theta^*)^2$. Direct measurement of this angle is difficult and requires the use of the missing E_T vector with very limited resolution. How-

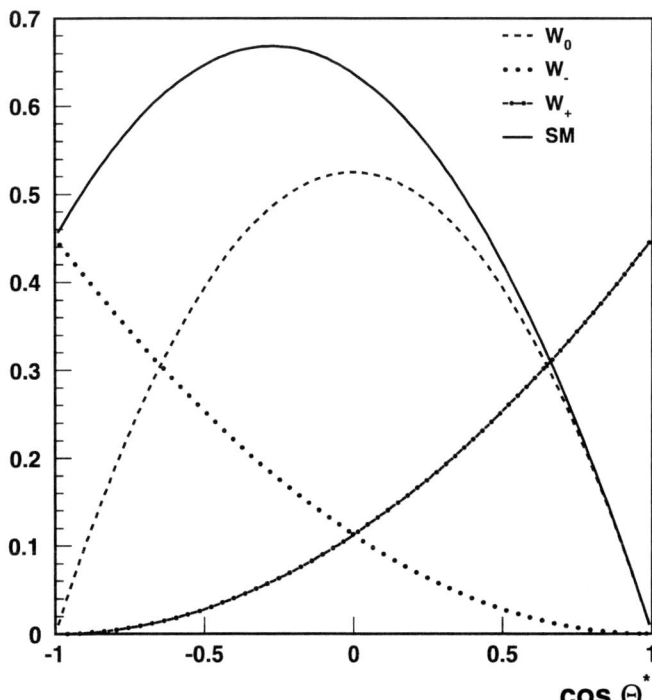

Fig. 88. The $\cos\theta^*$ distribution for the left-handed W-boson (*dotted line*), longitudinal W-boson (*dashed line*), right-handed W-boson (*dash-dotted line*) and the Standard Model expectation (*full line*)

ever, one can approximate $\cos\theta^*$ by relating it to the invariant mass of the b-quark and the charged lepton ($M_{\ell b}^2$):

$$\cos\theta^* = \frac{p_\ell p_b - E_\ell E_b}{|p_\ell||p_b|} \simeq \frac{2M_{\ell b}^2}{m_t^2 - M_W^2} - 1 \,, \quad (126)$$

a variable that depends only on lab-frame momenta and is independent of the measured $\not{E}_T$. The second observable exploits the fact that charged leptons from left-handed W decays are preferentially emitted in the background direction with respect to the W direction of motion, leading to softer lepton transverse momentum p_T in the lab frame, while the leptons from right-handed W's are preferentially emitted forward and thus have a harder p_T spectrum. Longitudinal W decays represent an intermediate case. Figure 89 shows the predicted cos* and lepton p_T distributions for $m_t = 175\,\text{GeV}/c^2$, after the event selection and reconstruction. The observed cos* distribution extends somewhat beyond the physical range $-1 \leq \cos\theta^* \leq 1$ because the world-average top and W masses are used in (126), rather than the event-by-event reconstructed masses which have larger uncertainties. In this analysis, $\mathcal{F}_0$ and $\mathcal{F}_+$ are measured from a combination of the $M_{\ell b}^2$ and the p_T technique.

The selection, reconstruction and background estimates for the dilepton events is taken from the corresponding cross section analysis (Sect. 4.2.1), yielding 13 events in this sample with an expected background of 2.7 ± 0.7 events. Similarly, the selection, reconstruction and background estimates in the lepton + jets channel with

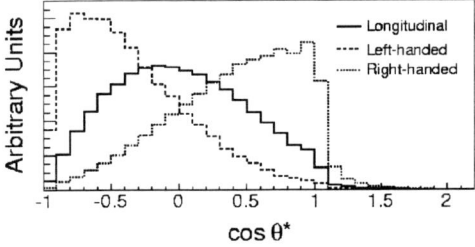
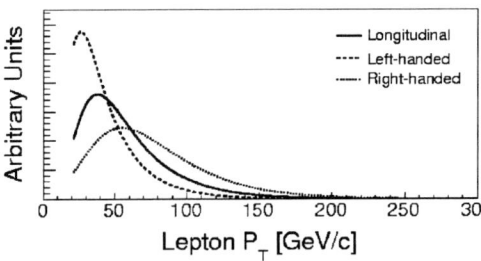

Fig. 89. Distribution of $\cos\theta^*$ and lepton p_T for top-quark decays to left-handed, right-handed, and longitudinally polarised W-bosons

one or more b-tagged jets using the SecVtx algorithm is taken from the corresponding cross section analysis (Sect. 4.2.3), yielding 57 events, of which approximately $2/3$ are $t\bar{t}$ events. The p_T analysis uses both samples, while the $M_{\ell b}^2$ analysis uses the lepton + jets sample only. In addition to those selection requirements, events selected for the $M_{\ell b}^2$ analysis are required to have a fourth jet with $E_T > 8$ GeV and $|\eta| < 2$. 37 events pass this cut, of which 9.9 ± 1.7 are estimated to be background. The presence of four jets allows the event to be kinematically reconstructed as a $t\bar{t}$ event with the top mass constrained to 175 GeV/c^2, and to associate the appropriate jet to the lepton in (126). 31 of the 37 events pass a $\chi^2_{\min}$ cut on the fit quality.

To create $\cos\theta^*$ (lepton p_T) templates for $t\bar{t}$ signal events, the Monte Carlo programs MADEVENT [236, 237] and HERWIG [226, 227] are used. The helicity of one W-boson in the top rest frame is fixed, while the other W takes on values according to the Standard Model prediction. Hadronisation and fragmentation are carried out using PYTHIA [223].

The data are fit separately to the $\cos\theta^*$ and p_T templates using likelihood functions that include a Gaussian constraint on the background, as well as corrections for trigger and event selection cuts that have helicity-dependent biases, such as those on the lepton p_T. The result of the fits to the various subsamples are shown in Table 41. The reconstructed $\cos\theta^*$ distribution from the data and the best-fit templates are shown in Fig. 90 (left). Figure 90 (right) shows the lepton p_T distribution for the two samples and the results of the fit.

Table 41. Summary of the results for the $\cos\theta^*(M_{\ell b}^2)$, p_T, and combined measurements of $\mathcal{F}_0$ and $\mathcal{F}_+$. N is the number of events or leptons used in the measurement. Where two uncertainties are given the first is statistical and the second is systematic. Uncertainties on the combined measurement are the total statistical and systematic uncertainty

Analysis	N	$\mathcal{F}_0$	$\mathcal{F}_+$
$\cos\theta^*(M_{\ell b}^2)$	31	$0.99^{+0.29}_{-0.35} \pm 0.19$	$0.23 \pm 0.16 \pm 0.08$
$p_T\,(\ell\ell)$	26	$-0.54^{+0.35}_{-0.25} \pm 0.16$	$-0.47 \pm 0.10 \pm 0.09$
$p_T\,(\ell+\text{jets})$	57	$0.95^{+0.35}_{-0.42} \pm 0.17$	$0.11^{+0.21}_{-0.19} \pm 0.10$
$p_T\,(\text{combined})$	83	$0.31^{+0.37}_{-0.23} \pm 0.17$	$-0.18^{+0.14}_{-0.12} \pm 0.12$
Combined		$0.74^{+0.22}_{-0.34}$	$0.00^{+0.20}_{-0.19}$
95% CL limit		$< 0.95, > 0.18$	< 0.27

The dominant systematic uncertainties in the $\cos\theta^*$ $(M_{\ell b}^2)$ and p_T analyses, estimated using Monte Carlo pseudo-experiments and summarised in Table 42, arise from uncertainties in the top-quark mass, the background shape and normalisation, the effects of initial- and final-state radiation (ISR/FSR), and uncertainty in the parton distribution functions (PDFs). The quadratic sum of all sources of systematic uncertainty leads to a final result of $\mathcal{F}_0 = 0.99^{+0.29}_{-0.35}(\text{stat.}) \pm 0.19(\text{syst.})$ for the $M_{\ell b}^2$ analysis and $\mathcal{F}_0 = 0.31^{+0.37}_{-0.23}(\text{stat.}) \pm 0.17(\text{syst.})$ for the p_T analysis. The results of the two analyses are combined taking into account both the statistical and systematic correlations between the

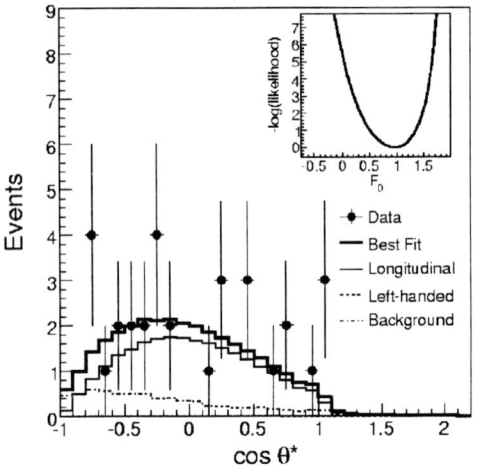
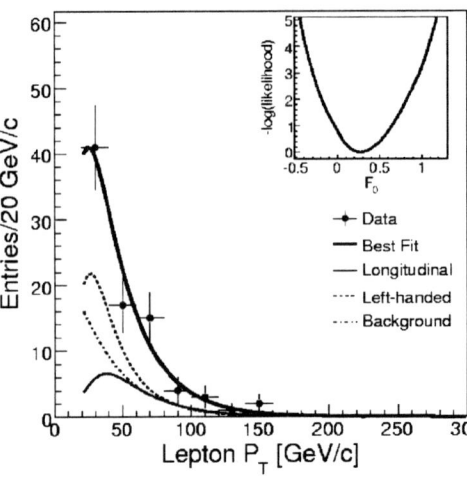

Fig. 90. The $\cos\theta^*$ distribution for the lepton + jets sample (*left*) and lepton p_T distribution for the lepton + jets and dilepton samples (*right*), overlaid with signal and background templates superimposed according to their best-fit values. The *inset* shows the projection of the negative log-likelihood along the $\mathcal{F}_0$ axis for the fit to the data

Table 42. Summary of systematic uncertainties for the CDF measurements of $\mathcal{F}_0$ and $\mathcal{F}_+$

Systematic source	p_T Method		$M_{\ell b}^2$ Method	
	$\Delta\mathcal{F}_0$	$\Delta\mathcal{F}_+$	$\Delta\mathcal{F}_0$	$\Delta\mathcal{F}_+$
Top mass	0.11	0.09	0.08	0.04
Bkg. Modelling	0.10	0.06	0.13	0.05
ISR/FSR	0.04	0.04	0.03	0.02
PDF	0.03	0.03	0.04	0.01
MC statistics	0.01	< 0.01	0.01	0.01
Accept. correction	0.02	0.01	< 0.005	< 0.005
Trigger correction	0.02	0.02	...	...
Jet energy scale	...	...	0.09	0.04
MC Modelling	...	...	0.04	0.02
b-tagging	...	...	0.01	< 0.005
Total	0.17	0.12	0.19	0.08

two techniques. Statistical correlations arise because the two analyses share the subset of the lepton + jets sample that passes the $\chi^2_{\min}$ cut on the top mass reconstruction, while common sources of systematic uncertainty include the top mass uncertainty and background normalisation. The correlation coefficients are determined via pseudo-experiments. The combined result is $\mathcal{F}_0 = 0.74^{+0.22}_{-0.34}$(stat. + syst.). In addition, $\mathcal{F}_+ = 0.00^{+0.20}_{-0.19}$(stat. + syst.) and $\mathcal{F}_+ < 0.27$ at the 95% CL. These results are consistent with the Standard Model predictions of $\mathcal{F}_0 = 0.70$, $\mathcal{F}_+ = 0$.

6.4.2 DØ analyses

Measurement of $\mathcal{F}_0$ in the lepton + jets channel. DØ has recently measured the longitudinal component $\mathcal{F}_0$ of the helicity of W-bosons in $t \to Wb$ decays in $125\,\mathrm{pb}^{-1}$ of Run-I data at $\sqrt{s} = 1.8\,\mathrm{TeV}$, using the matrix element method [391]. This analysis is based on the same lepton + jets sample that is used to extract the mass of the top quark in [396,397] (ℓ+jets analyses in Sect. 7.1.2). Making use of information contained in the events and comparing each individual event with the differential cross section for $t\bar{t}$ production and decay, the fraction $\mathcal{F}_0$ of longitudinal W-boson production in the data is extracted, assuming no contribution from right-handed W-bosons. In particular, this procedure relies on a direct comparison of data to the LO matrix element for the production and decay of $t\bar{t}$ states [396,397]. This method offers the possibility of increased statistical precision by using the decay of both W-bosons in these events, and is similar to that suggested for $t\bar{t}$ dilepton decay channels, and used in previous mass analyses of dilepton events [398–400].

The probability density for $t\bar{t}$ production and decay in the e + jets final state, for a given value of $\mathcal{F}_0$, is defined as:

$$P_{t\bar{t}}(\mathcal{F}_0) = \frac{1}{12\sigma_{t\bar{t}}} \int \mathrm{d}\rho_1 \, \mathrm{d}m_1^2 \, \mathrm{d}M_1^2 \, \mathrm{d}m_2^2 \, \mathrm{d}M_2^2$$
$$\times \sum_{\mathrm{perm}.\nu} |\mathcal{M}_{t\bar{t}}(\mathcal{F}_0)|^2 \frac{f(q_1)\,f(q_2)}{|q_1||q_2|} \Phi_6$$
$$\times W_{\mathrm{jets}}(E_p, E_j), \qquad (127)$$

where the integration and kinematic variables are identical to the ones described for the mass measurement in the ℓ + jets analyses in Sect. 7.1.2. All processes that contribute to the observed final state must be included in the probability density. The final probability density is therefore written as:

$$P_M(x; \mathcal{F}_0) = c_1 P_{t\bar{t}}(x; \mathcal{F}_0) + c_2 P_{\mathrm{bgd}}(x), \qquad (128)$$

where c_1 and c_2 are the signal and background fractions, and x is the set of variables needed to specify the measured event. $P_{t\bar{t}}$ and P_{bgd} refer to the signal and background production and decay probabilities, respectively.

The event probabilities are inserted into a likelihood function for N observed events. The $t\bar{t}$ probability density contains contributions from both W_0 ($\mathcal{F}_0$) and W_- ($\mathcal{F}_-$) helicities, and the ratio of $\mathcal{F}_0/\mathcal{F}_-$ is allowed to vary. The best estimate of $\mathcal{F}_0$ is obtained by maximising the following likelihood with respect to $\mathcal{F}_0$, subject to the constraint that $\mathcal{F}_0$ must be physical, i.e. $0 \le \mathcal{F}_0 \le 1$, and $\mathcal{F}_- + \mathcal{F}_0 = 1$:

$$\mathcal{L}(\mathcal{F}_0) = e^{-N \int P_m(x, \mathcal{F}_0)\mathrm{d}x} \prod_{i=1}^{N} P_m(x_i; \mathcal{F}_0), \qquad (129)$$

where P_m is the probability density for observing a given event i. The best value of $\mathcal{F}_0$ and the parameters c_i are obtained from minimising the negative log-likelihood with respect to all three parameters. Results from analysing samples of PYTHIA [223] Monte Carlo events indicate that a response correction, resulting from un-modelled gluon radiation, must be applied to the data.

The current uncertainty in the top quark mass is large enough to affect the value of $\mathcal{F}_0$. For sufficiently high statistics, the likelihood can be maximised as a function of the two variables ($\mathcal{F}_0, m_t$), which can then correctly take into account any correlations between the two parameters and the fact that $\mathcal{F}_0$ is bound between 0 and 1. Given the presently limited statistics, the next best way to account for the uncertainty in m_t is by projecting the 2-dimensional likelihood onto the $\mathcal{F}_0$ axis. In this way, the systematic uncertainty in $\mathcal{F}_0$ from the uncertainty in m_{top} can be obtained by integrating the probability over the mass, which is done here from 165 to $190\,\mathrm{GeV}/c^2$, in steps of $2.5\,\mathrm{GeV}/c^2$, using no other prior knowledge of the mass. Figure 91 shows the 2-dimensional probability density as a function of $\mathcal{F}_0$ and m_t for the data, after applying the response correction.

The other systematic uncertainties are quite small, and are calculated by varying their impact in the Monte Carlo or data, and added in quadrature (see Table 43). The final result is:

$$\mathcal{F}_0 = 0.56 \pm 0.31(\mathrm{stat.} + m_t) \pm 0.07(\mathrm{syst.}). \qquad (130)$$

After combining the two errors in quadrature, the final result is $\mathcal{F}_0 = 0.56 \pm 0.31$, which is consistent with expectations of the Standard Model.

Measurement of $\mathcal{F}_+$ in the lepton + jets channel. DØ has measured the positive helicity fraction $\mathcal{F}_+$ of the W-boson

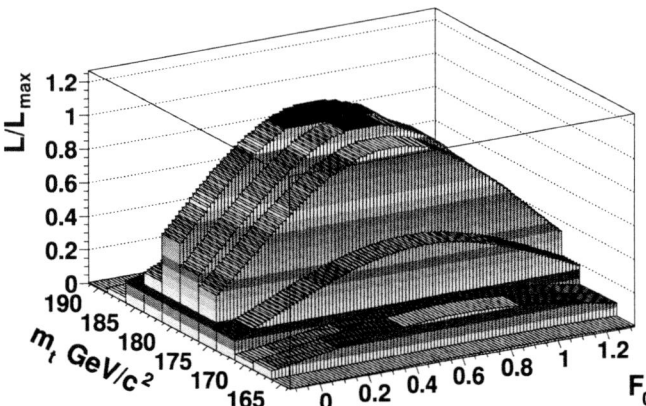

Fig. 91. Likelihood, normalised to its maximum value, as a function of m_t and $\mathcal{F}_0$

Table 43. Impact of systematic and statistical uncertainties on the measurement of $\mathcal{F}_0$

Source	$\Delta\mathcal{F}_0$
Acceptance and linearity response	0.055
Jet energy scale	0.014
Spin correlation in $t\bar{t}$ events	0.008
Parton distribution functions	0.008
Model for $t\bar{t}$ production	0.020
Multiple interactions	0.006
Multijet background	0.024
Total systematic uncertainty, except for m_t	0.070
Statistics and uncertainty in m_t	0.306
Total uncertainty	0.314

in top quark decays in the lepton + jets decay mode of $t\bar{t}$ events, using 230 pb^{-1} of Run-II data [401]. In this analysis, $\mathcal{F}_0$ is fixed at 0.7 and the positive helicity fraction $\mathcal{F}_+$ is measured. Selecting a $t\bar{t}$ event sample, the four vectors of the two top quarks and their decay products are reconstructed using a kinematic fit, from which $\cos\theta^*$ is calculated. The distribution in $\cos\theta^*$ is compared to templates for different $\mathcal{F}_+$ values using a binned maximum likelihood method.

Two separate analyses are performed and the results are combined. One analysis uses kinematic information to select $t\bar{t}$ events ("kinematic analysis") and the other uses secondary vertex b-jet identification (Sect. 3.3.7) as well as kinematic information in order to improve the signal to background ratio ("b-tagged analysis"). The kinematic analysis vetoes b-tagged events to simplify the combination of results with the b-tagged analysis.

The event selection follows closely that of the corresponding kinematic and b-tagged cross section measurement analyses (Sects. 4.3.2 and 4.3.3). The top quark and the W-boson four-momenta are reconstructed using a kinematic fit which is subject to the following constraints: two jets must form the invariant mass of the W-boson, the lepton and the $\not{E}_T$ together with the neutrino p_z component must form the invariant mass of the W-boson, and

the masses of the two reconstructed top quarks must be 175 GeV/c^2. Among the twelve possible jet combinations, the solution with the minimal χ^2 from the kinematic fit is chosen, yielding the correct solution in about 60% of all cases.

The $t\bar{t}$ signal events for seven different values of $\mathcal{F}_+$, $\mathcal{F}_+ = 0.00, \ldots 0.30$ in steps of 0.05, are generated with the ALPGEN [224] Monte Carlo program, followed by PYTHIA [223] for the hadronisation with a chosen top quark mass of $m_t = 175$ GeV/c^2. As the interference between $V - A$ and $V + A$ is suppressed by the small mass of the b-quark and is therefore negligible, these samples can be used to create $\cos\theta^*$ templates for any $\mathcal{F}_+$ value by a linear interpolation of the templates. All seven templates from these samples are normalised to unit area and a linear fit to the contents of each $\cos\theta^*$ bin as a function of $\mathcal{F}_+$ is performed. The ALPGEN Monte Carlo is also used to model the W + jets background. Using the matrix method (Sect. 4.3.2), the kinematic analysis calculates the number of multijet (QCD) background events for each bin in the $\cos\theta^*$ distribution from the data sample to obtain the shape of the multijet $\cos\theta^*$ templates. For the b-tagged analysis, the multijet template is formed from the data events after the event selection except that the leptons are required to satisfy the loose and to fail the tight criteria.

To discriminate between $t\bar{t}$ pair production and background, a discriminant $\mathcal{D}$, similar to that in the kinematic/topological cross section measurement in the lepton + jets channel (Sect. 4.3.2), is built using variables which exploit the difference in event topology: H_T, the minimum dijet mass of the jet pairs, the χ^2 from the kinematic fit, the centrality, $K'_{T,\min}$, and aplanarity and sphericity. All of these variables are used for the discriminant in the kinematic analysis. Only H_T, centrality, the minimum dijet mass, and χ^2 are used in the b-tagged analysis. The discriminant is built separately for the kinematic and b-tagged analyses, using the method described for the kinematic/topological $t\bar{t}$ cross section measurement in Sect. 4.3.2. Events are selected for which $\mathcal{D} > 0.6$ in the kinematic analysis, and $\mathcal{D} > 0.25$ in the b-tagged analysis. The number of $t\bar{t}$, W + jets, and multijet events in the sample is determined by a binned maximum likelihood fit, comparing the observed $\mathcal{D}$ distribution in the data to the sum of the distributions expected from $t\bar{t}$, W + jets and multijet events. Those numbers are multiplied by the efficiency for each type of event to pass the $\mathcal{D}$ selection, yielding the event composition of the sample used for measuring $\cos\theta^*$. The $\cos\theta^*$ distribution obtained in data after the full selection is shown in Fig. 92 (left) for the kinematic and in Fig. 92 (right) for the b-tagged analysis.

A binned maximum likelihood fit of signal and background $\cos\theta^*$ templates to the data is used to measure $\mathcal{F}_+$. The binned Poisson likelihood $\mathcal{L}(\mathcal{F}_+)$ of the data is computed to be consistent with the sum of signal and background templates, normalised to the number given by the above mentioned discriminant, $\mathcal{D}$, at each of the seven chosen $\mathcal{F}_+$ values. In both analyses, a parabola is fit to the negative log-likelihood points to determine the likelihood as a function of $\mathcal{F}_+$. These curves for the kinematic and b-tagged measurements are added for the combined analysis.

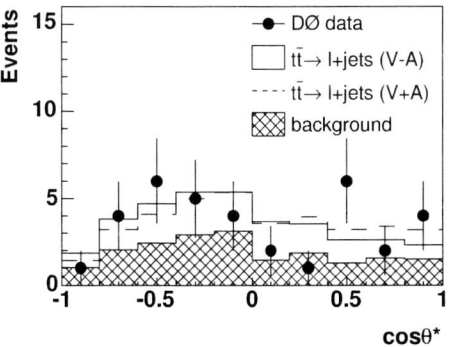 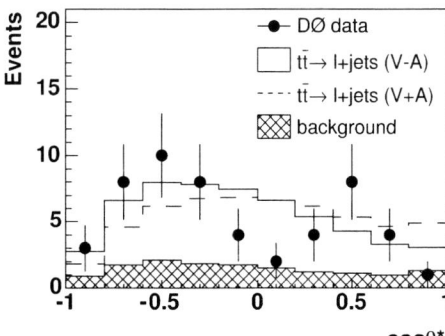

Fig. 92. $\cos\theta^*$ distribution observed in data along with the Standard Model prediction shown as the *solid line*, while a model with a pure $V + A$ interaction would result in the distribution given by the *dashed line*. *Left*: kinematic analysis. *Right*: b-tagged analysis

Table 44. Systematic uncertainties on $\mathcal{F}_+$ for the two independent analyses and for the combination

Source	Kinematic	b-tagged	Combined
Jet energy calibration	0.03	0.04	0.04
Top quark mass	0.04	0.04	0.04
Template statistics	0.05	0.02	0.03
b-tag	0.03	0.02	0.02
$t\bar{t}$ model	0.01	0.02	0.02
W + jets model	0.01	0.01	0.01
Sample composition	–	0.02	0.01
Calibration	0.01	0.01	0.01
Total	0.08	0.07	0.07

Systematic uncertainties, summarised in Table 44, are evaluated in ensemble tests which can affect the shape of the $\cos\theta^*$ distribution or the relative contribution from the three event sources ($t\bar{t}$, W + jets and multijets). The total systematic uncertainty is then taken into account in the likelihood by convoluting the average shift in the resulting $\mathcal{F}_+$ value with a Gaussian with a width that corresponds to the total systematic uncertainty.

Assuming a fixed value of 0.7 for $\mathcal{F}_0$, the combined result for $\mathcal{F}_+$ is:

$$\mathcal{F}_+ = 0.00 \pm 0.13 \,(\text{stat.}) \pm 0.07 \,(\text{syst.}) \,. \qquad (131)$$

Also a Bayesian confidence interval (using a flat prior distribution which is non-zero only in the physically allowed region of $\mathcal{F}_+ = 0.0 - 0.3$) is calculated which yields:

$$\mathcal{F}_+ < 0.25 \text{ at } 95\% \text{ CL} \,. \qquad (132)$$

Measurement of $\mathcal{F}_+$ in the dilepton channel. DØ has measured the W-boson positive helicity fraction $\mathcal{F}_+$ in $t\bar{t}$ decays in the dilepton final state (here lepton refers to electrons and muons, taus are only included via their subsequent decay to electrons or muons) with 370 pb^{-1} [402]. The presence of the two unmeasured neutrinos in the dilepton channels renders the $t\bar{t}$ system underconstrained, which means that one cannot reconstruct the W-boson rest frames without making additional assumptions. However, one can still measure $\mathcal{F}_+$ by noting charged leptons from right-handed W-bosons will tend to be emitted along the W-boson boost direction, and thus have larger p_T in the laboratory frame.

There, the lepton p_T is used as the measurement variable, yielding two measurements for each event. The expected distribution of lepton p_T for background and for signal with different $\mathcal{F}_+$ values, referred to as "templates", is estimated using Monte Carlo events. These templates are used in a binned likelihood fit to find the $V + A$ fraction $\mathcal{F}_+$ given by the data. The resulting log likelihood curves are interpreted using a Bayesian approach.

The dilepton event selection follows very closely the criteria applied in the corresponding cross section analysis (Sect. 4.3.1). Only in the $e\mu$ channel, the expected signal significance is improved by requiring the multivariate electron likelihood to be > 0.25. This selection yields 15 events in the $e\mu$ channel, 5 events in the ee channel, and 2 events in the $\mu\mu$ channel. The expected background contributions are determined using Monte Carlo and data events as described in the corresponding cross section analysis (Sect. 4.3.1). Figure 93 (left) shows the normalised lepton p_T distribution in $t\bar{t}$ Monte Carlo for a pure $V - A$ or pure $V + A$ coupling at the tWb vertex. Leptons with $p_T > 200$ GeV are included in the uppermost bin of each template (overflow bin).

The value of $\mathcal{F}_+$ favoured by the data is expected using a binned maximum likelihood fit. The ALPGEN [224] Monte Carlo program is used to simulate $t\bar{t}$ events with $\mathcal{F}_+ = 0.00$ and 0.30, and to simulate backgrounds from $Z\gamma^*$ and WW events. Models for intermediate values of $\mathcal{F}_+$ in increments of 0.05 are formed, performing a linear interpolation of the signal templates. For each $\mathcal{F}_+$ value, the likelihood of the data to be consistent with the sum of signal and background templates is computed by multiplying the Poisson probabilities of each template bin being consistent with the data. The prior expectation for the normalisation of the background, estimated as described in the cross section analysis, is expressed with a Gaussian term in the likelihood:

$$\mathcal{L}(\mathcal{F}_+) = \prod_{i=1}^{N_{\text{bgd}}} e^{(n_{b,i} - \overline{n_{b,i}})^2 / 2\sigma_{b,i}^2} \times \prod_{j=1}^{N_{\text{bins}}} P(d_j; n_j) \,. \qquad (133)$$

In the Gaussian term, N_{bkg} is the number of background sources, $\overline{n_{b,i}}$ is the nominal number of events for the i-th background, $\sigma_{b,i}$ is the uncertainty on $\overline{n_{b,i}}$ and $n_{b,i}$ is the fitted number of events for the i-th background. In the Poisson term, n_j is the total number of events expected in the j-th

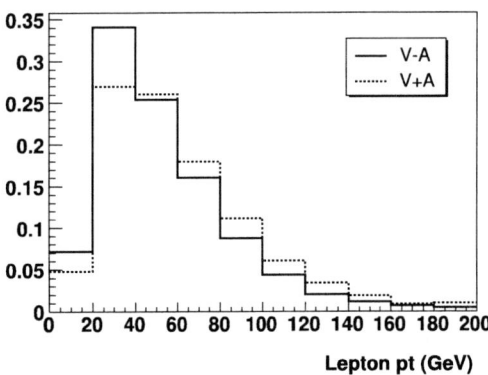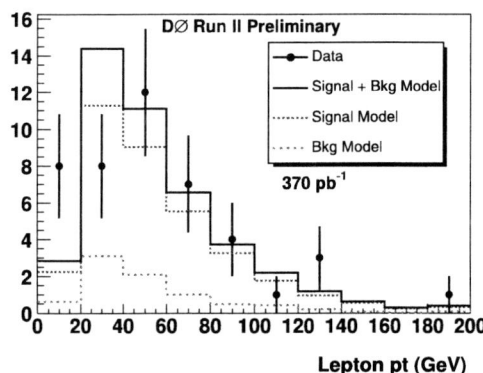

Fig. 93. *Left*: Normalised lepton p_T distribution for $m_{\text{top}} = 175\,\text{GeV}/c^2$ $t\bar{t}$ Monte Carlo events. The *solid (dashed) histogram* is for events with a pure $V - A$ (pure $V + A$) coupling at the tWb vertex. *Right*: Comparison of the sum of $e\mu$, ee, and $\mu\mu$ data (*points with error bars*) to the sum of the best-fit templates of signal and background (*solid histogram*). The signal and background contributions are shown separately as the *dashed* and *dotted histograms*

Table 45. Systematic uncertainties on $\mathcal{F}_+$

Source	Uncertainty
Monte Carlo statistics	0.046
Analysis self-consistency	0.010
Top quark mass	0.008
Jet energy calibration	0.013
$t\bar{t}$ model	0.030
Fake lepton model	0.013
Lepton p_T resolution	0.010
Trigger	0.008
Total	0.061

bin ($n_j = n_{s,j} + \sum_{i=1}^{N_{\text{bkg}}} n_{b,ik}$, where n_s is the fitted number of signal events), and d_j is the number of data events in the j-th bin. The negative log-likelihood curves are calculated for each channel separately and summed to arrive at a combined result. A parabola is fit to the summed points to determine the overall likelihood as a function of $\mathcal{F}_+$.

A Bayesian technique is used to determine a confidence level (CL) interval for the true value of $\mathcal{F}_+$. The prior probability is chosen to be flat in the physically-allowed range $0 < \mathcal{F}_+ < 0.30$ and zero elsewhere. With this choice, finding a Bayesian confidence interval is equivalent to integrating the likelihood curve. The performance of the maximum

likelihood fit is tested and confirmed by means of Monte Carlo ensemble tests. Systematic uncertainties can change the measurement in two ways: by altering the estimate of the background in the final sample and by modifying the shape of the lepton p_T templates. The effect of the systematic uncertainties on $\mathcal{F}_+$ are also estimated using Monte Carlo ensemble testing and summarised in Table 45.

In Fig. 93, the data is shown as the points with error bars, the best fit signal template as the dashed histogram, the best fit background template as the dotted histogram, and the sum as the solid histogram. The systematic uncertainties are included in the fit by convoluting a Gaussian function with a width given by the total systematic uncertainty with the Gaussian resulting from the maximum likelihood fit. The result of applying the maximum likelihood fit (Fig. 93, right) to the lepton p_T distribution observed in data is

$$\mathcal{F}_+ = 0.13 \pm 0.20\,(\text{stat.}) \pm 0.06\,(\text{syst.})\,, \qquad (134)$$
$$\mathcal{F}_+ < 0.28 \text{ at } 95\%\,\text{CL}\,. \qquad (135)$$

The statistical combination of the result from this analysis and the measurement in the lepton + jets channel reported above is performed by summing the negative log-likelihood curves from each analysis. Systematic uncertainties are combined using error propagation on the values of

Table 46. Measurement and 95% CL upper limits of the W helicity in top quark decays from CDF and DØ. † Preliminary result, not yet submitted for publication or not yet published as of October 2005

W helicity	Source	$\int \mathcal{L} dt\ (pb^{-1})$	Ref.	Method
$\mathcal{F}_0 = 0.91 \pm 0.39$	CDF Run I	106	[390]	p_T^ℓ
$\mathcal{F}_0 = 0.56 \pm 0.32$	DØ Run I	125	[391]	ME
$\mathcal{F}_0 = 0.74^{+0.22}_{-0.34}$	CDF Run II	200	[395]†	$M_{\ell b}^2 + p_T^\ell$
$\mathcal{F}_+ < 0.18$	CDF Run I	110	[403]	$M_{\ell b}^2 + p_T^\ell$
$\mathcal{F}_+ < 0.27$	CDF Run II	200	[395]†	$M_{\ell b}^2 + p_T^\ell$
$\mathcal{F}_+ < 0.25$	DØ Run II	230–370	[?, 402]†	$\cos\theta^* + p_T^\ell$

$\mathcal{F}_+$, with the correlations between the two analyses taken into account. The result of combining the two analyses is:

$$\mathcal{F}_+^{\text{comb.}} = 0.04 \pm 0.11\,(\text{stat.}) \pm 0.06\,(\text{syst.}), \quad (136)$$

$$\mathcal{F}_+^{\text{comb.}} < 0.25 \text{ at } 95\% \text{ CL}. \quad (137)$$

The W-boson positive helicity fraction $\mathcal{F}_+$, measured in $t\bar{t}$ decays, is consistent with the Standard Model prediction of $\mathcal{F}_+ = 3.6 \times 10^{-4}$.

6.4.3 Summary

At present, all Run I and Run II studies of the helicity of the W-boson in top quark decay are limited by statistics. However, with the increasing data set becoming available for analysis in CDF and DØ, the limits or measurements of $\mathcal{F}_0$, $\mathcal{F}_-$ and $\mathcal{F}_+$ are improving rapidly. Eventually, when sufficient data is available, simultaneous measurements of $\mathcal{F}_0$, $\mathcal{F}_-$ and $\mathcal{F}_+$ will be performed. At present, only measurements or limits from studies of single quantities are available. These results are summarised in Table 46.

6.5 Flavour changing neutral current decays of the top quark

Physics beyond the Standard Model can manifest itself by altering the expected rate of flavour-changing neutral-current (FCNC) interactions. As an analogous historical example within the Standard Model, the presence of the charm quark can be inferred from its effect on the branching fraction $B(K_L^0 \to \mu^+\mu^-)$ [11]. FCNC decays of the top quark are of particular interest [404, 405]. The large mass of the top quark suggests a strong connection with the electroweak symmetry breaking sector. Evidence for unusual decays of the top quark might provide insights into the mechanism of electroweak symmetry breaking. For the top quark, the FCNC decays $t \to qZ$ and $t \to q\gamma$ (where q denotes either a c- or a u-quark) are expected to be exceedingly rare (branching fractions of 10^{-10} or smaller) [209, 406–408], since they are suppressed by the GIM mechanism [11] and any observation of these decays in the available data sample would indicate new physics. In general, FCNC interactions are present in models which contain an extended Higgs sector [409], Supersymmetry [410], dynamical breaking of the electroweak symmetry [411, 412], or an additional symmetry [413]. For a general overview over top couplings in exotic models see [414, 415].

6.5.1 Search for FCNC top quark decays at the TEVATRON

CDF has performed a search for the flavour-changing neutral current decays of the top quark $t \to q\gamma$ and $t \to qZ$ in 110 pb^{-1} of Run-I data at $\sqrt{s} = 1.8$ TeV [416]. In this CDF Run-I analysis, electrons, muons, jets and missing E_T are essentially reconstructed as described in Sect. 3.3. A photon is identified as an energy cluster in the electromagnetic calorimeter with no track pointing to it. To improve the

identification efficiency, additionally one single soft track (presumably from a random overlap) with less than 10% of the energy of the photon is permitted to point at the cluster.

If the branching fraction of a particle into a particular final state (e.g. a FCNC decay) is x, the branching fraction into another final state (e.g. the decay $t \to Wb$) can be no longer larger than $(1-x)$. Therefore, the ratio r of the number of events detected in a rare decay mode normalised to a common decay mode is at least $r = x/(1-x)$, after corrections for efficiency and acceptance. Measuring r allows to calculate an upper limit on the branching fraction $x \leq r/(1+r)$, after corrections for misidentification, efficiency and acceptance, and for the fact that in $t\bar{t}$ events there are two top quarks that can decay into a particular final state.

In both the $t \to q\gamma$ and $t \to qZ$ searches, limits on branching fractions are calculated by comparing the number of candidate events in the FCNC candidate sample to the number of $t\bar{t}$ events observed in the normalisation sample. The latter consists of events consistent with the hypothesis where both top quarks decay via $t \to Wb$ decays, and one W decays leptonically and the other W decays hadronically (lepton + jets mode). In the selection of the data sample, at least one of the jets must be identified as a b-jet via the secondary vertex reconstruction technique. This yields in total 34 $t\bar{t}$ candidate events over an estimated background of 9 ± 1.5 events.

At TEVATRON energies, the dominant source of top quarks is $t\bar{t}$ pair production from $q\bar{q}$ annihilation. In the search for $t \to q\gamma$, it is assumed that the other top quark in the pair decays via the decay $t \to Wb$. Two event signatures are considered, depending on whether the W decays leptonically or hadronically. If the W decays leptonically, a search for a high p_T lepton, $\not{E}_T$, a moderately high E_T photon and at least two jets is performed. If the W decays hadronically, a search for events with at least 4 jets and a high E_T photon is performed. In both cases, there must be a photon and jet combination with mass between 140 and 210 GeV/c^2, consistent with the mass of the top quark. In the non-leptonic case, the remaining jets must have $\sum E_T \geq 140$ GeV, consistent with the decay of a second top quark in the event. 40% of the $t \to q\gamma$ acceptance is in the photon plus multijet mode, and 60% is in the lepton plus photon mode.

The background in the leptonic mode is expected to be dominated by $W + \gamma + 2$ or more jets. It is estimated from the calculated rate of $W + \gamma + 2$ jets production and $W + \gamma$ production according to [417], and the observed $W + \gamma$ event rate. The number of background events is estimated to be less than half an event. In the hadronic mode, it is estimated from a control data sample to be less than half an event. To set a conservative limit, it is assumed that any event passing the selection requirements is a signal event and no background is subtracted.

Two samples are defined. one in which the event topology is consistent with both top quarks decaying via $t \to Wb$ (the "Standard Model sample") and one in which the event topology is consistent with one top quark decaying via $t \to Wb$ and the other via $t \to c\gamma$ (the "FCNC sample"). Since the relative number of events in these two signatures is of interest, the ratio of acceptances and efficiencies in the two decay modes is calculated using the ISAJET [233] Monte Carlo event generator. To be conservative, the expected number of FCNC candidates from misidentified Standard Model decays is not subtracted off. One event is observed in the leptonic channel and no events are seen in the non-leptonic (i.e. photon plus multijet) channel. The one observed event is kinematically also consistent with the decay $t \to W^+b, \bar{t} \to W^-\bar{b}\gamma$ followed by $W^+ \to \mu^+\nu$ and $W^- \to$ jets. The observation of one event passing the selection requirements implies a 95% confidence limit of fewer than 6.45 events (including systematic uncertainties) which translates into a branching fraction limit of

$$B(t \to c\gamma) + B(t \to u\gamma) < 3.2\% . \qquad (138)$$

Also the search for $t \to qZ$ events is performed, using the channel where the Z decays to e^+e^- or $\mu^+\mu^-$, and the other t quark decays to 3 jets. The expected signature is therefore an event with four jets and with two leptons with an invariant mass consistent with a Z-boson. Because the Z branching fraction to charged leptons is small, this analysis is less sensitive than the $t \to q\gamma$ search. Z-bosons are identified as opposite-charge same-flavour lepton pairs inside the range $75 < M_{\ell^+\ell^-} < 105 \text{ GeV}/c^2$. The ISAJET [233] Monte Carlo is used to calculate the efficiencies and acceptances for top quark pairs to be identified as either Standard Model decay candidates or FCNC decay candidates. These acceptances and efficiencies are again determined relative to the Standard Model signal.

There are two comparable sources of background to the $t \to qZ$ signal. One is ordinary Z + multijet production; 0.5 background events from $Z + 4$ jet are expected. The second source of background is from $t\bar{t}$ events where both top quarks decay via $t \to Wb$, and the event topology is such that the event is accepted into this sample. The total expected background is approximately 1.2 events. As before, to set a conservative limit, any event passing the selection requirements is assumed to be signal and no subtraction of background is performed.

A single $Z \to \mu^+\mu^-$ event passes all the selection requirements. The event kinematics better fit the Z+multijet hypothesis than the FCNC decay hypothesis. Observation of one event passing the selection requirements implies a 95% confidence limit of fewer than 6.4 events (including systematic uncertainties), which translates to a branching fraction limit

$$B(t \to cZ) + B(t \to uZ) < 33\% . \qquad (139)$$

These limits on top quark decay branching ratios can be translated into limits on the flavour-changing neutral current couplings [418] κ_γ and κ_Z at 95% CL:

$$\kappa_\gamma < 0.42, \qquad (140)$$
$$\kappa_Z < 0.73 . \qquad (141)$$

6.5.2 Search for single top quark production at HERA

In ep collisions at the HERA collider, the production of single top quarks is kinematically possible due to the large centre-of-mass energy $\sqrt{s} \approx 318 \text{ GeV}$, which is well above the top production threshold. In the Standard Model, the dominant process for single top production at HERA is the charged current reaction $e^+p \to \bar{\nu}t\bar{b}X$ ($e^-p \to \nu t\bar{b}X$). This process has a tiny cross section of less than 1 fb [162, 419] and thus Standard Model top quark production is negligible. However, in several extensions of the Standard Model, the top quark is predicted to undergo flavour changing neutral current interactions, which could lead to a sizeable top quark production cross section. An observation of top quarks at HERA would thus be a clear indication of physics beyond the Standard Model.

The H1 Collaboration has reported [420, 421] the observation of events with energetic isolated electrons and muons together with missing transverse momentum in the positron-proton data collected between 1994 and 2000. The dominant Standard Model source is the production of real W-bosons [422, 423]. However, some of these events have a hadronic final state with large transverse momentum, which is atypical of W production. These outstanding events may indicate a production mechanism involving processes beyond the Standard Model. One such mechanism is the production of top quarks which predominantly decay into a b-quark and a W-boson. The lepton and the missing transverse momentum would then be associated with a leptonic decay of the W-boson, while the observed high p_T hadronic final state would be produced by the fragmentation of the b-quark.

H1 and ZEUS have both searched for the single top quark production via flavour-changing neutral current in ep collisions at HERA, using 118 pb^{-1} and 130 pb^{-1}, respectively [424, 425]. The searches cover the leptonic decay channel ($W \to \ell\nu$) and the hadronic decay channel ($W \to qq'$) of the W-boson that emerges from the top quark decay.

In ep collisions, due to the large Z mass, the contribution of the Z-boson and the $\gamma - Z$ interference are highly suppressed. Single top production is thus dominated by the t-channel exchange of a photon. H1 therefore neglects Z-exchange in the analysis and only considers the $\kappa_{tu\gamma}$, while ZEUS uses an improved signal simulation based on COMPHEP [426, 427], including the tuZ coupling in the top quark production and decay. The sensitivity of HERA is much larger to the coupling $\kappa_{tu\gamma}$ than to $\kappa_{tc\gamma}$ since the u-quark density in the proton is much larger than the c-quark density at the high Bjorken-x values needed to produce top quarks.

H1 selects 5 events as top quark candidate decays in the leptonic channel. The prediction for the Standard Model processes is 1.31 ± 0.22 events. The analysis of multi-jet production at high p_T, corresponding to a search for single top production in the hadronic channel, shows good agreement with the expectation for Standard Model processes within the uncertainties and, therefore, do not rule out a single top interpretation of the candidates observed in the electron and muon channels. For the combination of the electron, muon and hadronic channels a cross section for single top production of $\sigma = 0.29^{+0.15}_{-0.14}$ pb is obtained. This result is not in contradiction with limits obtained by other experiments. The addition of a contribution from a model of anomalous single top production yields a better description of the data than is obtained with the Standard Model alone. Assuming that the small number of top candidates are the result of a statistical fluctuation, exclusion limits for the single top cross section of $\sigma < 0.55$ pb and for the anomalous FCNC $tu\gamma$ coupling of $|\kappa_{tu\gamma}| < 0.27$ are also derived at 95% CL.

ZEUS selects no event in the leptonic channel and observes no excess over the Standard Model prediction in the hadronic channel. Therefore, limits are set on the FCNC couplings of the type tqV. The contribution of the charm quark, which has only a small density in the proton at high x, is ignored by setting $\kappa_{tc\gamma} = v_{tcZ} = 0$. Only the anomalous couplings involving a u-quark and $\kappa_{tu\gamma}$ and v_{tuZ}, are considered. By combining the results from both the leptonic and hadronic channels, an upper limit of $\kappa_{tu\gamma} < 0.174$ at 95% CL is derived assuming $m_{top} = 175$ GeV/c^2. The limit is $\kappa_{tu\gamma} < 0.158(0.210)$ for $m_{top} = 170(180)$ GeV/c^2. The above coupling limit corresponds to a limit on the cross section for single top production of $\sigma(ep \to etX, \sqrt{s} = 318$ GeV$) < 0.225$ pb at 95% CL.

The HERA bounds extend into a region of parameter space so far not covered by experiments at LEP and the TEVATRON.

6.5.3 Search for single top quark production via FCNC at LEP

Due to its large mass, top quarks may only be singly produced at the e^+e^- collider LEP. Single top quark production in the Standard Model process $e^+e^- \to e^- \bar{\nu}_e t\bar{b}$ has a cross section of about 10^{-4} fb at LEP 2 energies [428–430] and can not be seen with the available luminosities. Another possible process for single top production is the flavour-changing neutral current reaction $e^+e^- \to \bar{t}c(u)$. Such FCNC interactions are known to be absent at the tree level in the Standard Model but can naturally appear at the one-loop level due to CKM mixing which leads to cross sections of the order of 10^{-9} fb at LEP 2 energies [431].

The LEP experiments ALEPH, DELPHI, L3 and OPAL have searched for anomalous single top quark production via FCNC [432–435] in the leptonic and hadronic decay modes of the W-boson from the top quark decay, using approximately 214, 541, 634, 600 pb^{-1} at centre-of-mass energies between 189 and 209 GeV, respectively. In principle, a large FCNC coupling could not only lead to the associated production of a top plus a light quark at LEP 2, but also to sizable branching ratios of the top quark into $\gamma c(u)$ or $Zc(u)$. This analysis uses only the $t \to Wb$ channel. The reduction of the branching ratio $B(t \to Wb)$ due to possible FCNC decays is taken into account in the result calculation. ALEPH observes (expects) 2 (2.1) events in the semileptonic and 22 (18) events in the hadronic decay mode. DELPHI observes (expects) 10 (10.8) events in the semileptonic and 99 (100.1) events in the hadronic decay mode. L3 observes (expects) 346

Table 47. 95% CL upper limits on the flavour-changing neutral current couplings κ_γ and κ_Z from the LEP experiments for $m_t = 169, 174$, and 179 GeV/c^2. The DELPHI and L3 limits are given for $m_t = 170, 175$, and 180 GeV/c^2

Experiment	$\kappa_\gamma < \dots$ for $m_t =$			$\kappa_Z < \dots$ for $m_t =$		
	169	174	179	169	174	179
ALEPH	0.44	0.49	0.56	0.37	0.42	0.50
DELPHI	0.40	0.49	0.61	0.34	0.41	0.53
L3	0.43	0.43	0.49	0.38	0.37	0.43
OPAL	0.39	0.48	0.60	0.34	0.41	0.52

(357) events in the semileptonic and 321 (288) events in the hadronic decay mode. OPAL observes (expects) 85 (84) events in the semileptonic and hadronic decay mode combined

No evidence for single top quark production is observed in e^+e^- collisions at centre-of-mass energies between 189 and 209 GeV. Limits on the single top cross section have been derived at the 95% CL from the measurements on the number of observed events, the reconstruction efficiencies, and the integrated luminosities. The combination of all the data can be used to determine limits on the anomalous coupling parameters κ_γ and κ_Z. Each centre-of-mass energy for the leptonic and the hadronic channel is used as an independent channel. Taking statistical and systematic errors into account, limits on the anomalous coupling parameters in the $\kappa_\gamma - \kappa_Z$ plane are derived at the 95% CL, where the reduction of the branching ratio $B(t \to Wb)$ due

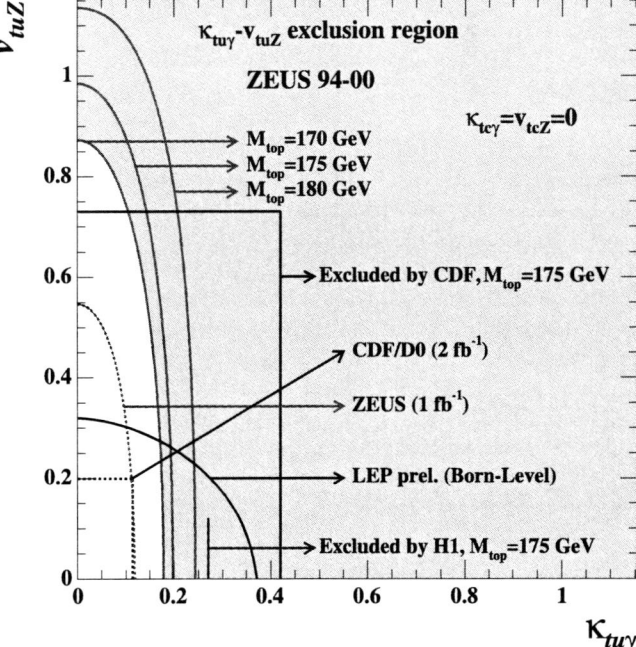

Fig. 94. Exclusion regions at 95% CL in the $\kappa_{tu\gamma} - v_{tuZ}$ plane of the FCNC couplings for ZEUS, H1, the LEP experiments and CDF along with the prospects for the HERA and TEVATRON experiments with higher luminosity samples (from [436])

to possible FCNC decays derived at each point in the $\kappa_\gamma - \kappa_W$ plane is taken into account in the limit calculation. The final upper limits on the flavour-changing neutral current coupling κ_γ and κ_Z at 95% CL are summarised in Table 47. The LEP experiments have approximately equal sensitivity to both couplings.

6.5.4 Summary

Figure 94 [436] shows the 95% CL limits on the FCNC couplings κ_γ and κ_Z of the different experiments along with high luminosity projections for HERA-II and the TEVATRON in Run II. HERA has the highest sensitivity to the $\kappa_{tu\gamma}$ coupling and will improve that even more during the coming years. LEP has the highest sensitivity to the κ_{tuZ} coupling, but the TEVATRON experiments will improve their sensitivity with $2\,\mathrm{fb}^{-1}$ per experiment to approximately equal strength.

7 Fundamental properties of the top quark

7.1 The mass of the top quark

The direct observation of the top quark in 1995 [42, 43, 437] was anticipated since the b-quark was expected to have an isospin partner to ensure the viability of the Standard Model, and therefore not a big surprise. What was a surprise is the very large mass of the top quark, almost 35 times the b-quark mass. The top quark mass is a fundamental parameter in the Standard Model, and plays an important role in electroweak radiative corrections, and therefore in constraining the mass of the Higgs boson. A large value of the top quark mass [46] indicates a strong Yukawa coupling to the Higgs, and could provide special insights in our understanding of electroweak symmetry breaking [438]. The top quark mass could have a different origin than the masses of the other light quarks. Thus, precise measurements of the top quark mass provide a crucial test of the consistency of the Standard Model and could indicate a strong hint for physics beyond the Standard Model. In doing that, it is important to measure and compare the top quark mass in the different decay channels. Since all top mass measurements assume a sample composition of $t\bar{t}$ and Standard Model background events, any discrepancy among the measured top masses could indicate the presence of non-Standard Model events in the samples.

The top mass has been measured in the lepton + jets, dilepton and the all-jets channel by both CDF and DØ. At present, the most precise measurements come from the lepton + jets channel containing four or more jets and large missing E_T. In this channel, three basic techniques are employed to extract the top mass. In the first, the so-called "template method" (TM) [42, 43, 87], a two-constraint kinematic fit is performed to the hypothesis $t\bar{t} \to W^+ b W^- \bar{b} \to \ell\nu_\ell b q\bar{q}' \bar{b}$ for each event, assuming that the four jets of highest E_T originate from the four quarks in $t\bar{t}$ decay. There are 24 possible solutions reflecting the allowed assignment of the final-state quarks to jets and two

possible solutions for the longitudinal momentum, p_z, of the neutrino when the W mass constraint is imposed on the leptonic W decay. The number of solutions is reduced to 12 when a jet with an identified secondary vertex is assigned as one of the b-quarks, and to 4 when the event has two such secondary vertices. A χ^2 variable is built based on the agreement of each possible solution with the $t\bar{t}$ hypothesis and the solution with the lowest χ^2 is defined as the best choice. The shape of the distribution of top masses from these fitted events is compared to templates modelled from a mixture of signal and background distributions for a series of assumed top masses. This comparison yields values of a likelihood as a function of the top mass hypothesis, from which a best value of the top mass and its uncertainty are obtained. In the second method, the "matrix element/dynamic likelihood method" (ME/DLM), similar to that originally suggested by Kondo et al. [400, 439–442] and Dalitz and Goldstein [367, 399, 443], for each event a probability is calculated as a function of the top mass, using a leading order matrix element. All possible assignments of reconstructed jets to final-state quarks are used, each weighted by a probability determined from the matrix element. The correspondence between measured four-vectors and parton-level four-vectors is taken into account using probabilistic transfer functions. CDF (TM) and DØ (ME/DLM) reduce the jet energy scale uncertainty by performing a simultaneous, *in situ* fit to the $W \to jj$ hypothesis using the jets without identified secondary vertices. In a third method, the "ideogram method" [444, 445], which combines some of the features of the above two techniques, each event is compared to the signal and background mass spectrum, weighted by the χ^2 probability of the kinematic fit for all 24 jet-quark combinations and an event probability. The latter is determined from the signal fraction in the sample and the event-by-event purity, as determined from a topological discriminant in Monte Carlo events.

Less precise determinations of the top mass come from the dilepton channel with two or more jets and large missing E_T, and from the all-jets channel. In the dilepton channel, a kinematically constrained fit is not possible because there are two neutrinos, so experiments must employ additional information in order to extract the mass. The general idea is based on the fact that, assuming a value for m_t, the $t\bar{t}$ system can be reconstructed up to an eight-fold ambiguity from the choice of associating leptons and quarks to jets and due to the two solutions to the p_z of each neutrino. Two basic techniques are employed, one based on matrix elements and one using templates, as in the lepton + jets channel. The first, ME/DLM, uses weights based on the SM matrix element for an assumed mass given the measured four-vectors (and integrating over the unknowns) to form a joint likelihood as a function of the top mass for the ensemble of fitted events. The second class of techniques incorporates additional information to render the kinematic system solvable. In this class, there are two techniques that assign a weight as a function of top mass for each event based on solving for either the azimuth, ϕ, of each neutrino given an assumed η, $(\eta(\nu))$ [370, 398, 446], or for η of each neutrino given an assumed ϕ, $(\phi(\nu))$ [447]. A modification of the latter method, ($\mathcal{M}\mathbf{WT}$) [370, 398], solves for η of each neutrino requiring the sum of the neutrino $\mathbf{p}_T$'s to equal the measured missing E_T vector. In another technique, $(\mathbf{p_z(t\bar{t})})$ [448], the kinematic system is rendered solvable by the addition of the requirement that the p_z of the $t\bar{t}$ system, equal to the sum of the p_z of the t and $\bar{t}$, be zero within a Gaussian uncertainty of $180\,\mathrm{GeV/c}$. In each of the techniques in this second class, a single mass per event is extracted and a top mass value found using a Monte Carlo template fit to the single-event masses in a manner similar to that employed in the lepton + jets TM technique.

In the all-jets channel there is no unknown neutrino momentum to deal with, but the S/B is the poorest. Both CDF and DØ use events with 6 or more jets, of which at least one is b-tagged. In addition, DØ uses a neural network selection based on eight kinematic variables, and a top-quark mass is reconstructed from the jet-quark combination that best fits the hadronic W-mass constraint and the equal-mass constraint for the two top quarks. At CDF, events with one b-tagged jet are required to pass a strict set of kinematic criteria, while events with two b-tagged jets are required to exceed a minimum total energy. The top quark mass for each event is then reconstructed applying the same fitting technique used in the ℓ + jets mode. At both, CDF and DØ the resulting mass distribution is compared to Monte Carlo templates for various top quark masses and the background distribution, and a maximum likelihood technique is used to extract the final measured value of m_t and its uncertainty.

It should be noted that the different techniques make assumptions about the Standard Model-like production and decay of the top quark at different levels. In general, methods which make stronger assumptions about the top quark production and decay mechanism, such as the matrix element/dynamic likelihood method, have the highest sensitivity to the top quark mass. Simple template methods in the lepton + jets channel, which only rely on energy and momentum conservation in the kinematic reconstruction and detector resolution functions, are on the one hand less sensitive to the top quark mass but on the other hand more stable with respect to possible modification in the details of the production and decay mechanisms involved.

In this chapter, the different analyses are described in turn, including prospects for scenarios with integrated luminosity up to $8\,\mathrm{fb}^{-1}$ in some of the lepton + jets analyses and the combination of the top quark mass measurements by the TEVATRON Electroweak Working Group. A summary table and plot is shown at the end of this chapter.

7.1.1 CDF analyses

Lepton + jets channels.

1-dimensional template analysis. CDF has measured the top quark mass in $162\,\mathrm{pb}^{-1}$ of Run-II data, using lepton + jets events with one or more secondary vertex b-tags and applying the 1-dimensional template method [449]. Lepton+jets events are selected according to the standard event selection (Sect. 4.2.3). In addition, out of the

four or more jets, the first three should have a measured $E_T > 15\,\mathrm{GeV}$ and $\eta < 2.0$ and the fourth jet passes a relaxed $E_T > 8\,\mathrm{GeV}$ cut. One of the jets with $E_T > 15\,\mathrm{GeV}$ should be identified as a b-jet using the SecVtx algorithm (Sect. 3.3.7). Events with $8\,\mathrm{GeV} < E_T^{j4} < 15\,\mathrm{GeV}$ are called "3.5-jet events". In this analysis, a total of 10 such 3.5-jet events and 27 4-jet events are selected, of which one is a double b-tagged 3.5-jet event, and four are double-tagged 4-jet events. For each lepton + jet event, an invariant mass of the top quark is reconstructed using the lepton, four jets, and missing transverse energy. The invariant mass distribution for the $t\bar{t}$ candidates in data is compared to equivalent distributions for $t\bar{t}$ Monte Carlo with different top quark pole masses (called top mass templates). This comparison is quantified and the best estimate for the true top pole mass in the data is determined using a maximum likelihood.

In the event reconstruction, the jet energies are corrected for different hadronisation effects of b and light quark jets and for semi-leptonic b-decays. In the lepton+ jets events with one b-tagged jet, there are six ways to assign four leading jets to the four quarks in a $t\bar{t}$ lepton + jets event, while there are two solutions for the longitudinal momentum of the neutrinos, obtained from a quadratic equation in the kinematics. In this reconstruction, 5 and $0.5\,\mathrm{GeV/c^2}$ are assigned for the masses of the b-quark and the light quarks, respectively. A kinematic fit is performed on each event, where MINUIT [450] is used to minimise the following χ^2:

$$
\begin{aligned}
\chi^2 = &\sum_{i=l,4\mathrm{jets}} \frac{\left(p_T^{i,\mathrm{fit}} - p_T^{i,\mathrm{meas}}\right)^2}{\sigma_i^2} \\
&+ \sum_{j=x,y} \frac{\left(p_j^{\mathrm{UE,fit}} - p_j^{\mathrm{UE,meas}}\right)^2}{\sigma_i^2} \\
&+ \frac{(M_{jj} - M_W)^2}{\Gamma_W^2} + \frac{(M_{\ell\nu} - M_W)^2}{\Gamma_W^2} \\
&+ \frac{(M_{bjj} - M_t)^2}{\Gamma_t^2} + \frac{(M_{b\ell\nu} - M_t)^2}{\Gamma_t^2}, \quad (142)
\end{aligned}
$$

where σ_ℓ and σ_{jet} are the transverse momentum or energy resolutions of the lepton and four leading jets, and $p_{x,y}^{\mathrm{UE}}$ and $\sigma_{x,y}$ are the unclustered energy (not clustered into jets) and its resolution. The mass of the t and $\bar{t}$ quark are constrained to be the same, and the two W masses are both constrained to the PDG value of $M_W = 80.41\,\mathrm{GeV/c^2}$ [167]. In each event, from 12 combinations (only 4 for double-tags) the one with the lowest χ^2 is chosen as the best jet-parton assignment. An additional requirement of $\chi_{\mathrm{min}}^2 < 9$ rejects badly reconstructed $t\bar{t}$ events. This cut has $\approx 80\%$ efficiency for signal, and gives $\approx 30\%$ rejection power for the background. 28 out of the 37 events pass this cut.

In order to extract the top quark mass, top mass templates need to be constructed for signal and backgrounds. Signal templates are generated using HERWIG Monte Carlo [226, 227] with the input top quark mass value in

2.5–5 GeV/c^2 intervals from 130 to 230 GeV/c^2. Smooth functions are fitted to these normalised distributions, yielding the probability density function $P_{\text{sig}}(m_{\text{rec}}; m_{\text{top}})$ to reconstruct a mass m_{rec} for a given true top quark mass, m_{top}. The background is mainly due to W-boson production associated with a b-jet or associated gluon jets with a misidentified b-jet (mistags), and QCD multijet background due to fake leptons. Templates for this background are constructed using Monte Carlo samples generated with ALPGEN [224] and PYTHIA [223] and fitted to a smooth function, yielding the probability density function, $P_{\text{b}}(m_{\text{rec}})$ to reconstruct a mass m_{rec} from background events.

The mass of the top quark is extracted using a maximum likelihood method:

$$\mathcal{L} = \mathcal{L}_{\text{shape}} \times \mathcal{L}_{\text{bg}} \times \mathcal{L}_{\text{params}} , \qquad (143)$$

$$\mathcal{L}_{\text{shape}} = \frac{e^{-(n_s+n_{\text{b}})}(n_s + n_{\text{b}})^N}{N!}$$
$$\times \prod_{i=1}^{N} \frac{n_s P_{\text{sig}}(m_i; m_{\text{top}}) + n_{\text{b}} P_{\text{b}}(m_i)}{n_s + n_{\text{b}}} , \qquad (144)$$

$$-\ln \mathcal{L}_{\text{bg}} = \frac{(n_{\text{b}} - n_{\text{b}}^{\text{expect}})^2}{2\sigma_{n_{\text{b}}}^2} , \qquad (145)$$

$$-\ln \mathcal{L}_{\text{params}} = \sum_{i=0}^{17} \sum_{j=0}^{17} \frac{1}{2} \left(\delta p_i^{\text{sig}} C_{ij}^{\text{sig}} \delta p_j^{\text{sig}} \right)$$
$$+ \sum_{i=0}^{2} \sum_{j=0}^{2} \frac{1}{2} \left(\delta p_i^b C_{ij}^b \delta p_j^b \right) . \qquad (146)$$

The likelihood $\mathcal{L}_{\text{shape}}$ is the joint probability density for a sample of N reconstructed masses m_i to come from a parent distribution with a background fraction, $n_{\text{b}}/(n_{\text{b}} + n_s)$. The shapes of the previously described templates (probability densities) enter here. The background fraction is constrained by the background likelihood $\mathcal{L}_{\text{bg}}$ to be consistent with the expected background $n_{\text{b}}^{\text{expect}}$. Because of the finite statistics of the Monte Carlo sample, the shapes of the signal and background templates are constrained to agree with the input parameter values using the likelihood $\mathcal{L}_{\text{params}}$. C is the inverted covariance matrix, and δp_i is the deviation from the input parameter values p_i, describing the probability density functions. With these constraints, the likelihood is maximised with respect to the true top quark mass, m_{top}. The resulting statistical error is scaled up by 1.065 to satisfy 68% coverage of the true value, as the pull width is found in Monte Carlo ensemble tests to be slightly larger than one. Each systematic uncertainty is estimated by performing a series of pseudo-experiments with Monte Carlo samples modified by $\pm 1\sigma$ of the respective source of systematic effects. A summary of the systematic uncertainties is given in Table 48.

The application of the likelihood fit to the 28 selected events yields a top quark mass of:

$$m_{\text{top}} = 174.9^{+7.1}_{-7.7}(\text{stat.}) \pm 6.5(\text{syst.}) \, \text{GeV}/c^2 . \qquad (147)$$

Table 48. Summary of the systematic uncertainties for $W + \geq 3.5$ jets and $W + \geq 4$ jets events in the 1-dimensional CDF template analysis

Source of systematics	Δm_{top} (GeV/c^2)	
	≥ 3.5 jets	≥ 4 jets
Jet energy calibration	6.3	6.6
ISR	0.4	0.6
FSR	0.9	1.0
PDF's	0.2	0.2
Generators	0.4	0.4
Background shape	0.8	0.8
Other MC modelling (jet reso., $p_{\text{T}}^{\text{top}}$)	0.7	0.7
b-tagging	0.1	0.1
Total	6.5	6.8

2-dimensional template analysis. CDF has developed the template method further and now also employs the reconstructed mass of the hadronic W-boson decays $W \to jj$ to constrain *in situ* the largest systematic uncertainty of the top quark mass measurement: the jet energy scale. Monte Carlo templates of the reconstructed top quark and W-boson mass are produced as a function of the true top quark mass and the jet energy scale and compare to the data using a likelihood fit. CDF applies this technique to 138 $t\bar{t}$ lepton + jets events, selected in 318 pb^{-1} of Run-II data [176]. The lepton + jets events are selected as in the 1-dimensional template analysis described before, including the $\chi^2 < 9$ cut on the χ^2 of the kinematic fit. This latter requirement is not applied for the W-boson mass reconstruction since it reduces the sensitivity of this observable to the jet energy scale. To improve the statistical power of the method, the lepton + jets sample is divided into four subsamples with different sensitivity to the mass of the top quark. Events with 2-, 1-, and 0-tag (using the SecVtx algorithm – Sect. 3.3.7) are considered separately since the mass resolution, which is dominated by incorrect jet-quark combinations, improves and the background level decreases with increasing number of b-tags. Events with 1-tag are separated further. Events in the 1-tag(T) ('tight') category have 4 jets with $E_{\text{T}} > 15$ GeV, while events in the 1-tag(L) ('loose') category have 3 jets with $E_{\text{T}} > 15$ GeV and the 4th jet with $8 < E_{\text{T}} < 15$ GeV. Events in the 1-tag(T) sample are less contaminated by background.

The *a priori* determination of the jet energy scale, using information from external, independent data sets, with all the different corrections is described in Sect. 3.3.2. The resulting $\pm 1\sigma$ total uncertainty on the jet energy scale is used as the unit of jet energy scale in this analysis. Furthermore, this *a priori* information on the jet energy scale is used in the likelihood fit as an additional constraint on the jet energy scale (JES). $t\bar{t}$ events are simulated using the HERWIG Monte Carlo program [226,227] for various values of the true top quark mass and jet energy scale hypothesis. A kinematic fit, identical to the one described in the 1-dimensional template analysis before, is employed. The χ^2 of the fit is calculated according to (142). The $t\bar{t}$

hypothesis is fit to the 12 possible assignments of jets to quarks. The number of jet-quark assignments is reduced to 6 and 2 when 1- and 2-tags are available, respectively. There is always an additional combination due to the 2 solutions for the p_z of the neutrino arising from solving a quadratic equation. The jet-quark assignment yielding the lowest χ^2 is chosen for the top mass reconstruction. The efficiency for the additional requirement of $\chi^2 < 9$ ranges from 65% (38%) for 2-tag events to 91% (83%) for 0-tag events for signal (background) events.

The dijet mass from the hadronic W-boson decay m_{jj} is sensitive to the jet energy scale but is relatively insensitive to the true top quark mass. It can thus be used to determine fully *in situ* the jet energy scale with little uncertainty on m_{top}. In this analysis, the jet energy scale is determined using both the m_{jj} templates and the *a priori* determination of the jet energy scale, providing an optimal constraint on this parameter (JES). m_{jj} is reconstructed from the measured three-jet momenta of the jets without using a χ^2 fitter. All possible assignments of the four highest-E_{T} jets, that are not b-tagged, to the quarks from the W decay are considered yielding 1, 3, and 6 m_{jj} values per event for the 2-tag, 1-tag and 0-tag subsample, respectively, and therefore providing optimal sensitivity of m_{jj} to JES.

Distributions of $m_{\text{t}}^{\text{reco}}$ and m_{jj} are constructed from HERWIG $t\bar{t}$ Monte Carlo for m_{top} values from 130 to 230 GeV/c^2 and JES values varying from -3 to $+3\sigma$. Smooth probability density functions ($P_{\text{sig}}(m_{\text{t}}^{\text{reco}}; m_{\text{top}}$, JES) and $P_{\text{sig}}(m_j; m_{\text{top}}$, JES)) are obtained by fitting the mass distributions as a function of m_{top} and JES. The parameters of the fit functions depend linearly on each of these two parameters. Figure 95 shows the reconstructed top quark mass distribution for various true top quark masses (JES fixed at 0) for the 2-tag subsample (left). Also shown in Fig. 95 is the m_{jj} distribution for various jet energy scale values (m_{top} fixed at 175 GeV/c^2) for the 2-tag subsample (right). Background templates for the W + jets background with heavy flavour production and mistagged jets are reconstructed using the ALPGEN [224] Monte Carlo samples. The mistag template is also used for the QCD multijet background template, as they are found to have very similar shape. The resulting probability density functions for background events do not depend on m_{top} and JES.

The reconstructed mass distributions from data are compared to the signal and background templates using an unbinned likelihood fit. The likelihood involves parameters for the expectation values of the number of signal and background events in each subsample, and for the true top quark pole mass and jet energy scale. For each subsample, the likelihood is given by:

$$\mathcal{L}_{\text{sample}} = \mathcal{L}_{\text{shape}}^{m_{\text{t}}^{\text{reco}}} \times \mathcal{L}_{\text{shape}}^{m_{jj}} \times \mathcal{L}_{\text{nev}} \times \mathcal{L}_{\text{bg}}, \quad (148)$$

where

$$\mathcal{L}_{\text{shape}}^{m_{\text{t}}^{\text{reco}}} = \prod_{k=1}^{r^W} \frac{\epsilon_{\text{s}} n_{\text{s}}^W P_{\text{s}}\left(m_{\text{t}}^k; m_{\text{top}}, \text{JES}\right) + \epsilon_{\text{b}} n_{\text{b}}^W P_{\text{b}}\left(m_{\text{t}}^k\right)}{\epsilon_{\text{s}} n_{\text{s}}^W + \epsilon_{\text{b}} n_{\text{b}}^W}, \quad (149)$$

$$\mathcal{L}_{\text{shape}}^{m_{jj}} = \prod_{k=1}^{r^W n_i^c} \frac{n_{\text{s}}^W P_{\text{s}}\left(m_{jj}^k; m_{\text{top}}, \text{JES}\right) + n_{\text{b}}^W P_{\text{b}}\left(m_{jj}^k\right)}{n_{\text{s}}^W + n_{\text{b}}^W}, \quad (150)$$

$$\mathcal{L}_{\text{nev}} = \sum_{r_{\text{s}}^W + r_{\text{b}}^W = r^W} P_{\text{Pois}}\left(r_{\text{s}}^W; n_{\text{s}}^W\right) P_{\text{Pois}}\left(R_b^W; n_{\text{b}}^W\right) \quad (151)$$

$$\times \left[\sum_{r_{\text{s}}^t + r_{\text{b}}^t = r_t}^{r_{s,b}^t \leq r_{s,b}^W} P_{\text{bin}}\left(r_{\text{s}}^t; r_{\text{s}}^W, \epsilon_{\text{s}}\right) P_{\text{bin}}\left(r_{\text{b}}^t; r_{\text{b}}^W, \epsilon_{\text{b}}\right)\right], \quad (152)$$

$$\mathcal{L}_{\text{bg}} = \exp\left(-\frac{\left(n_{\text{b}}^W - n_{\text{b}}^W(\text{const})\right)^2}{2\sigma_{n_{\text{b}}^W}^2}\right). \quad (153)$$

The most information on the true top quark mass is provided by the products in $\mathcal{L}_{\text{shape}}^{m_{\text{t}}^{\text{reco}}}$, the k-th term of which gives the probability of observing the k-th data event with reconstructed mass m_k, given the background template, $P_{\text{b}}(m_k)$, and the signal template with a true top quark mass of m_{top} and energy scale shift JES, $P_{\text{s}}(m_k; m_{\text{top}}, \text{JES})$. The third term represents the information arising from the number of signal and background events in the top quark mass and dijet mass samples,

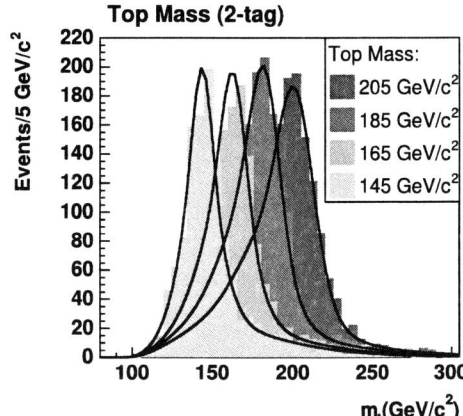

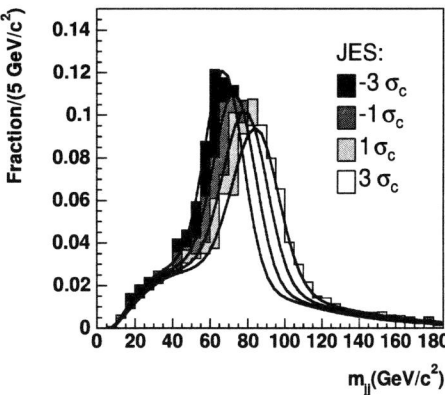

Fig. 95. *Left*: Signal m_t^{reco} templates with top quark masses ranging from 145 GeV/c^2 to 205 GeV/c^2 (JES set to 0) for the 2-tag subsample. *Right*: Signal m_{jj} for JES values ranging from -3σ to $+3\sigma$ (m_{top} set to 175 GeV/c^2) for the 2-tag subsamples. Overlaid are the fitted parameterisations at each generated mass (*left*) and JES (*right*)

which are correlated. The number of expected signal and background events in the $W \to jj$ sample is denoted n_s^W and n_b^W, respectively. The expected numbers of signal and background events in the m_t^{reco} sample are given by $\epsilon_s n_s^W$ and $\epsilon_b n_b^W$, respectively, where the two parameters ϵ_s and ϵ_b represent the efficiency of the χ^2 cut for signal and background events. The third term in the likelihood, $\mathcal{L}_{\text{nev}}$, expresses the likelihood associated with observing r^W and r^t events in the two samples given the expected number of events and the expected efficiencies. The first sum expresses the Poisson probability to observe r_s^W signal and r_b^W background events given Poisson means of n_s^W and n_b^W, respectively. The sum in the third term runs over those signal and background events that equal the observed number of events in the m_{jj} sample: $r_s^W + r_b^W = r^W$. For each pair in this sum, the binomial probability to observe r_s^t signal events and r_b^t background events in the m_t sample gives the number of observed events in the m_{jj} sample and the χ^2_{min} cut efficiency is included. The second sum in the $\mathcal{L}_{\text{nev}}$ runs over the pairs of signal and background events in the m_t sample that equal the observed number of events: $r_s^t + r_b^t = r^t$. When independent estimates of background are available, the background normalisations are constrained in the likelihood fit by Gaussian terms with the form of $\mathcal{L}_{\text{bg}}$. The background normalisations are constrained for the 2-tag, 1-tag(T), and 1-tag(L) samples. Both, n_s and n_b are required to be greater than zero.

The *a priori* constraint on the jet energy scale is used in the likelihood in the form of a Gaussian constraint:

$$\mathcal{L}_{\text{JES}} = \exp\left(-\frac{(\text{JES} - \text{JES}^{\text{expected}})^2}{2\sigma_{\text{JES}}^2} \right)$$
$$= \exp\left(-\frac{\text{JES}^2}{2} \right), \qquad (154)$$

where the simplification arises by the definition of the measured energy scale, $\text{JES}^{\text{expected}} = 0$ and the uncertainty $\sigma_{\text{JES}} = 1.0$.

The total likelihood is given by the product of the likelihoods for the four subsamples and the jet energy scale constraint:

$$\mathcal{L} = \mathcal{L}_{\text{2-tag}} \times \mathcal{L}_{\text{1-tag(T)}} \times \mathcal{L}_{\text{1-tag(L)}}$$
$$\times \mathcal{L}_{\text{0-tag}} \times \mathcal{L}_{\text{JES}}. \qquad (155)$$

Table 49. Summary of the systematic uncertainties for the 2-dimensional CDF template analysis

Source of systematics	Δm_{top} (GeV/c^2)	$\Delta\text{JES}(\sigma)$
Jet energy scale	N/A	N/A
b-jet energy scale	0.6	0.25
Fit method	0.5	0.02
ISR modelling	0.4	0.08
FSR modelling	0.6	0.06
PDF's	0.3	0.04
Monte Carlo generators	0.2	0.15
Bgd shape (Q^2 scale)	1.1	0.17
b-tagging	0.1	0.01
MC statistics	0.3	0.05
Total	1.7	0.36

The true top quark mass m_{top} and the jet energy scale JES are shared between the four likelihoods and are free parameters in the fit. The likelihood is maximised with respect to all ten parameters (n_s and n_b for four subsamples, JES, and m_{top}).

Ensemble tests of the procedure show that the pull width as a function of m_{top} is slightly larger than one: 1.027. The statistical uncertainties obtained in the data are scaled by that factor to guarantee 68% coverage of the 1σ uncertainties. Various sources of systematic uncertainties are considered for this measurement, apart from the jet energy scale that is given from the fit. Table 49 summarises the estimated impact of those sources on the measurement of the top quark mass.

As a result, the mass of the top quark has been measured to be

$$m_{\text{top}} = 173.5^{+3.7}_{-3.6} \,(\text{stat.} + \text{JES}) \pm 1.7 \,(\text{syst.})\, \text{GeV/c}^2$$
$$= 173.5^{+2.7}_{-2.6} \,(\text{stat.}) \pm 2.5 \,(\text{JES}) \pm 1.7 \,(\text{syst.})\, \text{GeV/c}^2 \qquad (156)$$

and the measurement of the jet energy scale is $-0.10^{+0.78}_{-0.80}$ (stat.) ± 0.36 (syst.)σ, i.e. fully consistent with the result of the *a priori* jet energy scale. Figure 96 (left) shows the confidence level contour of the likelihood in the $(m_{\text{top}}, \text{JES})$ plane for the combined fit to the four subsamples. The cross-hair shows the best fit point. For demonstration purposes, Fig. 96 also shows the 1-dimensional distribution of

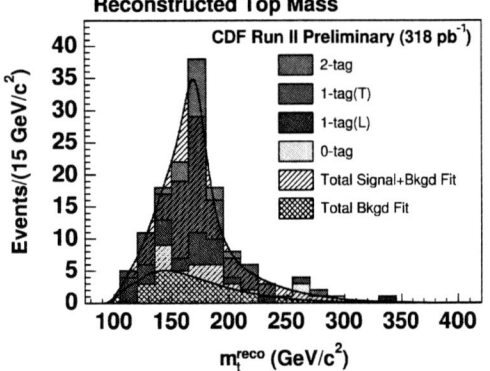

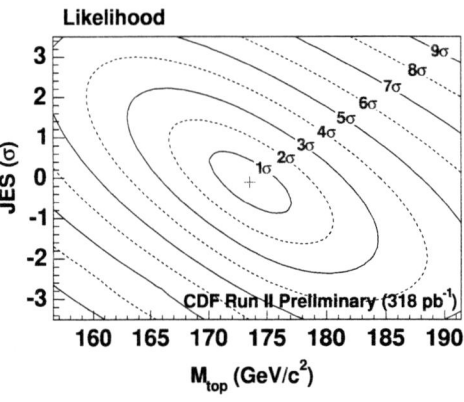

Fig. 96. *Left*: 1-dimensional distribution of the reconstructed top quark masses, m_{top}, for each subsample with the MC templates overlaid corresponding to the best fit. *Right*: Confidence level contour of the likelihood in the m_{top}, JES plane for the combined fit to the four subsamples

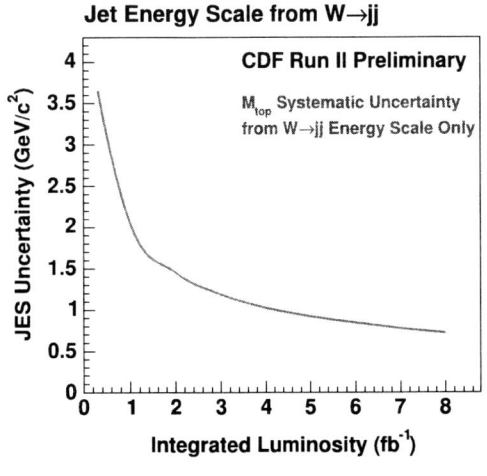

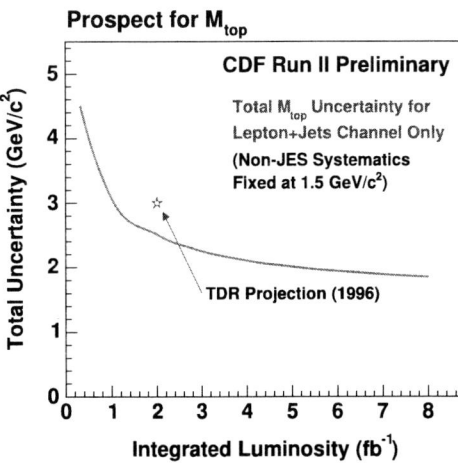

Fig. 97. Prospects for the jet energy scale uncertainty from $W \to jj$ calibration (*left*) and the total top mass uncertainty (*right*) as a function of integrated luminosity. Note the significant improvement over the projection made in the technical design report in 1996

the reconstructed top quark masses, m_{top}, for each subsample with the MC templates overlaid corresponding to the best fit.

The jet energy scale has also been estimated for the previously described 1-dimensional template analysis, which s also used as a cross check, yielding $3.1 \,\text{GeV}/c^2$ instead of $2.5 \,\text{GeV}/c^2$ in the 2-dimensional template analysis. This clearly demonstrates that including the *in situ* calibration in addition to the external calibration of the jet energy scale improves the result further. In comparison to the 1-dimensional template analysis described above, however, it is also clear that most of the improvement in the jet energy scale uncertainty does originate from numerous, detailed studies and significant improvements in the understanding of the external jet energy scale calibration, rather than from the $W \to jj$ constraint. In the future, however, the jet energy scale uncertainty is expected to improve down to $\approx 1 \,\text{GeV}/c^2$ as more data is available to perform the $W \to jj$ calibration, as shown in Fig. 97. The total top mass uncertainty in the lepton + jets channel measured in CDF is therefore expected to decrease to $\approx 2 \,\text{GeV}/c^2$.

1-dimensional template analysis using the jet probability algorithm. In another 1-dimensional template analysis of $318 \,\text{pb}^{-1}$, CDF measures the mass of the top quark in lepton + jets events, in which the $t\bar{t}$ content is enhanced by using the jet probability tagger in addition to the SecVtx b-tagging algorithm [451].

The lepton + jets events are selected according to the standard selection criteria (Sect. 4.2.3). Four or more jets within $|\eta| < 2.0$ are required with $E_{\text{T}} > 15 \,\text{GeV}$ for the first three, and $E_{\text{T}} > 8 \,\text{GeV}$ for the fourth jet in tagged events. In 0-tag events, a tighter cut is applied on jets in order to reduce the background: $E_{\text{T}} > 21 \,\text{GeV}$ for the first four jets. The jet probability b-tagging algorithm (Sect. 3.3.7), has a looser tagging condition than SecVtx (Sect. 3.3.7), resulting in a higher efficiency for heavy flavour jets (per jet on average 28% for SecVtx and 33% for jet probability) at the price of a slightly higher mistag rate (per jet on average 0.34% for SecVtx and 4.1% for jet probability). Since SecVtx is better at rejecting light flavour jets than jet probability, at least one jet in the event is required to

be tagged by SecVtx. Jet probability is used as a second b-tag requirement to enhance the double tagging capability in the analysis, aiming at tagging the second b-jet in an event. The jet probability cut value is optimised for the best expected statistical precision of this analysis, using Monte Carlo ensemble testing. The candidate event sample is subdivided into five categories, labelled 0-tag, 1-tag(L), 1-tag(T), 2-tag(S+S), and 2-tag(S+J), with 20, 21, 43, 16 and 18 candidate events respectively. The definition of the categories is equivalent to those in 2-dimensional template analysis, except that the double tagged events are labelled S+S if they have two SecVtx tags, and S+J if they have one SecVtx and one jet probability tag.

Using the same kinematic fit, χ^2 definition and cut, and the same Monte Carlo samples as in the 1-dimensional and 2-dimensional template methods described above, the 1-dimensional signal and background templates are derived and fitted to smooth functions, yielding the probability densities $f_s(m_{\text{t}}^{\text{reco}}; m_{\text{top}})$ and $f_b(m_{\text{t}}^{\text{reco}})$. The number of allowed jet-quark assignments is 24 for 0-tag, 12 for 1-tag, and 4 for 2-tag events.

The top quark mass is determined by fitting the reconstructed top mass distribution in the data sample to a sum of templates for the $t\bar{t}$ signal and background using an unbinned likelihood method. The likelihood $\mathcal{L}$ for each tagged sample is the product of two likelihoods: $\mathcal{L} = \mathcal{L}_{\text{shape}} \times \mathcal{L}_{\text{bkg}}$, where $\mathcal{L}_{\text{shape}}$ and $\mathcal{L}_{\text{bkg}}$ describe the likelihoods for the top mass distribution and for the number of background events:

$$\mathcal{L}_{\text{shape}} = \frac{e^{-(N_{\text{s}}+N_{\text{b}})}(N_{\text{s}}+N_{\text{b}})^N}{N!}$$
$$\times \prod_{i=1}^{N} \frac{N_{\text{s}}f_s(m_{\text{t}}^i; m_{\text{top}}, \alpha) + N_{\text{b}}f_b(m_{\text{t}}^i; \beta)}{N_{\text{s}} + N_{\text{b}}}, \tag{157}$$

$$\mathcal{L}_{\text{bkg}} = \exp\left(-\frac{1}{2}\frac{(N_{\text{b}} - N_{\text{b}}^{\text{pred.}})^2}{\sigma_{n_{\text{b}}^{\text{pred.}}}^2}\right). \tag{158}$$

Here, m_{t}^i is the reconstructed top mass of event i in the data sample. In this fit, m_{top}, N_{s} and N_{b} are free parameters, describing the true top mass, and the number of signal and background events, respectively. N is the total number of events in the sample, and α and β are the parameters of the smooth probability density functions describing the signal and background template shapes. The quantities $N_b^{\text{pred.}}$ and $\sigma_{N_b^{\text{pred.}}}$ refer to the estimated number of background event and its error. In the 0-tag sample, the number of background events is not constrained, so that: $\mathcal{L}^{0-tag} = \mathcal{L}^{0-\text{tag}}_{\text{shape}}$. Since the subsamples are statistically independent of each other, they are combined by multiplying the likelihood functions:

$$\mathcal{L}^{\text{combine}} = \mathcal{L}^{0-tag} \times \mathcal{L}^{1-tag(L)} \times \mathcal{L}^{1-tag(T)}$$
$$\times \mathcal{L}^{2-tag(S+S)} \times \mathcal{L}^{2-\text{tag}(S+J)} . \quad (159)$$

The true top quark mass is a free parameter that is common to the five likelihood functions. The statistical errors are scaled by a factor 1.037, due to the pull distributions observed to be slightly larger than 1.0.

Table 50 summarises the systematic uncertainties in this top quark mass measurement. The jet energy scale is clearly the dominant component. Figure 98 shows the reconstructed top mass distributions along with the likelihood fit to data for the five subsamples separately. The resulting combined measurement of the top quark mass is:

$$m_{\text{top}} = 173.0^{+2.9}_{-2.8} \, (\text{stat.}) \pm 3.3 \, (\text{syst.}) \, \text{GeV}/c^2 . \quad (160)$$

Dynamical likelihood method. CDF has measured the mass of the top quark in lepton + jets events, applying the dynamical likelihood method (DLM) to 318 pb^{-1} of Run-II data [452]. The analysis is described and compared to an earlier version using the same technique [453], but based

Table 50. Summary of the systematic uncertainties for the jet probability template analysis

Source of systematics	Δm_{top} (GeV/c^2)
ISR modelling	0.3
FSR modelling	0.6
Jet energy scale	3.0
PDF's	0.4
MC generators	0.2
b-jet modelling	0.6
Background shape	0.7
b-tagging	0.3
MC statistics	0.4
Total	3.3

on 162 pb^{-1} and using an earlier version of the CDF-II jet energy calibration.

The lepton + jets $t\bar{t}$ candidates events are selected according to the standard event selection (Sect. 4.2.3). In order to minimise the contamination by initial and final state radiation, exactly four jets with $E_T \geq 15$ GeV and $|\eta| < 2.0$ are required. At least one of the four jets has to be identified as a b-jet using the SecVtx algorithm (Sect. 3.3.7). This selection yields 63 (22) events in the 318(162) pb^{-1} analysis.

The dynamic likelihood method (DLM) [400, 439–442] assumes the Standard Model and uses the differential cross section per unit phase space to infer the parton kinematics and the mass of the top quark. Each event is given an individual probability to be a $t\bar{t}$ event of a certain, hypothesised top quark mass m_{top}, taking the observed kinematics of the final state jets and leptons and, most im-

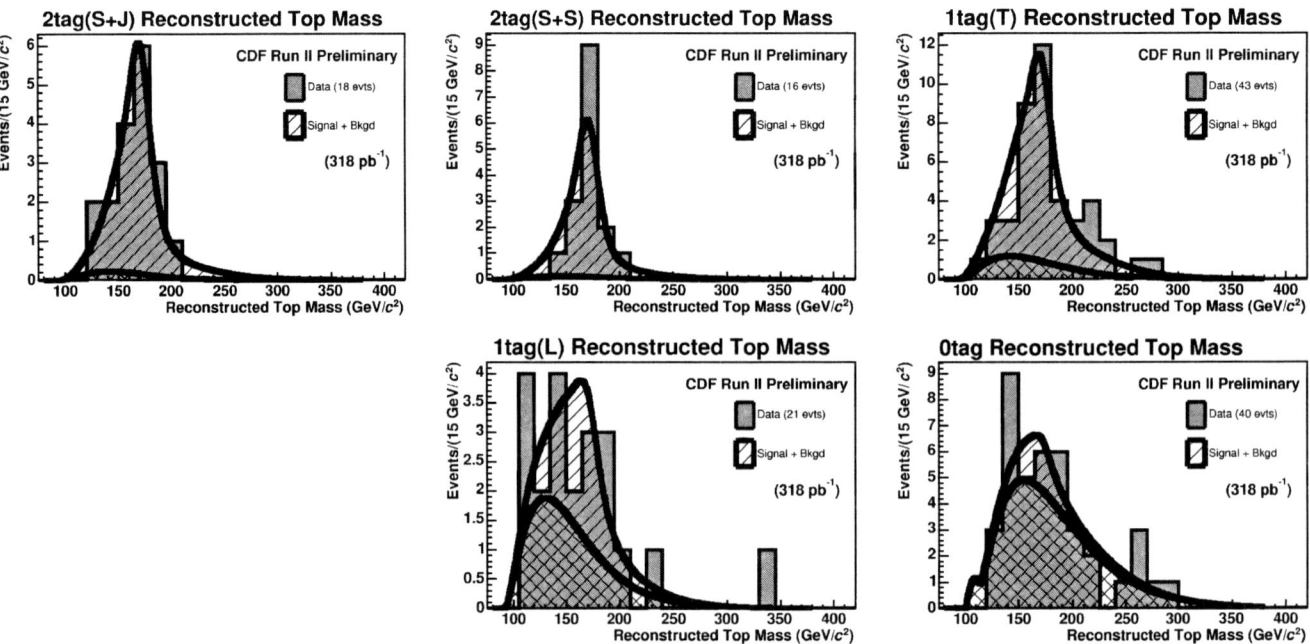

Fig. 98. Results of the likelihood fit to data in the jet probability analysis: The *histograms* are distributions of the reconstructed top mass in data. The *lines* show the best fit and the background component in the five different subsamples

portantly, their individual resolutions into account. Well measured events enter the overall likelihood with a higher weight than not so well measured events, while in the previously described template analyses, all events enter 1- or 2-dimensional histograms with the same weight. Any information on individual events is lost in those analyses.

The dynamic likelihood method is very similar to DØ's matrix element method, which was the first method applied to the TEVATRON data, that constructs individual event likelihoods. It was applied to DØ's Run-I data in the lepton + jets channel ([396, 397], Sect. 7.1.2). The DØ matrix element method is strongly inspired by the publications of Dalitz and Goldstein [367, 399, 443].

The parton process in a $p\bar{p}$ collision can generally be written as

$$a/p + b/\bar{p} \to \ldots \to C, C \equiv \sum_{i=1}^{n} c_i, \qquad (161)$$

where a and b are the initial partons, a quark, antiquark or gluon, in the proton and anti-proton, respectively, and $c_1, c_2, \ldots, c_n$ are final state partons. In the case of the $t\bar{t}$ lepton + jets channel, the initial parton set (a, b) is $(q, \bar{q})$ or (g, g), and the final partons are $\ell^+, \nu, q, b, \bar{q}', \bar{b}$ or their anti-particles, where $\ell = e$ or μ and $q, \bar{q}'$ are quarks from the W-boson decay. The final partons are assumed to have their pole masses so that the 3-momenta define their states unambiguously.

The hadronic cross section for the $t\bar{t}$ production is given by

$$d\sigma = dz_a \, dz_b \, d^2\mathbf{p}_T f^*_{a/p}(z_a) f^*_{b/\bar{p}}(z_b) f_T(p_T)$$
$$\times \hat{\sigma}(a + b \to C; m_{\text{top}}), \qquad (162)$$

where m_{top} is the mass of the top quark, and $d\hat{\sigma}$ is the parton level cross section,

$$d\hat{\sigma}(a + b \to C; m_{\text{top}}) = \frac{(2\pi)^4 \delta^4(a + b - C)}{4\sqrt{(a\,b)^2 - m_a^2 m_b^2}}$$
$$\times |\mathcal{M}(a + b \to C; m_{\text{top}})|^2$$
$$\times d\Phi_n^{(f)}(a + b; C). \qquad (163)$$

Variables z_a and z_b are the energy fractions of a and b in hadrons p and $\bar{p}$, respectively, and $\mathbf{p}_T$ is the total transverse 3-momentum of the initial and final systems. Functions $f^*_{a/p}(z_a)$ and $f^*_{b/\bar{p}}(z_b)$ denote the effective parton (longitudinal momentum) distribution functions (PDF's), while $f_T(\mathbf{p}_T)$ is the probability function for the total transverse momentum of the system acquired by initial state radiation. $\mathcal{M}$ is the matrix element of process (161), which includes the strong production mechanism of the top quark pair, their decays to a W-bosons and b-quarks, and the subsequent decay of the W-boson. t, $\bar{t}$, W^+ and W^- are treated as resonances on their mass shell. $d\Phi_n^{(f)}$ is the Lorentz invariant phase space factor,

$$d\Phi_n^{(f)} \equiv \prod_{i=1}^{n} \frac{d^3\mathbf{c}_i}{(2\pi)^3 2E_i}. \qquad (164)$$

The basic postulate in the dynamic likelihood method is that final partons in a single event occupy an n-dimensional unit phase space in the neighbourhood of $\mathbf{c}(c_1, \ldots, c_n)$. The total probability for this final state to occur is obtained by integrating (162) with initial state variables, z_a, z_b and $\mathbf{p}_T$, as

$$\frac{d\sigma}{d\Phi_n^{(f)}} = I(a, b)|\mathcal{M}(a + b \to C; m_{\text{top}})|^2, \text{ where} \qquad (165)$$

$$I(a, b) = \frac{(2\pi)^4}{4\sqrt{(a\,b)^2 - m_a^2 m_b^2}}$$
$$\times f^*_{a/p}(z_a) f^*_{b/\bar{p}}(z_b) f_T(\mathbf{p}_T), \qquad (166)$$

is the integration factor for the initial state. Because of the δ-function in (163), the initial parton momenta a and b are uniquely defined by that of C.

When the parton kinematics are given, the likelihood is defined as

$$\mathcal{L}_1(m_{\text{top}}|\mathbf{c}) = l_0 \frac{d\sigma}{d\Phi_n^{(f)}}, \qquad (167)$$

where l_0 stands for the integrated luminosity required to generate the event. Since the beam state is unknown, it is set to a constant (according to Bayes's postulate).

Final partons are not directly observed; they undergo parton-shower/hadronisation and are observed by detectors with finite resolution. Quarks and gluons are observed as jets. The correlation between the final parton state and observed quantities (observables) are described by transfer functions (TF, equivalent to resolution functions), denoted by $w(\mathbf{y}|\mathbf{x}||m_{\text{top}})$, which represents an m_{top} dependent probability density function for the observable $\mathbf{y}$ when the corresponding parton variable $\mathbf{x}$ is given. In this analysis, transfer functions are only used for jets, as the lepton kinematics are relatively well measured. The transfer functions are determined in $t\bar{t}$ Monte Carlo, separately for b-jets and light-quark jets.

For a single event I_i, the likelihood is determined in the following way: (i) Jets are assigned to quarks and each assignment is given a topology number ($I_t = 1, \ldots, N_t$, up to 24). (ii) The 3-momenta of the corresponding partons b, $\bar{b}$, q and $\bar{q}'$ are generated from the TFs. (iii) The neutrino p_T is identified with the measured $\not\!\!E_T$ and the (up to two) solutions ($I_s = 1, 2$) for the neutrino p_z are calculated from the invariant mass constraint of the W-boson. (iv) For a set of variables thus reconstructed, called a "path" $k(1, \ldots K)$, the value of m_{top} is scanned in its search region. The resulting single-path likelihood for the k-th path of event i with observables $\mathbf{y}^{(i)}$ is given by

$$\mathcal{L}_1^{(i)}(m_{\text{top}}; I_t, I_s, k|\mathbf{y}^{(i)}) = l_0 \frac{d\sigma}{d\Phi}(I_t, I_s, k; m_{\text{top}}), \qquad (168)$$

where the subscript 1 stands for a single path. The single event likelihood for the i-th event is then given by the average over all the single path likelihoods for that event,

$$\mathcal{L}^{(i)}(m_{\text{top}}) = \frac{1}{2KN_t} \sum_{I_t=1}^{N_t} \sum_{k=1}^{K} \sum_{I_s=1}^{2}$$
$$\times \mathcal{L}_1^{(i)}(m_{\text{top}}; I_t, I_s, k|\mathbf{y}^{(i)}). \quad (169)$$

By this definition, the event likelihood for the true values of I_t and I_s is always included in $\mathcal{L}(i)(m_{\text{top}})$, if the event is signal. The event likelihoods for individual events are mutually independent probability functions for m_{top}. To obtain the top quark mass from a total of N_{ev} events, the single event likelihoods are thus multiplied, and the $-2\ln$ (likelihood) of the product is taken,

$$\Lambda(M_{\text{top}}) = -2\ln\left(\prod_{i=1}^{N_{\text{ev}}} \mathcal{L}^{(i)}(m_{\text{top}})\right). \quad (170)$$

The mass of the top quark is determined from the maximum likelihood estimator of m_{top}, i.e. from the minimum of $\Lambda(m_{\text{top}})$.

Note that this procedure only calculates the probability for an event to be signal, but no integration over the corresponding background matrix element is carried out to calculate the probability for an event to be background. Also, the estimated fraction of background contained in the event sample does not enter the likelihood combination. The background contamination needs to be considered afterwards. These aspects are handled differently in DØ's matrix element method in the $\ell+$jets channel ([396, 397], Sect. 7.1.2).

In this analysis, CDF estimates the total number of background events to be 9.2 ± 1.8 (4.2 ± 0.7) for the 318 (162) pb^{-1} analysis.

A mass-dependent correction factor, called 'mapping function', is obtained with a larger number of sets of pseudo-experiments. There are two sources that cause a non-unit slope between the input top mass and the reconstructed top mass. One is the top mass dependence of the transfer function, and the other is the effect of background. Both effects are integrated into a single mapping function. Figure 99 (left) shows the joint likelihood for the 22 selected $t\bar{t}$ candidate events in the 318 pb^{-1} sample. This likelihood is fitted with a second order polynomial, from

which $m_{\text{top}} = 171.8^{+2.2}_{-2.0}$ (stat.) GeV/c^2 is obtained, assuming there is no background. The application of the mapping function removes the mass-pulling effect of the background. Figure 99 (right) shows the extracted top mass as a function of the background fraction. The top mass changes by less than 2.0 GeV/c^2 in the background fraction range 0%-20%. For the final result, the estimated 14.5% background fraction is used. The statistical errors are scaled due to the slope of the mapping function as a function of the top mass and due to the pull width of 1.04. The systematic uncertainties are summarised in Table 51. Both versions of the analysis are strongly dominated by the uncertainty on the jet energy calibration. However, it can be clearly seen, that significant improvements in the understanding of the jet energy calibration have been made between the two analyses, almost reducing the jet energy scale uncertainty by a factor of 2. The latest version of the jet energy scale calibration and its corresponding uncertainty is described in Sect. 3.3.2.

The early measurement of the top quark mass, using the dynamic likelihood method in 162 pb^{-1} of data, yields:

$$m_{\text{top}} = 177.8^{+4.5}_{-5.0}\,(\text{stat.}) \pm 6.2\,(\text{syst.})\,\text{GeV/c}^2. \quad (171)$$

Table 51. Summary of the systematic uncertainties for the dynamic likelihood analysis

Source of systematics	Δm_{top} (GeV/c^2)	
	in 162 pb^{-1}	in 318 pb^{-1}
Jet energy calibration	5.3	3.0
Transfer function	2.0	0.2
ISR	0.5	0.4
FSR	0.5	0.5
PDFs	2.0	0.5
MC generator	0.6	0.3
Spin correlation	0.4	–
NLO effect	0.4	–
Background fraction ($\pm 5\%$)	0.5	0.4
Background modelling	0.5	0.4
Monte Carlo modelling	0.6	–
b-jet energy modelling	–	0.6
b tagging	–	0.2
Total	6.2	3.2

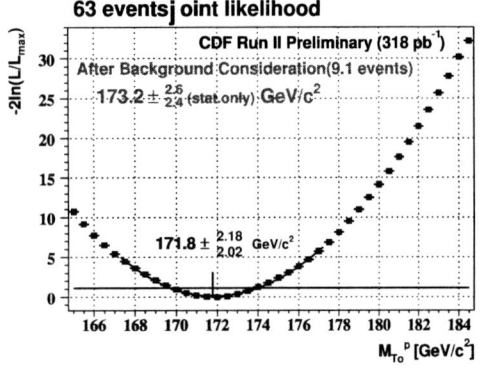

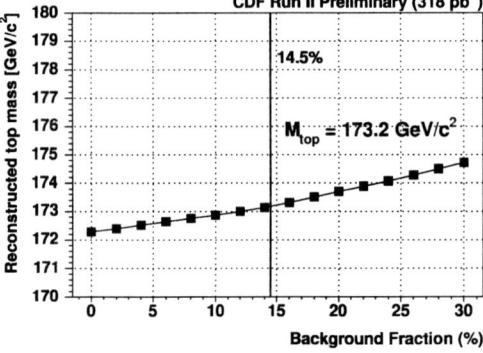

Fig. 99. *Left*: The joint negative log-likelihood of the 63 events in the 318 pb^{-1} analysis gives $m_{\text{top}} = 171.8^{+2.2}_{-2.0}$ GeV/c^2. *Right*: Extracted top quark mass using the mapping function as a function of the background fraction

The update analysis, using the dynamic likelihood method in $318\,\mathrm{pb}^{-1}$ of data, yields:

$$m_{\mathrm{top}} = 173.2^{+2.6}_{-2.4}\,(\mathrm{stat.}) \pm 3.2\,(\mathrm{syst.})\,\mathrm{GeV}/c^2\,. \quad (172)$$

Matrix element method. CDF has measured the top quark mass in $318\,\mathrm{pb}^{-1}$ of CDF-II data, using the matrix element method in the lepton + jets channel [454]. The lepton + jets events are selected according to the standard criteria, where exactly four reconstructed jets are required, of which at least one has to be identified as a b-jet. The matrix element method is described in more detail in Sect. 7.1.2.

This CDF-II measurement yields a value for the top quark mass of

$$m_{\mathrm{top}} = 172.0 \pm 2.6\,(\mathrm{stat.}) \pm 3.3\,(\mathrm{syst.})\,\mathrm{GeV}/c^2\,. \quad (173)$$

Multivariate technique. CDF has measured the top quark mass in the lepton + jets channel using a multivariate template method, applied to $162\,\mathrm{pb}^{-1}$ of data [455]. The method used here differs in several ways from the traditional template method. Using the mass of the W-boson for jet energy calibration on event-by-event basis, the systematic uncertainty from the jet energy scale is reduced. Estimating the probability to pick the correct jet-quark assignment on an event-by-event basis reduces the statistical uncertainty. Using other kinematic variables in addition to the reconstructed top mass improves the signal to background separation.

The lepton + jets events are selected according to the standard criteria (Sect. 4.2.3). At least one jet is required to be tagged by the SecVtx algorithm, allowing at most six jet-quark assignments, and in the kinematic fit only the four highest-E_T jets are considered. For each parton assignment, an integration over the W mass Breit–Wigner distribution is performed, using seven m_W^2 values equidistant in the cumulative Breit–Wigner probability. A jet energy scale factor, JES, is included in the W mass kinematic fit with a Gaussian width constraint, $\sigma_{\mathrm{JES}} = 0.07\%$. The momenta of all jets in the event are multiplied by the jet energy scale obtained from the mass fit of the hadronically decaying W-boson. The masses of the t and $\bar{t}$ quarks, m_{t} and $m_{\bar{t}}$ are then determined together with their errors, $\sigma_{m_{\mathrm{t}}}$ and $\sigma_{m_{\bar{t}}}$, and the correlation coefficient, ρ, for each of the seven mass assignments of the leptonically decaying W-boson. Out of the two neutrino solutions, the one with the smaller difference $|m_{\mathrm{t}} - m_{\bar{t}}|$ is chosen. Finally, the equal mass constraint $m_{\mathrm{t}} = m_{\bar{t}}$ is imposed and the top quark mass $m_{\mathrm{t},W}$ and its error $\sigma_{m_{\mathrm{t},W}}$ are calculated according to:

$$m_{\mathrm{t},W} = \frac{\sigma_{m_{\bar{t}}}^2 m_{\mathrm{t}} + \sigma_{m_{\mathrm{t}}}^2 m_{\bar{t}} - \rho m_{\mathrm{t}} m_{\bar{t}}(m_{\mathrm{t}} + m_{\bar{t}})}{\sigma_{m_{\bar{t}}}^2 + \sigma_{m_{\mathrm{t}}}^2 - 2\rho\sigma_{m_{\bar{t}}}\sigma_{m_{\mathrm{t}}}}\,, \quad (174)$$

$$\sigma_{m_{\mathrm{t},W}}^2 = \frac{\sigma_{m_{\mathrm{t}}}^2 \sigma_{m_{\bar{t}}}^2 (1 - \rho)}{\sigma_{m_{\bar{t}}}^2 + \sigma_{m_{\mathrm{t}}}^2 - 2\rho\sigma_{m_{\bar{t}}}\sigma_{m_{\mathrm{t}}}}\,, \quad (175)$$

$$\chi_{\mathrm{t},W}^2 = \frac{(m_{\mathrm{t}} - m_{\bar{t}})^2}{\sigma_{m_{\bar{t}}}^2 + \sigma_{m_{\mathrm{t}}}^2 - 2\rho\sigma_{m_{\bar{t}}}\sigma_{m_{\mathrm{t}}}}\,. \quad (176)$$

The top mass for each jet to quark assignment is determined by summing over the points in the W mass integration grid weighted by the combined χ^2 of the mass constraints:

$$m_{\mathrm{t}} = \frac{\sum e^{-\left(\chi_{t,W}^2 + \chi_W^2\right)/2} m_{t,W}}{\sum e^{-\left(\chi_{t,W}^2 + \chi_W^2\right)/2}}\,. \quad (177)$$

The overall jet permutation χ^2 is defined as $\chi^2 = -2\log\left(1/N \sum e^{-(\chi_{t,W}^2 + \chi_W^2)/2}\right)$, where N is the total number of points in the W mass integration grid (here 49). The jet permutation with the lowest overall χ^2 is chosen as the reconstructed top mass for each event.

Three jet-quark combinatorics scenarios are differentiated: (i) The four leading jets correspond to the four quarks, and the chosen jet permutation is the correct one. This subsample is called 'Good Permutation – GP'. (ii) The four leading jets correspond to the four quarks, but the chosen jet permutation is not correct. This subsample is called 'bad permutation – BP'. (iii) One or more of the four leading jets do not originate from any of the four daughter quarks of the $t\bar{t}$ decay. This subsample is called 'incorrect jets – IJ'. Since these three subsamples are expected to have significantly different top mass resolutions, a probability, p_{CJ} is calculated to pick correct jets using $t\bar{t}$ Monte Carlo. Furthermore, in the correct jets (CJ) subsample, a conditional probability, $P(\mathrm{GP}|\mathrm{CJ})$, is calculated to pick the good permutation. The fraction of incorrectly assigned permutations appears to decrease exponentially as a function of the difference between permutation χ^2 values. Therefore, an initial GP probability is calculated assuming the jets are correctly chosen as

$$P(\mathrm{GP}|\mathrm{CJ}) = \frac{a_b}{\sum_{i=1}^{12} a_i \exp\left(-\frac{\chi_i^2 - \chi_b^2}{w_e(\chi_i^2 + \chi_b^2)}\right)}\,,$$
$$\text{where} \quad w_e(y) = \exp\left(b_e + c_e y + d_e y^2\right)\,. \quad (178)$$

Coefficient a_i models the event density in the correct region for jet permutation i. Index b denotes the permutation with the smallest χ^2. Index e in the function $w_e(y)$ refers to the error type assigned to the permutation with index b when the correct permutation is the one with index i. The constants in the above formula (a_i, b_e, c_e, d_e) are determined from the $t\bar{t}$ signal Monte Carlo. Additional kinematic information is included in this probability calculation in order to improve the separation between the correct permutation sample and the other subsamples. This information is included by sequential application of the "approximate Bayesian update" formula. Let X be some kinematic variable whose value may be used to improve the separation between the GP and the BP subsamples. Then

$$P(GP|CJ, X) = \frac{\kappa P(\mathrm{GP}|\mathrm{J})}{\kappa P(\mathrm{GP}|\mathrm{CJ}) + (1 - P(\mathrm{GP}|\mathrm{CJ}))}\,, \quad (179)$$

follows from Bayes theorem when $\kappa = \frac{P(X|\mathrm{GP})}{P(X|\mathrm{BP})}$ and $P(\mathrm{GP}|\mathrm{CJ}) + P(\mathrm{BP}|\mathrm{CJ}) = 1$. The parameter κ is chosen according to an optimisation procedure. The first kinematic variables used to update $P(\text{GP}|\text{CJ})$ is $\cos\phi$, where ϕ is the angle between the lepton ℓ and b_ℓ in the rest frame of the leptonically decaying W-boson (using information from the W helicity). The second kinematic variable is $\cos\theta_1\cos\theta_2$, where θ_1 is the angle between the lepton momentum and the beam axis in the rest frame of the corresponding top quark. θ_2 is the angle between the direction of the light quark and the beam axis in the rest frame of the corresponding top quark. The product $\cos\theta_1\cos\theta_2$ is sensitive to the $t\bar{t}$ spin correlation. Studies in the $t\bar{t}$ Monte Carlo demonstrate that $P(GP|CJ,X)$ is a good estimate of the fraction of good permutations in the subsample containing correct jets.

For the multivariate template method, a set of good variables is needed, which both increase the sensitivity of the likelihood to the top quark mass and improve the discrimination between signal and background. The set of variables should also be optimal with respect to the systematic uncertainties of the measurement. Based on ensemble tests with several kinematic variables, CDF has found the choice of 2-dimensional templates with m_{top} and the sum of the transverse momenta of the four leading jets as variables to be optimal. The 2-dimensional templates are constructed using a non-parametric multivariate density reconstruction method called 'kernel density estimation' [456]. Figure 100 shows the 2-dimensional templates for the example of good permutations in $t\bar{t}$ events with $m_{\text{top}} = 150\,\text{GeV}/c^2$ and $200\,\text{GeV}/c^2$ and for the non-W background.

The likelihood for the observed data sample is defined as

$$\mathcal{L}(m_{\text{t}}) = \prod_{i=1}^{N}(f_{\text{b}}P_{\text{b}}(m_i, \mathbf{x}_i) + (1 - f_{\text{b}})P_{\text{s}}(m_i, \mathbf{x}_i, m_{\text{t}})), \quad (180)$$

where N is the number of observed events, m_i is the top mass in the i-th event, $\mathbf{x_i}$ symbolises all template variables besides the measured mass, P_{s} and P_{b} are the signal and background densities in the $(m, \mathbf{x})$ space, f_{b} is the background fraction, which is allowed to float freely. The signal density is composed as follows:

$$P_{\text{s}}(m, \mathbf{x}, m_{\text{t}}) = p_{\text{CJ}}(p_{\text{GP}}S_{0,m_{\text{t}}}(m, \mathbf{x}) + (1 - p_{\text{GP}})S_{1,m_{\text{t}}}(m, \mathbf{x})) + (1 - p_{\text{CJ}})S_{2,m_{\text{t}}}(m, \mathbf{x}), \quad (181)$$

where $S_{0,m_{\text{t}}}$, $S_{1,m_{\text{t}}}$ and $S_{2,m_{\text{t}}}$ are the three signal templates for the given generated m_{t}. p_{GP} is a short notation for $P(\text{GP}|\text{CJ}, X)$. The background densities are modelled as

$$P_{\text{b}}(m, \mathbf{x}) = \sum_{\text{bg types}} a_j B_j(m, \mathbf{x}), \quad \text{with} \quad \sum_{\text{bg types}} a_j = 1, \quad (182)$$

where B_j are the templates for different background types, and a_j are the background composition coefficients obtained from the $t\bar{t}$ cross section studies. The likelihood in (180) is not normalised, so that a calibration procedure for the width and the mean of the pull distribution is applied in $t\bar{t}$ Monte Carlo events.

Systematic uncertainties are estimated from many pseudo-experiments of 33 events each in Monte Carlo. Their impact on the top mass determination is summarised in Table 52.

Applying this multivariate technique to $162\,\text{pb}^{-1}$ of data and scaling the statistical error by 1.10 according to the observed pull width, the mass of the top quark is measured to be:

$$m_{\text{top}} = 179.6^{+6.4}_{-6.3}\,(\text{stat.}) \pm 6.8\,(\text{syst.})\,\text{GeV}/c^2, \quad (183)$$

where the result for the most probable background fraction and its statistical error is $f_{\text{b}} = 0.34 \pm 0.14\%$.

Decay length technique. CDF has measured the mass of the top quark in lepton + jets events, using the transverse decay length of b-hadrons from top decays in $318\,\text{pb}^{-1}$ [457]. All currently employed techniques to measure the top quark mass are limited by the same systematic uncertainty, the calorimeter jet energy scale. CDF has developed and studied a novel method to measure the top quark mass, using the transverse decay length of b-hadrons from top decays [458]. The method exploits the fact that top quarks at the TEVATRON are produced nearly at rest. In the rest frame of the top quark, the boost given to the bottom quark as a consequence of the top's decay and can simply

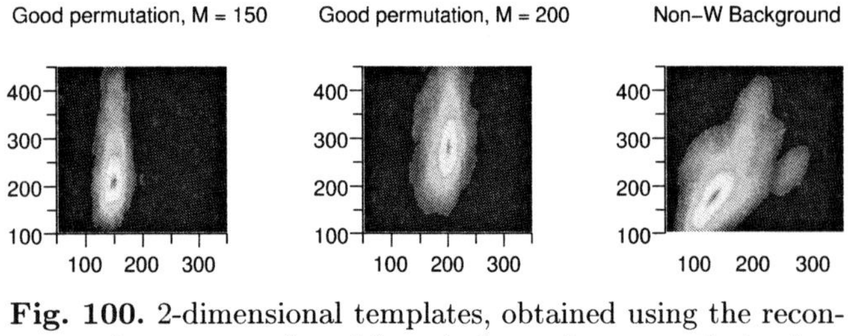

Good permutation, M = 150   Good permutation, M = 200   Non–W Background

Fig. 100. 2-dimensional templates, obtained using the reconstructed top mass (x-axis) and the scalar sum of the four leading jet p_{T} values (y-axis). Shown are examples for good permutations in $t\bar{t}$ events with $m_{\text{top}} = 150\,\text{GeV}/c^2$ (*left*), and $200\,\text{GeV}/c^2$ (*middle*) and for the non-W background (*right*)

Table 52. Summary of the systematic uncertainties for the multivariate analysis

Source of systematics	Δm_{top} (GeV/c^2)
Jet energy calibration	6.7
MC generator	0.2
ISR	0.2
FSR	0.6
PDFs	0.6
Background fraction	0.4
b tagging	0.3
Fitting procedure	0.7
Total	6.8

be written:

$$\gamma_{\rm b} = \frac{m_{\rm t}^2 + m_{\rm b}^2 - m_W^2}{2 m_{\rm t} m_{\rm b}} \approx 0.4 \frac{m_{\rm t}}{m_{\rm b}} \,, \qquad (184)$$

where the approximation makes use of the fact that $m_{\rm t} \gg m_{\rm b}$. The top quark's mass therefore (to the extent that the threshold approximation holds) is strongly correlated with the boost given to the b-quark and the subsequent b-hadron after fragmentation. Thus, the average lifetime of the b-hadrons resulting from top decays can be used to statistically infer the mass of the top quark. In this analysis, rather than measuring the average lifetime, the experimentally more accessible average transverse decay length of the b-hadrons, denoted $\langle L_{xy} \rangle$, is measured. This technique relies on tracking to precisely determine the decay length. It does not use any calorimeter information and thus avoids any jet energy scale uncertainty. It is therefore complementary to the other measurements of the top quark mass, and hence adds information to the overall measurement in combination with other results, despite its presently large statistical uncertainty.

The selection of the lepton + jets events follows the standard selection criteria, detailed in Sect. 4.2.3. At least one jet is required to be b-tagged using the SecVtx algorithm (see Sect. 3.3.7).

The method requires an accurate simulation of L_{xy}. This is checked in triple-tagged (2 SecVtx tags with one jet, in addition, being tagged by the SLT tagger) dijet events. The mean of the fitted decay length distribution in data and Monte Carlo are found to be: $\langle L_{xy}^{\rm data} \rangle = 0.3813 \pm 0.0031$ cm and $\langle L_{xy}^{\rm MC} \rangle = 0.3815 \pm 0.0045$ cm, respectively. The data/MC mean ratio is calculated to be $\mathrm{SF}_{L_{xy}} = 1.000 \pm 0.014$, concluding that the simulation models the transverse decay length of b-hadron in data with sufficient accuracy, with a 1.4% systematic uncertainty.

The tagged lepton + jets sample selected has an expected signal-to-background ratio of about 2.5 : 1. The rate of the different background components is derived as described in the $t\bar{t}$ cross section measurement (Sect. 4.2.3). The decay length distributions for signal and background are determined from Monte Carlo simulations. $t\bar{t}$ events are

simulated using HERWIG [226, 227] with top quark masses ranging from 130–230 GeV/c^2 in 5 GeV/c^2 intervals. The W + heavy flavour backgrounds are simulated using ALPGEN [224] + HERWIG $Wb\bar{b}$ + parton, $Wc\bar{c}$ + parton, and Wc + parton Monte Carlo. The W + mistag background is modelled using ALPGEN + HERWIG W + parton Monte Carlo. For the non-W (QCD) background, the L_{xy} distribution is obtained directly from CDF data. The single-top distribution is modelled using PYTHIA [223] Monte Carlo.

The modelling of L_{xy} distributions for the various background processes is cross-checked in W + 1 and 2 jet events in the W + jet sample. The combination of the L_{xy} background distributions according to their estimated rates describes the observed distribution in data reasonably well, yielding a Kolmogorov–Smirnov probability of 15.3%.

The signal and background L_{xy} distributions are treated as probability density functions from which ensembles of pseudo-experiments are formed. In forming each ensemble, the number of events is obtained by separately Poisson fluctuating the signal and each background about their expected contributions. The number of events for each process is converted to a number of tags by multiplying by the double-tag probability for that process. This process is repeated 1000 times for each mass point over the full mass range 130–230 GeV/c^2. The mean L_{xy} resulting from each pseudo-experiment is histogrammed from which the mean and $\pm 1\sigma$ variance are extracted as a function of mass. These points are fitted to third order polynomials. The fit to the mean establishes the most probable value for a true top mass given a measured mean L_{xy} and is the function used to make the top mass measurement from the L_{xy} extracted from data. Similarly, the fits to the variance form $\pm 1\sigma$ Neyman confidence intervals used to give the statistical uncertainty of the measurement. Those fits as a function of the mean decay length $\langle L_{xy} \rangle$ are shown in Fig. 101 (left).

Systematic uncertainties for this measurement, summarised in Table 53, can be classified according to three types of sources. The first arises from the accuracy of modelling factors that affect the top (or subsequent bottom) quark's momentum such as radiation, fragmentation and

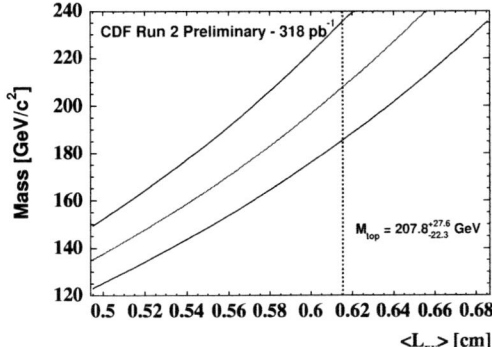 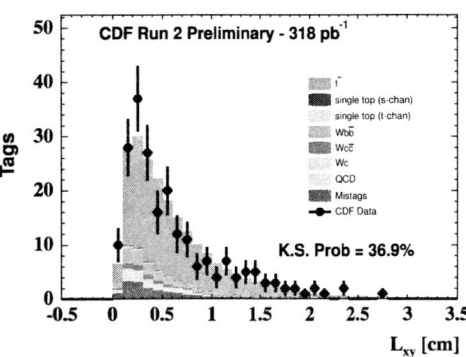

Fig. 101. *Left*: Most-probable (*central red*) and 1σ (*outer blue*) $m_{\rm t}$ curves as a function of the mean decay length L_{xy}. The mean transverse decay length measured in data is overlaid as *green, vertical dashed line*. *Right*: Transverse decay length L_{xy} distribution of positive tags in selected events in the $W + \geq 3$ jets data (*black points*). The expected contributions from signal and background Monte Carlo are overlaid in the *solid stacked histogram*

Table 53. Summary of the systematic uncertainties for the decay length analysis

Source of systematics	Δm_{top} (GeV/c^2)
Generator/fragmentation	0.7
Gluon fraction	0.2
QED radiation	0.6
ISR	1.0
FSR	0.9
PDFs	0.5
Top p_{T} spectrum	1.6
Jet energy scale	0.3
Background shape	2.3
Background normalisation	2.3
Data/MC $\langle L_{xy} \rangle$ SF	5.1
Total	6.5

PDF's. The second type of systematic uncertainty comes from potential inaccuracies in the size or shape of the background L_{xy} distributions. The third type of systematic uncertainty arises from imperfections of the detector simulation of L_{xy} (or other experimentally indistinguishable disagreements between Monte Carlo and data, for example from the imprecise knowledge of b-hadron lifetimes).

For the 216 positive SecVtx tags in 178 events, the L_{xy} distribution of positive tags, is plotted in Fig. 101 (right) together with the expected contributions from signal and background Monte Carlo overlaid. From those distributions, the mean decay length is measured to be

$$\langle L_{xy} \rangle = 0.6153 \pm 0.0356 \,\text{cm} \,. \qquad (185)$$

This value is translated into a measured top quark mass using the fits described above and shown in Fig. 101. This way, the top quark mass is measured to be:

$$m_{\text{top}} = 207.8^{+27.6}_{-22.3} \,(\text{stat.}) \pm 6.5 \,(\text{syst.}) \,\text{GeV}/c^2 \,. \qquad (186)$$

Dilepton channels.

Matrix element analysis. CDF has measured the top quark mass in the dilepton channel, using the matrix element technique in 340 pb^{-1} [459]. The dilepton channel has the lowest expected signal yield amongst the different $t\bar{t}$ decay channels, but it also has the least background contamination. In order to extract maximal information on the top quark mass from these dilepton events, the matrix element method [396, 397] is, for the first time, applied to dilepton events.

The general idea, as described in more detail for the ℓ + jets analyses in Sect. 7.1.2, is the expression of the top mass information in an event as the conditional probability $P(\mathbf{x}|m_{\text{t}})$, where m_{t} is the top quark pole mass and $\mathbf{x}$ is the vector of measured event quantities. The posterior probability density is calculated using the theoretical description of the $t\bar{t}$ production process, expressed with respect to the measured event quantities:

$$P(\mathbf{x}|m_{\text{t}}) = \frac{1}{\sigma(m_{\text{t}})} \frac{\text{d}\sigma(m_{\text{t}})}{\text{d}\mathbf{x}} \,, \qquad (187)$$

where $\text{d}\sigma/\text{d}\mathbf{x}$ is the per-event differential cross section. This expression is evaluated by integrating over all unmeasured quantities where the jet energy resolutions are convoluted using the transfer functions, $f(p, j)$, giving the probability of measuring jet energy j to give parton energy p. The total, resulting expression for the probability of a given pole mass for a specific event is then:

$$P(\mathbf{x}|m_{\text{t}}) = \frac{1}{N} \int \text{d}\Phi_6 |\mathcal{M}_{t\bar{t}}(p; m_{\text{t}})|^2$$
$$\times \prod_{\text{jets}} f(p_i, j_i) f_{\text{PDF}}(q_1) f_{\text{PDF}}(q_2) \,, \qquad (188)$$

where the integral is over the entire six-particle phase space, q_1 and q_2 are the incoming parton momenta, p is the vector of resulting parton-level quantities: lepton and quark momenta, and $|\mathcal{M}(p; m_{\text{t}})|$ is the $t\bar{t}$ production matrix element as defined in [193, 194]. The constant term in front of the integral ensures the normalisation for the probability: $\int \text{d}\mathbf{x} P(\mathbf{x}|m_{\text{t}}) = 1$.

In a similar fashion the probability densities for a given event to be background are calculated. This calculation is performed for the Drell–Yan process with associated jets, for W pair production with associated jets and $W + 3$ jets production where one jet is incorrectly identified as a lepton. The overall event likelihood is then given by

$$P(\mathbf{x}|m_{\text{t}}) = P_s(\mathbf{x}|m_{\text{t}}) p_s + P_{\text{bg1}}(\mathbf{x}) p_{\text{bg1}}$$
$$+ P_{\text{bg2}}(\mathbf{x}) p_{\text{bg2}} + \dots \,, \qquad (189)$$

where the weights $p_s, p_{\text{bg1}}, p_{\text{bg2}}, \dots$ for each term are determined from the number of expected background events in each category.

The method is tested with a large number of Monte Carlo pseudo-experiments, resulting in a small correction on the top mass as a function of the measured mass. Also the statistical uncertainty is slightly scaled up due to the width of the observed pull distribution in Monte Carlo events to be slightly larger than one. The largest contributing effects are yet coming from radiation rather than from b-quark hadronisation (20%), imperfect resolution of the lepton momenta ($\approx 10\%$) and imperfect resolution of jet angles ($\approx 10\%$).

A summary of the systematic uncertainties is given in Table 54. The jet energy scale uncertainty is the largest uncertainty, followed by the signal and background Monte Carlo, background modelling and the overall sample composition.

Using the matrix element method in dilepton events, CDF has measured the top quark mass to be:

$$m_{\text{top}} = 165.3 \pm 6.3 \,(\text{stat.}) \pm 3.6 \,(\text{syst.}) \,\text{GeV}/c^2 \,. \qquad (190)$$

Neutrino weighting analysis ($\nu(\eta)$). CDF measures the top quark mass in dilepton events using the neutrino weighting algorithm in 197 pb^{-1} [460] and in 360 pb^{-1} [461] of data.

Table 54. Summary of the systematic uncertainties for the matrix element method in dilepton events

Source of systematics	Δm_{top} (GeV/c^2)
Jet energy scale	2.6
Generator	1.0
Method	0.6
Sample composition uncertainty	0.7
Background MC	1.5
Background modelling	0.8
FSR modelling	0.5
ISR modelling	0.5
PDF's	1.1
Total	6.5

The neutrino weighting algorithm (NWA) was first introduced by DØ in Run-I [370], and later used by CDF [446] and DØ [398] for their final top mass measurements in the dilepton channel in Run-I. Here, it is used to reconstruct each dilepton event according to the $t\bar{t}$ decay hypothesis, with each event yielding a most probable top mass. Subsequently, an unbinned likelihood fit is applied to find the top mass hypothesis which best explains the observed data values as a mixture of background and $t\bar{t}$ signal events.

CDF selects the 46 (19) dilepton events in the 360 (197) pb^{-1} sample according to the lepton+track (LTRK) selection criteria (see Sect. 4.2.1) rather than the dilepton (DIL) criteria as the former has a larger efficiency and hence gives a smaller, albeit still dominant, statistical uncertainty.

The neutrino weighting method assumes: (i) the top mass, (ii) the W-boson mass, (iii) the η's of the two neutrinos, and (iv) the lepton-jet pair which originated from the top quark decay, e.g. $\ell^+ - jet_1$. Then, energy-momentum conservation is applied on the t and the $\bar{t}$ side, yielding up to two possible solutions for the 4-vector (ν) of the neutrino and another two solutions for the 4-vector ($\bar{\nu}$) of the antineutrino. Each of the resulting four neutrino-antineutrino solutions ($\nu, \bar{\nu}$) is assigned a probability (weight, w_i), that it describes the observed missing transverse energies, E_x and E_y, within their uncertainties σ_x and σ_y, respectively:

$$w_i = \exp\left(-\frac{(\not{E}_x - P_x^\nu - P_x^{\bar{\nu}})^2}{2\sigma_x^2}\right)$$
$$\exp\left(-\frac{(\not{E}_y - P_y^\nu - P_y^{\bar{\nu}})^2}{2\sigma_y^2}\right). \quad (191)$$

CDF uses $\sigma_x = \sigma_y = 15\,\text{GeV}$, which is obtained from $t\bar{t}$ Monte Carlo with $m_{\text{top}} = 175\,\text{GeV/c}^2$, and, independently, by using the individual resolutions of the observed objects in data. Given the assumed top mass and the neutrino η values, any of the four solution pairs ($\nu, \bar{\nu}$) could have occurred in nature. Therefore the four weights are added up:

$$w(m_{\text{t}}, \eta_\nu, \eta_{\bar{\nu}}, \ell\text{-jet}) = \sum_{i=1}^{4} w_i. \quad (192)$$

Not knowing which are the true neutrino η's in the event, the above steps are repeated for many possible ($\nu, \bar{\nu}$) pairs.

Monte Carlo $t\bar{t}$ simulations indicate that the neutrino and antineutrino η's are uncorrelated and distributed according to a Gaussian around 0 with width of 1. The neutrino η distributions, obtained in $t\bar{t}$ Monte Carlo assuming the Standard Model to describe nature, are scanned from -3 to 3 in steps of 0.1 and each ($\nu, \bar{\nu}$) pair is assigned a probability of occurrence $P(\eta_\nu, \eta_{\bar{\nu}})$, derived from the aforementioned Gaussian. Each trial ($\nu, \bar{\nu}$) pair contributes to the event weight according to its weight (192) and probability of occurrence $P(\eta_\nu, \eta_{\bar{\nu}})$:

$$w(m_{\text{t}}, \ell\text{-jet}) = \sum_{\eta_\nu, \eta_{\bar{\nu}}} P(\eta_\nu, \eta_{\bar{\nu}}) w(m_{\text{t}}, \eta_\nu, \eta_{\bar{\nu}}, \ell\text{-jet}). \quad (193)$$

Since b and $\bar{b}$ jets are not distinguished, the problem is solved by adding up both possible lepton-jet pairings and their respective weights $w(m_{\text{t}}, \ell\text{-jet})$. Thus, the final weight is only a function of the top mass, with all other unknowns integrated out:

$$W(m_{\text{t}}) = \sum_{\ell^+\text{-jet}_1}^{\ell^+\text{-jet}_2} w(m_{\text{t}}, \ell\text{-jet}). \quad (194)$$

CDF tries top masses from $100\,\text{GeV/c}^2$ to $500\,\text{GeV/c}^2$ in $1\,\text{GeV/c}^2$ steps. Finally, the weight distribution from each event is normalised to one. Each event is represented by one indicative top mass, which is chosen to be the top mass value which best explains the event as a $t\bar{t}$ dilepton decay, i.e. the maximum of the weight distribution. In a more sophisticated approach, the shape of the weight distribution for each individual event could also be used.

Applying this procedure to simulated Monte Carlo events of $t\bar{t}$ and background processes, template distributions of the top masses reconstructed as described above are built. The $t\bar{t}$ dilepton events are generated with PYTHIA [223] at top masses from $130\,\text{GeV/c}^2$ to $230\,\text{GeV/c}^2$ in $5\,\text{GeV/c}^2$ steps. The Drell–Yan production is also simulated using PYTHIA, while diboson production and fakes are simulated using ALPGEN [224] + HERWIG [226, 227]. Parameterising the resulting templates, probability density functions (pdf's) are derived. $P_{\text{s}}(m; m_{\text{top}})$ is the probability of reconstructing a top mass m when the true top mass is m_{top}. $P_{\text{b}}(m)$ is the corresponding probability to reconstruct a mass m in background events, including the NWA acceptance for background events of 96% (the NWA signal acceptance is 99.8%).

The overall probability that the data is described as an admixture of background events and dilepton $t\bar{t}$ events with top mass m_{top}, is found using the following likelihood function:

$$\mathcal{L}(m_{\text{top}}) = \mathcal{L}_{\text{shape}}(m_{\text{top}}) \times \mathcal{L}_{n_{\text{b}}} \times \mathcal{L}_{(n_{\text{s}}+n_{\text{b}})}, \quad (195)$$

with

$$\mathcal{L}_{\text{shape}} = \prod_{i=1}^{N} \frac{n_{\text{s}} P_{\text{s}}(m_i; m_{\text{top}}) + n_{\text{b}} P_{\text{b}}(m_i)}{n_{\text{s}} + n_{\text{b}}} \quad (196)$$

Table 55. Summary of the systematic uncertainties for the neutrino weighting $\nu(\eta)$ analysis in dilepton events

Source of systematics	Δm_{top} (GeV/c^2)	
	$(197\,\text{pb}^{-1})$	$(360\,\text{pb}^{-1})$
Jet energy scale	7.4	2.6
Background shape	2.8	3.0
Signal templates	0.3	0.2
Background templates	1.3	1.3
Signal MC generators	0.6	0.5
PDFs	0.8	0.4
ISR	2.5	0.8
FSR	1.3	0.6
$E\!\!\!/_T$ resolution	0.3	0.2
Total	8.6	4.4

$$-\ln \mathcal{L}_{n_{\text{b}}} = \frac{(n_{\text{b}} - n_{\text{b}}^{\text{exp}})^2}{2\sigma_{n_{\text{b}}}^2}, \quad \text{and} \tag{197}$$

$$\mathcal{L}_{(n_{\text{s}}+n_{\text{b}})} = \frac{e^{-(n_{\text{s}}+n_{\text{b}})}(n_{\text{s}}+n_{\text{b}})^N}{N!}, \tag{198}$$

where the term $\mathcal{L}_{\text{shape}}$ determines the relative abundance of signal and background events, n_{s} and n_{b}, respectively, by comparing the distribution of top masses m_i in the data with the probability distribution functions for signal and background, $P_{\text{s}}(m_i; m_{\text{top}})$ and $P_{\text{b}}(m_i)$, respectively. The term $\mathcal{L}_{n_{\text{b}}}$ constrains (within uncertainty $\sigma_{n_{\text{b}}}$) the number of background events to the a priori estimate of $n_{\text{b}}^{\text{exp}}$ events. The term $\mathcal{L}_{(n_{\text{s}}+n_{\text{b}})}$ imposes that the total number of signal and background events $(n_{\text{s}}+n_{\text{b}})$ be in agreement with the event count, N, in the data sample. The top mass hypothesis which minimises $-\ln(\mathcal{L})$ is retained.

The method is tested in sets of 10 000 pseudo-experiments. From this a small correction to the central value and the statistical uncertainty of the extracted top quark masses is derived. Several sources of systematic uncertainties are considered and evaluated from pseudo-experiments in the Monte Carlo. Table 55 summarises the contribution of the different systematic uncertainties, amounting to 6.8 (8.6) GeV/c^2 in total for the 360(197) pb^{-1} analysis. Figure 102 shows the reconstructed top mass for the 45 data events compared to the probability distribution function from signal and signal + background.

Using the neutrino weighting method in dilepton events, CDF has measured the top quark mass in 197 pb^{-1} to be:

$$m_{\text{top}} = 168.1^{+11.0}_{-9.8}\,(\text{stat.}) \pm 8.6\,(\text{syst.})\,\text{GeV/c}^2. \tag{199}$$

This analysis has recently been updated with the analysis of 360 pb^{-1}, yielding:

$$m_{\text{top}} = 170.0^{+7.1}_{-6.6}\,(\text{stat.}) \pm 4.4(\text{syst.})\,\text{GeV/c}^2. \tag{200}$$

Neutrino weighting analysis $(\nu(\phi))$. CDF has measured the top quark mass in dilepton events using the neutrino ϕ weighting method in 193 pb^{-1} [462]. This analysis has recently been updated using 340 pb^{-1} [447]. 33 (13) dilepton events are selected in the 340 (193) pb^{-1} sample using

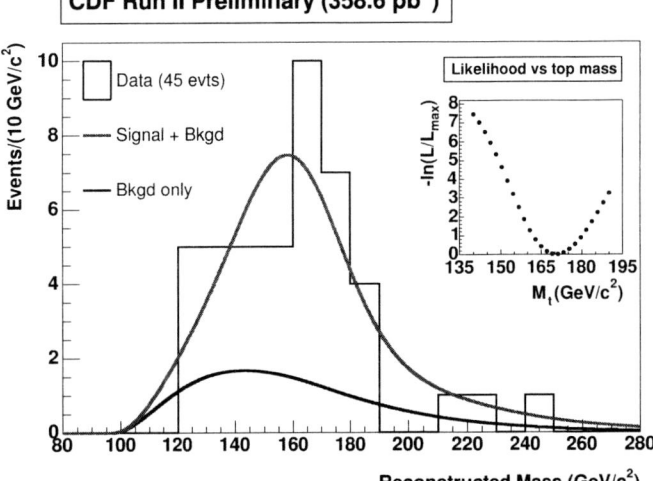

Fig. 102. Reconstructed top mass for the 45 data events in the 360 pb^{-1} analysis (*histogram*). The normalised shapes of the probability distribution functions for background and signal plus background are shown as *hatched curves*. The shape of the likelihood function is shown in the *inset*

the standard event selection criteria ('DIL' selection in Sect. 4.2.1).

In contrast to the lepton + jets mode, for the dilepton case the kinematics are underconstrained due to the presence of two neutrinos in the final state. The number of independent variables is one more than the number of kinematic constraints ($-1C$ kinematics). One must assume some of the event parameters ($\mathbf{R}$) to be known in order to constrain the kinematics and then vary $\mathbf{R}$ to determine a set of solutions. Each solution then needs to be assigned a weight. In this analysis, $\mathbf{R}$ is chosen to be the azimuthal angles of the neutrino momenta $\mathbf{R} = (\phi_{\nu_1}, \phi_{\nu_2})$ and a net of solutions, solving for the neutrino η, is created. This approach is complementary to the previously described neutrino weighting method in η, where η values are assumed for the two neutrinos and the ϕ values are calculated from the mass constraints. Taking into account the symmetry of the problem, it is sufficient to only consider the quadrant ($0 < \phi_{\nu_1} < \pi$, $0 < \phi_{\nu_2} < \pi$), which is split into 12×12 grid points. As in the $\nu(\eta)$ analysis, every point in this $(\phi_{\nu_1}, \phi_{\nu_2})$ plane has 8 solutions. This includes 2 solutions for the charged lepton to leading jet association, and two η solutions for each neutrino satisfying the $t\bar{t}$ kinematics. For every event, 1152 $1C$ minimisations are carried out with an output χ^2_{ijk} of the corresponding kinematic fit and a reconstructed top mass m^{rec}_{ijk} ($i = 1, \ldots, 12; j + 1, \ldots, 12; k = 1, \ldots, 8$). The indices i and j run over the $(\phi_{\nu_1}, \phi_{\nu_2})$ grid points and k runs over the 8 kinematic solutions. For every grid point (i, j fixed; $k = 1, \ldots, 8$) the minimal χ^2_{ijk} is chosen. The final output from this procedure is an array of 144 χ^2_{ij} and m^{rec}_{ij} ($i, j = 1, \ldots, 12$). The overall normalisation of the weight distribution is chosen to be one. The expression for the weights is:

$$w_{ij} = \frac{\exp(-\chi^2_{ij}/2)}{\sum_{i=1}^{12}\sum_{j=1}^{12}\exp(-\chi^2_{ij}/2)}. \tag{201}$$

These sets of weights w_{ij} as a function of the corresponding reconstructed masses $m_{ij}^{\rm rec}$ are sampled from a large number of Monte Carlo events to yield signal and background templates. In this procedure, only masses $m_{ij}^{\rm rec}$ with weights $w_{ij} > 30\% w_{ij}^{\rm max}$, the maximal weight in the event, are considered. The signal templates are created from HERWIG [226, 227] $t\bar{t}$ Monte Carlo with top masses in the range 140–230 GeV/c^2 with 5 GeV/c^2 steps. The background templates for diboson production, Drell–Yan production, and for $Z \to \tau\tau$ and "fake" lepton are also created from Monte Carlo samples and added according to the expected number of background events derived in the corresponding cross section analysis (Sect. 4.2.1). The signal and background templates are fitted to smooth functions, yielding $f_s(m_{\rm t}|m_{\rm top})$, the probability density function to reconstruct a mass $m_{\rm t}$ give a true top mass $m_{\rm top}$, and $f_{\rm b}(m_{\rm t})$, the probability density function to reconstruct a mass $m_{\rm t}$ in background events.

A maximum likelihood is used to extract the top quark mass by comparing the reconstructed top mass distribution of the data with the superposition of signal and background. The following likelihood is used:

$$\mathcal{L} = \mathcal{L}_{\rm shape} \times \mathcal{L}_{\rm bgd} \times \mathcal{L}_{\rm param}, \quad \text{where} \quad (202)$$

$$\mathcal{L}_{\rm shape} = \frac{e^{-(n_{\rm s}+n_{\rm b})}(n_{\rm s}+n_{\rm b})^N}{N!}$$

$$\times \prod_{n=1}^{N} \frac{n_{\rm s}f_s(m_n|m_{\rm top})+n_{\rm b}f_{\rm b}(m_n)}{n_{\rm s}+n_{\rm b}}, \quad (203)$$

$$\mathcal{L}_{\rm bgd} = \exp\left(-\frac{(n_{\rm b}-n_{\rm b}^{\rm exp})^2}{2\sigma_{n_{\rm b}}^2}\right), \quad (204)$$

$$\mathcal{L}_{\rm param} = \exp\left(-0.5\left[(\alpha-\alpha_0)^T U^{-1}(\alpha-\alpha_0) + (\beta-\beta_0)^T V^{-1}(\beta-\beta_0)\right]\right), \quad (205)$$

where the parameters $n_{\rm s}$ and $n_{\rm b}$ are the expected number of signal and background events in the dilepton sample, N is the total number of events observed in the data. $\mathcal{L}_{\rm bgd}$ is included to constrain the number of background events to the expected value. $\mathcal{L}_{\rm param}$ serves to constrain the α and β parameters, which describe the signal and background template parameterisation, respectively. U and V are the covariance matrices for α_0 and β_0, respectively. The $-\ln \mathcal{L}$ is maximised with respect to the two free fit parameters $m_{\rm top}$ and $n_{\rm s}$.

Using a large number of Monte Carlo pseudo-experiments, the validity of the method is tested. From the resulting distribution of the pull width, a scale factor for the statistical uncertainties of 1.051 (1.058) for the 340 (193) pb^{-1} sample is extracted. The systematic uncertainties, studied using a large number of pseudo-experiments, are summarised in Table 56. The jet energy scale clearly dominates the uncertainties even after the update with the improved jet energy calibration.

Analysing 193 pb^{-1} of data, the two-component background constrained fit (with 2.7 ± 0.7 expected background events) for the selected 13 dilepton events results in a top

Table 56. Summary of the systematic uncertainties for the neutrino weighting $\nu(\phi)$ analysis in dilepton events

Source of systematics	$\Delta m_{\rm top}$ (GeV/c^2)	
	(193 pb^{-1})	(340 pb^{-1})
Jet energy scale	6.7	3.3
ISR	1.8	0.7
FSR	0.7	0.8
PDFs	2.2	0.9
MC generators	0.7	1.0
Background shape	0.7	0.7
b-jet energy scale	–	0.7
Total	7.4	3.8

mass measurement of

$$m_{\rm top} = 170.0 \pm 16.6\,({\rm stat.}) \pm 7.4\,({\rm syst.})\,{\rm GeV}/c^2, \quad (206)$$

with 10.5 ± 3.6 signal and 2.7 ± 0.7 background events.

In the updated analysis, using 340 pb^{-1} of data, the two-component background constrained fit (with 11.6 ± 2.1 expected background events) for the selected 33 dilepton events results in a top mass measurement of

$$m_{\rm top} = 169.8^{+9.2}_{-9.3}\,({\rm stat.}) \pm 3.8\,({\rm syst.})\,{\rm GeV}/c^2, \quad (207)$$

with $23.4^{+6.3}_{-5.7}$ signal and 11.0 ± 2.1 background events.

Kinematic template analysis ($p_z(t\bar{t})$). CDF has measured the mass of the top quark in dilepton events, using the full kinematic template method in 193 pb^{-1} [463]. This analysis has recently been updated using 340 pb^{-1} [448]. This analysis is based on 30 (13) events selected in the 340 (193) pb^{-1} sample according to the standard dilepton analysis ('DIL') criteria (Sect. 4.2.1). As explained above, the kinematics of dilepton events are underconstrained even after imposing the mass of the W-boson and the equal-mass constraint, $m_{\rm t} = m_{\bar{\rm t}}$. One additional constraint is needed to solve the kinematics of the events. Given that the $t\bar{t}$ production at the TEVATRON is dominated by the top quark production at rest, the quantity $P_{t\bar{t}_z} = P_{t_z} + P_{\bar{t}_z}$, i.e. the z-component of the momentum of the $t\bar{t}$ system which is equal to the sum of the z-components of the top and the anti-top quarks, is studied. As naively expected and confirmed in $t\bar{t}$ Monte Carlo events, the most probable value of $P_{t\bar{t}_z}$ is zero with a Gaussian width of 180 GeV/c. Taking this constraint and its uncertainty into account, a kinematic fit is applied to each event. The measured event quantities such as the momentum of the leptons, the energy of the jets, and $\not{E}_{\rm T}$ have experimental uncertainties. For a given event, these quantities are smeared and the changes propagated to $\not{E}_{\rm T}$ and $P_{t\bar{t}_z}$ about 1000 times, so that a distribution of possible top masses is obtained for each event. The most probable value of such a distribution is considered the 'raw top mass' for each event. In principle, this method can give up to 4 solutions for each event, two for the kinematics of the two undetected neutrinos, and two from the possible assignments of the leptons and the b-jets. Only one out of the four solutions is considered according to the following criteria:

- from the two kinematically allowed solutions, the one yielding the smaller effective mass of the $t\bar{t}$ system is taken,
- from the two solutions due to the assignment of the leptons to the b-jets, the one with the higher reconstruction probability is chosen.

The normalised reconstructed 'raw mass' distributions are called templates. The signal templates are created using HERWIG $t\bar{t}$ Monte Carlo for top masses at 140–210 GeV/c^2 in 10 GeV/c^2 steps. The background templates are created from the simulation of diboson (WW, WZ, ZZ) production, Drell–Yan production and 'fake' leptons in W + jets events with the relative fractions as determined in the corresponding cross section analysis. The signal and background templates are fitted by smooth functions, $f_s(m^{\rm rec}, m_{\rm top})$ and $f_b(m^{\rm rec})$, respectively, describing the probability densities to observe the reconstructed mass $m^{\rm rec}$ for a true top mass $m_{\rm top}$ in signal events or for reconstructing a mass $m^{\rm rec}$ in background events.

The top quark mass is obtained by comparing the distribution of reconstructed masses in data with those in the combination of $t\bar{t}$ signal at different top masses and the background, using an unbinned maximum likelihood method. The following likelihood function is used:

$$\mathcal{L} = \mathcal{L}_{\rm shape} \times \mathcal{L}_{\rm bg}, \quad (208)$$

$$\mathcal{L}_{\rm shape} = \prod_{i=1}^{n} \left[b f_b(m_i^{\rm rec}) + (1-b) f_s(m_i^{\rm rec}, m_{\rm top}) \right], \quad (209)$$

$$-\ln \mathcal{L}_{\rm bg} = \frac{(b - b_{\rm exp})^2}{2\sigma_b^2}, \quad (210)$$

where $m_{\rm top}$ and b are free parameters of the fit. b represents the background fraction in the data sample and is varied within its uncertainties σ_b according to a Gaussian distribution ($\mathcal{L}_{\rm bg}$). As determined in the corresponding cross section analysis, 7.1 ± 1.8 (2.7 ± 0.7) background events are expected in this data sample and 30 out of the 33 (12 of the 13) selected events pass the kinematic reconstruction procedure in the 340(193) pb^{-1} sample and are fitted.

The systematic uncertainties, summarised in Table 57, are studied in a large number of Monte Carlo pseudo-experiments. By far the dominant systematic uncertainty is the jet energy scale calibration, which has been significantly reduced when going to the larger data set. Figure 103 shows the distribution of reconstructed top mass for the 12 events together with the probability density function for background and signal + background. Also shown is the negative log-likelihood curve.

The resulting top mass in 193 pb^{-1} is found to be:

$$m_{\rm top} = 176.5^{+17.2}_{-16.0} \,({\rm stat.}) \pm 6.9 \,({\rm syst.}) \,{\rm GeV}/c^2. \quad (211)$$

Table 57. Summary of the systematic uncertainties for the $p_z(t\bar{t})$ analysis in dilepton events

Source of systematics	$\Delta m_{\rm top}$ (GeV/c^2)	
	(193 pb^{-1})	(340 pb^{-1})
Jet energy scale	6.5	3.2
b-jet energy scale	–	0.6
MC generators	0.7	0.6
ISR	1.3	0.6
FSR	0.4	0.3
PDFs	0.9	0.5
Background shape	1.9	1.5
Background rate	–	0.3
Background statistics	–	0.8
Total	6.9	3.8

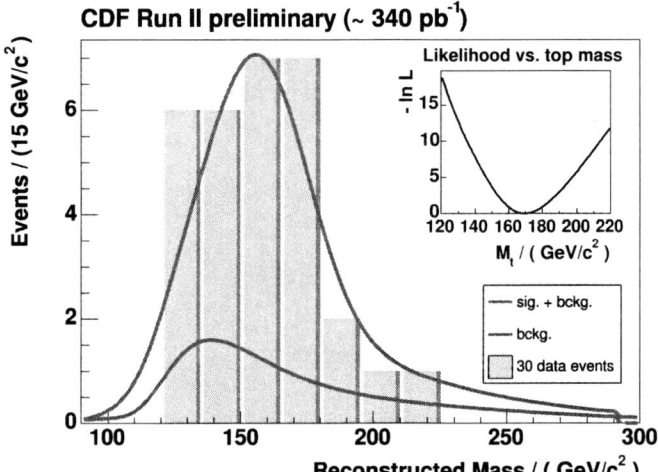

Fig. 103. Distribution of reconstructed top mass for the 30 data events in the 340 pb^{-1} data sample together with the probability density functions for background and signal + background. Also shown is the negative log-likelihood curve from the fit in the *inset*

The updated analysis, using 340 pb^{-1} of data, results in a top mass measurement of

$$m_{\rm top} = 170.2^{+7.8}_{-7.3} \,({\rm stat.}) \pm 3.8 \,({\rm syst.}) \,{\rm GeV}/c^2. \quad (212)$$

All jets channels. The most recent measurement of the top quark mass in the fully hadronic decay mode of the $t\bar{t}$ pair by CDF was made on 109 pb^{-1} of Run-I data [464]. This analysis, which yielded the first observation of top quark production in this decay mode, uses two separate approaches. In the first (Technique I), events with at least one identified b-jet are required to pass strict kinematic criteria that favour $t\bar{t}$ production and decay. Using this technique, CDF selects 187 events containing a total of 222 b-tags with an expected background of 164.8 ± 1.2 (parameterisation) ± 10.7 (syst.) tags. This excess has a significance of three standard deviations. In the second approach (Technique II), on events with two identified b-jets a minimum energy requirement is imposed, yielding

157 events with two or more b-tags with a predicted background of 122.7 ± 13.4 from QCD heavy flavour production and fake double tags. This excess corresponds to a significance of two standard deviations. In both cases, an excess of events with respect to the background prediction is observed, from which CDF measures the $t\bar{t}$ production cross section.

To determine the top quark mass, full kinematic reconstruction is applied to the sample of events with 6 or more jets, one or more tags, and the kinematic requirements of Technique I. Events are reconstructed according to the $t\bar{t} \to W^+ b W^- \bar{b}$ hypothesis, where both W-bosons decay into a quark pair, with each quark associated with one of the six highest E_T jets. This corresponds to 16 four-momentum conservation equations with 13 unknown variables, the three-momenta of the two top quarks and the two W bosons, and the unknown top quark mass. Since all events contain at least one b-tag, the tagged jets are required to be assigned to a b or $\bar{b}$ quark. A kinematic fit is applied, and the combination with the lowest χ^2 is chosen. In order to avoid threshold effects in the mass distribution, the $\sum E_T$ cut is lowered from 300 to 200 GeV, while keeping the other requirements unchanged. The 3-jet mass distribution for the 136 selected tagged events is shown in Fig. 104 along with the expected background and $t\bar{t}$ contributions. A maximum likelihood method is applied to extract the top quark mass. The experimental data are compared to HERWIG [226, 227] Monte Carlo samples of $t\bar{t}$ events, in a top quark mass range from 160 to 210 GeV/c^2, and a background sample from the untagged events.

The systematic uncertainties are summarised in Table 58. As a result, the top quark mass is measured to be

$$m_{\text{top}} = 186 \pm 10 \,(\text{stat.}) \pm 12 \,(\text{syst.}) \,\text{GeV/c}^2. \quad (213)$$

7.1.2 DØ analyses

Lepton + jets channel.

Low bias template analysis. DØ has measured the top quark mass in the lepton + jets channel using the low bias template method. This technique, which was already used for the first DØ measurement of the top quark mass in Run-I [465], is applied to lepton + jets events, selected based on a discriminant which utilises the unique topology of $t\bar{t}$ events (topological analysis) in 160 and 229 pb^{-1} of Run-II data, and to a sample in which at least one jet in the event is required to be b-tagged (b-tagged analysis, 229 pb^{-1}) [466, 467].

The event selection follows that of the corresponding cross section analyses (Sects. 4.3.2 and 4.3.3). To further reduce the background from mis-identified electrons in multijet events, $E_T^W = |p_T^\ell| + |\not{E}_T| > 65$ GeV is required. The events are reconstructed using a 2-C constrained kinematic fit, imposing the mass of the hadronically and the leptonically decaying W-boson and the equal mass constraint $m_t = m_{\bar{t}}$. This fit incorporates the measured object resolutions and considers all 12 possible jet-quark assignments. It returns a best fit top quark mass and a fit

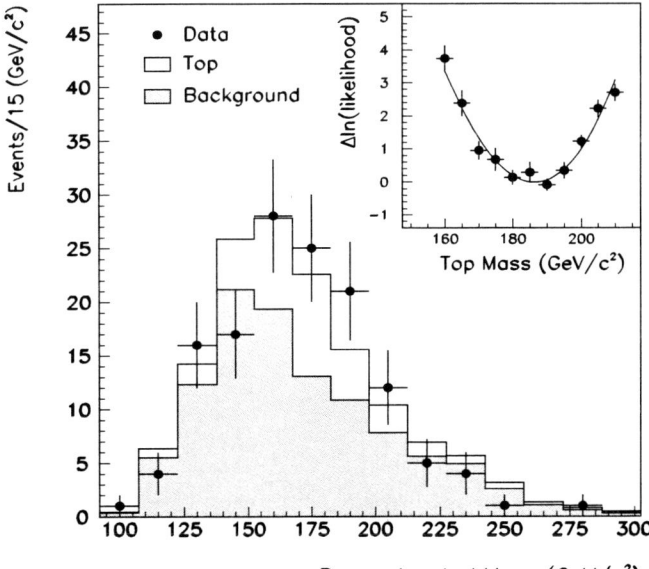

Fig. 104. Reconstructed mass distribution for events with at least one tag (•). Also shown are the background distribution (*shaded*) and the contribution from $t\bar{t}$ Monte Carlo events with $m_t = 175$ GeV/c^2 (*hollow*). The *inset* shows the difference in $-\ln$ (likelihood) with respect to its minimum and the fit used to determine the top mass

Table 58. Summary of the systematic uncertainties for the CDF alljets analysis in Run-I

Source of systematics	$\Delta m_{\text{top}}/m_{\text{top}}$ (%)
Jet energy scale	± 2.9
Gluon rad., fragmentation	± 4.6
Fitting procedure	± 2.8
Background estimate	± 0.9
Total	$\pm 6.2 \,(\Delta m_{\text{top}} \approx 12 \,\text{GeV/c}^2)$

χ^2 for each permutation. At least one permutation is required to have a $\chi^2 < 10$. This cut keeps 96% of the $t\bar{t}$ events and reduces the W + jets (multijet) background by 7% (10%). The fit with the lowest χ^2 is used in the top mass fit.

In order to get further discrimination between signal and background, a discriminant (low bias discriminant, D) is constructed from the topology of the events. It is designed to be uncorrelated with the top quark mass. It closely follows the procedure described in the topological/kinematic $t\bar{t}$ cross section measurement (Sect. 4.3.2) and uses the following kinematic variables: (i) The reconstructed $\not{E}_T$, (ii) the aplanarity $\mathcal{A}$, (iii) $H'_{T2} = H_{T2}/H_{||}$, and (iv) $K'_{T,\text{min}} = \Delta R_{jj}^{\text{min}} p_T^{\text{min}}/E_T^W$. $\mathcal{A}$ and $K'_{T,\text{min}}$ are defined as in the topological/kinematic cross section analysis. H_{T2} is the scalar sum of the $|p_T|$ of the jets excluding the leading jet, and $H_{||}$ is the scalar sum of $|p_z|$ of the jets, isolated lepton, and the neutrino. The selected events are required to all have $D > 0.4$ and $H_{T2} > 90$ GeV. This set of requirements selects 94 (87) events in the topological analysis using 229(160) pb^{-1} and 69 events in the b-tagged analysis.

The template analysis is based on comparing the fitted mass from the kinematic fit on data with the results obtained from fitting simulated Monte Carlo samples of known top masses. The top mass templates in the signal $t\bar{t}$ and the $W+4$-jets background Monte Carlo samples are constructed in 10 GeV/c^2 wide bins from 80 to 280 GeV/c^2. In order to extract the top quark mass from this comparison, a binned likelihood is used. The probability distribution function for the mass estimator is written in terms of the number of signal events, n_s, and the number of background events, n_b in the sample. The fraction of background events is constrained to the expected number using a Poisson probability term. For each hypothesised top quark mass, the likelihood is maximised as a function of the number of signal and background events. The mass with the largest likelihood is the measured top quark mass.

This procedure is tested on a large series of Monte Carlo ensembles. The method is found to be well calibrated for both, the topological and the b-tagged analysis. The uncertainties are consistent with the statistical spread seen in the ensembles.

The systematic uncertainties, determined from variations of the event selection, reconstruction, calibration in

Monte Carlo events, and the modelling of the underlying event (UE) and multiple interactions (MI) are summarised in Table 59.

Figure 105 shows the distributions of the fit masses in all three template analyses together with the fit mass distributions in signal and background. The gain from increasing statistics and from the use of b-tagging, improving the overall mass resolution by increasing the fraction of correct jet-quark assignments, is clearly visible.

The topological analysis of 160 pb^{-1} yields a top quark mass of

$$m_{\text{top}} = 170.0 \pm 6.5\,(\text{stat.})^{+10.5}_{-6.1}\,(\text{syst.})\,\text{GeV/c}^2\,,$$
(214)

with 38 ± 8 $t\bar{t}$ events (statistical uncertainties only), where 47 of the 87 events are expected to be $t\bar{t}$ according to the topological cross section analysis.

The updated topological analysis, using 229 pb^{-1} yields a top quark mass of

$$m_{\text{top}} = 169.9 \pm 5.8\,(\text{stat.})^{+7.8}_{-7.1}\,(\text{syst.})\,\text{GeV/c}^2\,,$$
(215)

Table 59. Summary of the systematic uncertainties for the DØ template analyses in the lepton + jets channel

Source of systematics	Δm_{top} (GeV/c^2)		
	(160 pb^{-1}, topo.)	(229 pb^{-1}, topo.)	(229 pb^{-1}, b-tag)
Jet energy scale	$+9.0$ -4.0	$+6.8$ -6.5	$+4.7$ -5.3
Jet energy resolution	± 2.3	± 0.9	
Gluon radiation	$-$	± 2.6	± 2.4
UE & MI	$+3.0$	$-$	$-$
MC statistics	± 0.5	± 0.5	± 0.5
Trigger uncertainty	± 1.0	± 0.5	± 0.5
Method calibration	± 0.5	± 0.5	± 0.5
$t\bar{t}$ modelling	± 3.8	$+2.3$	$+2.3$
Background modelling	$-$	$+0.7$	± 0.8
b-tagging	$-$	$-$	± 0.7
Total	$+10.5$ $-\ 6.1$	$+7.8$ -7.1	± 6.0

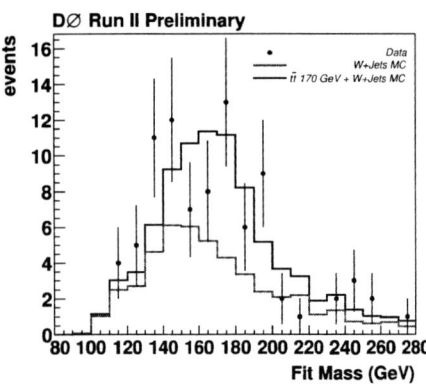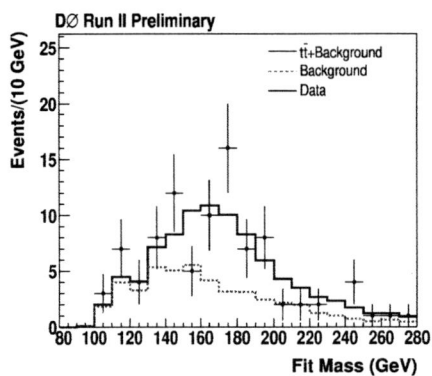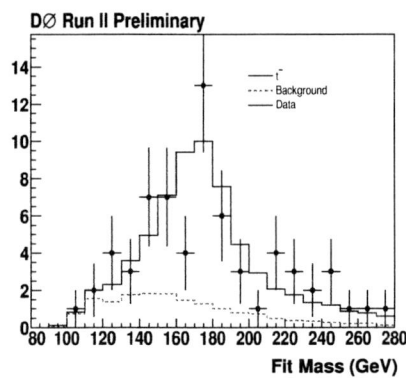

Fig. 105. Distribution of the fit masses in the topological analysis using 160 pb^{-1} (*left*), 229 pb^{-1} (*middle*) and in the b-tagging analysis (*right*). Also shown are the corresponding fit mass distributions in signal and background Monte Carlo

with 44.2 ± 6.6 $t\bar{t}$ events (statistical uncertainties only), where 47.9 ± 8.8 of the 94 events are expected to be $t\bar{t}$ according the topological cross section analysis.

The b-tagged analysis, using $229\,\mathrm{pb}^{-1}$ yields the most precise top quark mass:

$$m_{\mathrm{top}} = 170.6 \pm 4.2\,(\mathrm{stat.}) \pm 6.0\,(\mathrm{syst.})\,\mathrm{GeV}/c^2\,, \tag{216}$$

with 49.2 ± 6.3 $t\bar{t}$ events (statistical uncertainties only), where 52.4 ± 4.2 of the 69 events are expected to be $t\bar{t}$ according the b-tagging cross section analysis.

Ideogram method. DØ has measured the mass of the top quark in lepton + jets events, using the ideogram method in $160\,\mathrm{pb}^{-1}$ [466]. The ideogram method is very similar to the technique that was used by the DELPHI experiment to measure the mass of the W-boson at LEP [444, 445]. In this technique, the mass information from the constrained kinematic fit is used to construct an event likelihood taking into account all possible jet permutations. The low bias discriminant, D, introduced in the DØ template analysis described above, is used on event-by-event basis to estimate the probability that an event is background. Therefore no cut on D is necessary and is omitted to improve the statistical sensitivity. Similarly, no cut on $H_{\mathrm{T}2}$ is deemed necessary. Apart from this, the event selection and the kinematic fit are identical to the ones described above for the DØ template analysis, yielding here 101 events in the $e +$ jets and 90 events in the $\mu +$ jets channel. Finally, the overall fraction of signal events in the sample is allowed to float freely in the likelihood fit.

The kinematic fit used is identical to the one used in the DØ template analysis described above. Here, however, all information from this fit is taken into account. It is used to construct an event likelihood $\mathcal{L}_{\mathrm{evt}}(m_{\mathrm{t}}, P_{\mathrm{samp}})$ as a function of the top mass m_{t} and overall $t\bar{t}$ fraction in the sample, P_{samp}. For each event all 12 jet-quark combinations are considered, and for each combination up to two different starting guesses for the p_z of the neutrino are considered. In about 60% of parton-matched $e +$ jets $t\bar{t}$ events at $175\,\mathrm{GeV}/c^2$ (55% in the $\mu +$ jets channel), considering both starting guesses for the neutrino in the 'correct' jet-quark combination leads to two identical fitted masses. In 20% it leads to two top mass solutions differing by less than 5 GeV. In the remaining 20% of the cases (25% in the $\mu +$ jets channel) it yields two mass solutions that differ by more than 5 GeV. Whether or not the mass values are different, two masses per jet-quark combination are included in the likelihood. Thus, each event yields 24 masses m_i, uncertainties σ_i and χ_i^2 values indicating the quality of the fit. A relative probability of each jet assignment w_i is calculated as

$$w_i = \exp\left(-\frac{1}{2}\chi_i^2\right)\,. \tag{217}$$

If, for a particular jet-quark combination, both neutrino solutions i and $i+1$ fail to converge, the corresponding weights w_i and w_{i+1} are chosen to be zero. However, if for a particular jet-quark combination only one of the two neutrino solutions leads to a converging fit, that solution is used twice for consistency with other jet-quark combinations. Also the probability for the event to be signal P_{evt} is estimated (see below) and the likelihood to observe this event is calculated as follows:

$$\mathcal{L}_{\mathrm{evt}}(m_{\mathrm{t}}, P_{\mathrm{samp}}) =$$
$$P_{\mathrm{evt}}\left[\int_{100}^{300} \sum_{i=1}^{24} w_i \mathsf{G}(m_i, m', \sigma_i)\mathrm{BW}(m', m_{\mathrm{t}})\,\mathrm{d}m'\right]$$
$$+ (1 - P_{\mathrm{evt}}) \sum_{i=1}^{24} w_i \mathrm{BG}(m_i)\,. \tag{218}$$

The signal term consists of a convolution of the sum of the Gaussian resolution functions $\mathsf{G}(m_i, m', \sigma_i)$ describing the experimental resolution with a relativistic Breit–Wigner $\mathrm{BW}(m', m_{\mathrm{t}})$, representing the expected distribution of the average of the two invariant masses of the top and anti-top quark in the event, for a top mass m_{t}. The background term consists of the weighted sum $\mathrm{BG}(m_i)$, where $\mathrm{BG}(m)$ is the shape of the mass spectrum from $W + 4$-jet and multijet events observed in Monte Carlo simulation. The Breit–Wigner and background shape are both normalised to unity on the integration interval: 100 to 300 GeV. This interval is chosen large enough not to bias the mass in the region of interest. The sensitivity to the signal fraction P_{samp} in the sample enters through the estimated event purity P_{evt}, which depends on P_{samp} and on the value of the topological discriminant D for that event:

$$P_{\mathrm{evt}} = \left(\frac{\mathrm{S}}{\mathrm{S+B}}\right)_{\mathrm{evt}} = \frac{(\mathrm{S/B})_{\mathrm{evt}}}{(\mathrm{S/B})_{\mathrm{evt}} + 1}$$
$$= \frac{(\mathrm{S/B})_{\mathrm{samp}}(\mathrm{S/B})_D}{(\mathrm{S/B})_{\mathrm{samp}}(\mathrm{S/B})_D + 1}\,, \quad \text{where} \tag{219}$$
$$(\mathrm{S/B})_{\mathrm{samp}} = \frac{P_{\mathrm{samp}}}{1 - P_{\mathrm{samp}}}\,. \tag{220}$$

and $(\mathrm{S/B})_D$ is derived from the estimated event purity $P(D)$, parameterised as a function of the discriminant value D for a sample with a S/B ratio equal to unity,

$$(\mathrm{S/B})_D = \frac{P(D)}{1 - P(D)}\,. \tag{221}$$

Since each event is independent, the combined likelihood for the whole sample is calculated as the product of the single event likelihood curves:

$$\mathcal{L}_{\mathrm{shape}}(m_{\mathrm{t}}, P_{\mathrm{samp}}) = \prod_j \mathcal{L}_{\mathrm{evt},j}(m_{\mathrm{t}}, P_{\mathrm{samp}})\,. \tag{222}$$

This likelihood is maximised with respect to the top mass m_{t} and the estimated fraction of $t\bar{t}$ signal in the sample, P_{samp}.

The performance of the ideogram method is tested in large ensembles of Monte Carlo events. The method is reliable over the whole range of sample purities and top masses

Table 60. Summary of the systematic uncertainties for the DØ Ideogram analysis in the lepton + jets channel

Source of systematics	Δm_{top} (GeV/c^2)
Jet energy scale	$^{+4.6}_{-5.0}$
Jet energy resolution	± 1.0
Trigger uncertainty	± 0.5
UE & MI modelling	$+1.8$
MC statistics	± 0.3
Noise/MI	± 2.6
Background level	± 0.8
Background modelling	± 1.4
$t\bar{t}$ modelling	± 3.8

tested. Based on this calibration and the corresponding pull distributions, corrections for the top mass and its statistical uncertainty are derived. The systematic uncertainties, including variations of the modelling of the underlying event (UE) and multiple interactions (MI), are studied in large ensembles of Monte Carlo events and summarised in Table 60.

Figure 106 shows the mass likelihood curves for the $e +$ jets (left) and $\mu +$ jets channels (middle) separately, and the combined lepton + jets likelihood curve (right). The shoulder observed around 145 GeV/c^2 has been studied and is attributed to a detector-related class of background, which is not well modelled in the Monte Carlo. This shoulder, however, does not affect the top mass result from the minimum of the likelihood curve.

After the calibration correction, the top quark mass is found to be $m_{\mathrm{t}} = 177.6 \pm 11.9$ GeV/c^2 in the $e +$ jets channel and 177.895 GeV/c^2 in the $\mu +$ jets channel, where the uncertainties are statistical only. The combination of these two yields a top quark mass measurement of

$$m_{\mathrm{top}} = 177.5 \pm 5.8 \,(\mathrm{stat.}) \pm 7.1 \,(\mathrm{syst.}) \;\mathrm{GeV/c}^2 \,. \tag{223}$$

Matrix element method in Run-I. DØ has recently developed a new technique to measure the top quark mass, the matrix element method. This method was first applied to the 125 pb^{-1} of lepton + jets data recorded at $\sqrt{s} = 1.8$ TeV during Run-I for a re-analysis, resulting in significantly improved statistical and systematic uncertainties [396, 397] compared to the previous measurement [465]. The event selection of the lepton + jets events follows that in [465] and is very similar to that of the topological/kinematic $t\bar{t}$ cross section analysis in DØ Run-II (Sect. 4.3.2), i.e. it does not require any b-tagging. The new analysis involves the comparison of the lepton + jets events with a leading-order matrix element for $t\bar{t}$ production and decay. In order to minimise the effect of higher-order corrections, the analysis is restricted to events containing exactly four jets, yielding 71 selected events.

The matrix element method is similar to that suggested for $t\bar{t}$ dilepton decay channels [367, 399, 400, 439–443], and used in a previous mass analysis of dilepton events in Run-I [398]. A similar approach has also been suggested for the measurement of the mass of the W-boson at LEP [444, 468]. Given N events, the top quark mass is estimated by maximising the likelihood:

$$\mathcal{L}(\alpha) = e^{-N \int P_m(\mathbf{x},\alpha)\,\mathrm{d}\mathbf{x}} \prod_{i=1}^{N} P_m(\mathbf{x_i}, \alpha) \,, \tag{224}$$

where x_i is a set of variables needed to specify the i-th measured event, P_m is the probability density for observing that event, and α represents the parameters to be determined (in this case α is the mass of the top quark). Detector and reconstruction effects are taken into account in two ways. Geometric acceptance, trigger efficiencies, and event selection enter through a multiplicative function $A(\mathbf{x})$ that is independent of α, and relates the observed probability density $P_m(\mathbf{x}, \alpha)$ to the production probability $P(\mathbf{x}, \alpha)$: $P_m(\mathbf{x}, \alpha) = A(\mathbf{x})P(\mathbf{x}, \alpha)$. Energy resolution and merging and splitting of jets are taken into account in a "transfer" function, $W(\mathbf{y}, \mathbf{x})$, discussed below. The production probability density can be written as a convolution of the

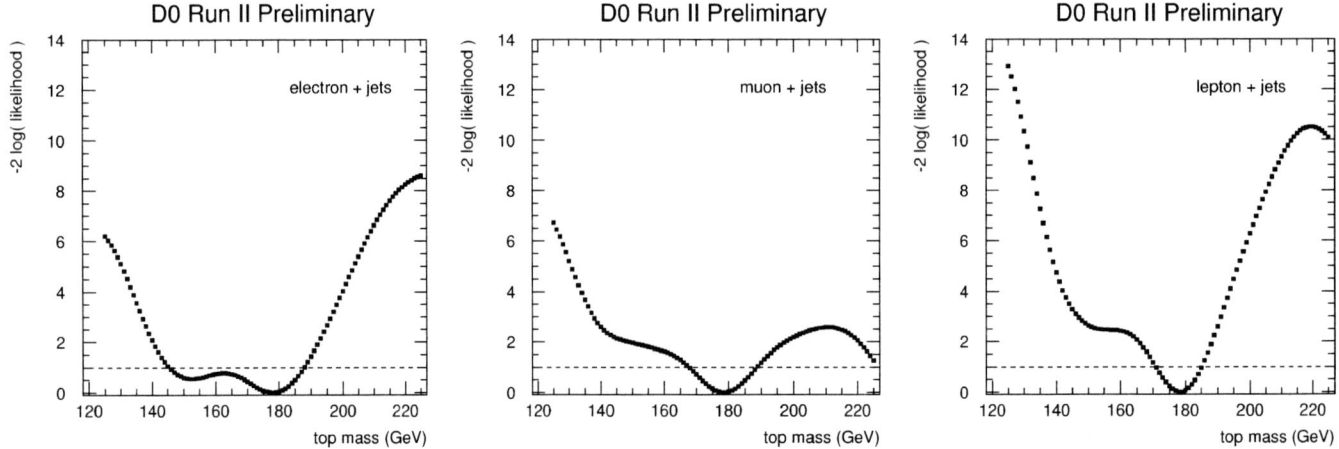

Fig. 106. Mass likelihood curves for the data events in the $e +$ jets channel (*left*), the $\mu +$ jets channel (*middle*), and both channels combined (*right*). These likelihood curves are shown before calibration

calculable cross section and $W(\mathbf{y}, \mathbf{x})$:

$$P(\mathbf{x}, \alpha) = \frac{1}{\sigma(\alpha)} \int d\sigma(\mathbf{y}, \alpha) \, dq_1 \, dq_2 f(q_1) f(q_2)$$
$$\times W(\mathbf{y}, \mathbf{x}), \qquad (225)$$

where $W(\mathbf{y}, \mathbf{x})$, the general transfer function, is the normalised probability density that the measured set of variables $\mathbf{x}$ arises from a set of partonic variables $\mathbf{y}$, $d\sigma(\mathbf{y}, \alpha)$ is the partonic differential cross section, and $f(q_i)$ are parton distribution functions for the incoming partons with longitudinal momentum q_i. Dividing by $\sigma(\alpha)$, the total cross section for the process, ensures that $P(\mathbf{x}, \alpha)$ is properly normalised. The integral in (225) sums over all possible parton states leading to what is observed in the detector.

For the $t\bar{t}$ production probability, the measured angles of the jets and of the charged leptons are assumed to be the angles of the partons in the final state. Given the detector resolutions, the electron energy is assumed to be exact, and the muon energy is described by its known resolution [469]. Evaluation of (225) for the $e +$ jets channel involves two incident parton energies (approximated by the quarks only, the $\approx 10\%$ contribution from gluon fusion is neglected), and six objects in the final state. The integration over the essentially fifteen sharp variables (three components of electron momentum, eight jet angles, and four equations of energy-momentum conservation), leave five integrals that must be performed to obtain the probability that an event represents $t\bar{t}$ production for some specified value of top quark mass M_t:

$$P_{t\bar{t}} = \frac{1}{12\sigma_{t\bar{t}}} \int d\rho_1 \, dm_1^2 \, dM_1^2 \, dm_2^2 \, dM_2^2$$
$$\times \sum_{\mathrm{perm}, \nu} |\mathcal{M}_{t\bar{t}}|^2 \frac{f(q_1) f(q_2)}{|q_1||q_2|} \Phi_6$$
$$\times W_{\mathrm{jets}}(E_{\mathrm{part}}, E_{\mathrm{jet}}). \qquad (226)$$

For $|\mathcal{M}_{t\bar{t}}|^2$, the leading-order matrix element [193, 194] is used, $f(q_1)$ and $f(q_2)$ are CTEQ 4M parton distribution functions for the incident quarks [470], Φ_6 is the phase-space factor for the six-object final state, and the sum is over all twelve permutations of the jets (the permutation of the jets from the W-boson decay is performed by symmetrising the matrix element), and the up-to-eight possible neutrino solutions. Conservation of transverse momentum is used to calculate the transverse momentum of the neutrino. $W_{\mathrm{jets}}(E_{\mathrm{part}}, W_{\mathrm{jet}})$ is the part of $W(y, x)$ that refers to the mapping between parton-level energies E_{part} and energies measured in the detector, E_{jet}. Four of the variables chosen for integration (m_1, M_1, m_2 and M_2), namely the masses of the W-bosons and of the top quarks in the event, are economical in computing time, because the value of $|\mathcal{M}_{t\bar{t}}|^2$ is essentially negligible except at the peaks of the four Breit–Wigner terms in the matrix element. ρ_1 is the energy of one of the quarks in the hadronic decay of one of the W-bosons. The narrow-width approximation is used to integrate over the top quark masses, and Gaussian adaptive quadrature [471] is used to perform the three remaining integrals. $W_{\mathrm{jets}}(E_{\mathrm{part}}, E_{\mathrm{jet}})$ is the product of four

functions $F(E_{\mathrm{part}}^i, E_{\mathrm{jet}}^i)$, one for each jet. The parameters used for b-quarks are different from those for the lighter quarks. For a final state with a muon, W_{jets} is expanded to include the muon momentum resolution, and an integration over the muon momentum is included in (226).

The $W + 4$-jets matrix element from VECBOS [472] is used in (225) to calculate the background probability, P_{bkg}. The integration is performed over the energy of the four partons leading to jets and the W-boson mass. The probability is summed over the twenty-four permutations and two neutrino solutions. The integration over parton energies is performed using Monte Carlo techniques, increasing the number of random points until the integral converges. Monte Carlo studies show that the 20% background from multijet events are represented satisfactorily by the jets in the $W +$ jets events.

After adding the probabilities for the non-interfering $t\bar{t}$ and $W + 4$-jets channels, the final likelihood as a function of M_t is written as:

$$-\ln \mathcal{L}(\alpha) = -\sum_{i=1}^{N} \ln \left[c_1 P_{t\bar{t}}(\mathbf{x_i}, \alpha) + c_2 P_{\mathrm{bkg}}(\mathbf{x_i}) \right]$$
$$+ N c_1 \int A(\mathbf{x}) P_{t\bar{t}}(\mathbf{x}, \alpha) \, d\mathbf{x}$$
$$+ N c_2 \int A(\mathbf{x}) P_{\mathrm{bkg}}(\mathbf{x}) \, d\mathbf{x}. \qquad (227)$$

The above integrals are calculated using Monte Carlo methods, for which the acceptance $A(\mathbf{x})$ is 1.0 or 0.0, depending on whether the event is accepted or rejected by the analysis criteria. The best value of α, representing the most likely M_t, and the parameters c_i are defined by minimising $-\ln \mathcal{L}(\alpha)$.

Studies of the performance of this method in ensembles of Monte Carlo events show a systematic shift in M_t that depends on the amount of background in the data sample. To minimise this bias, only events with a background probability $P_{\mathrm{bkg}} < 10^{-11}$ are considered, reducing the mass bias to 0.5 GeV/c^2. This event selection yields 22 events for the measurement of the top quark mass.

Figure 107 (left) shows the value of $-\ln \mathcal{L}(\alpha)$ as a function of M_t for the 22 events. $-\ln \mathcal{L}(\alpha)$ is minimised with respect to the parameters c_i at each mass point. Figure 107 (right) shows the likelihood normalised to its maximum value.

The systematic uncertainties are studied in ensembles of Monte Carlo events and summarised in Table 61. The by-far-dominant systematic uncertainty is the jet energy scale, followed by the modelling of the $t\bar{t}$ signal and the $W + 4$-jets background. The matrix element method makes explicit the assumption that the top quarks and the background events are produced via the Standard Model processes. The method is very sensitive to the top quark mass, but it is also very sensitive to the exact modelling of the Standard Model production mechanisms.

After applying the 0.5 GeV/c^2 correction, the new value of the top quark mass is measured to be:

$$m_{\mathrm{top}} = 180.1 \pm 3.6 \, (\mathrm{stat.}) \pm 3.9 \, (\mathrm{syst.}) \, \mathrm{GeV}/c^2. \qquad (228)$$

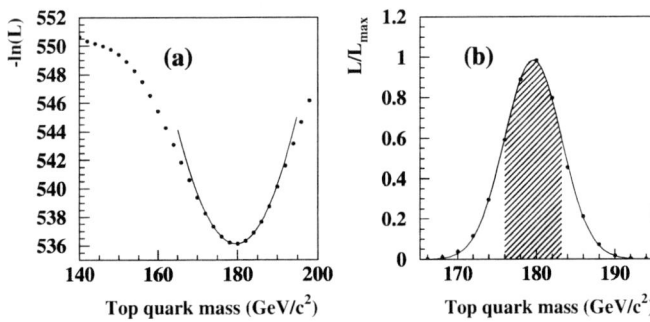

Fig. 107. *Left*: Negative of the log of the likelihood as a function of the top quark mass. *Right*: The likelihood normalised to its maximum value in the left plot. The *curve* is a Gaussian fit to the likelihood plot. The *hatched area* corresponds to the 68.27% probability interval

Table 61. Summary of the systematic uncertainties for the matrix element method applied to DØ Run-I data in the lepton + jets channel

Source of systematics	Δm_{top} (GeV/c^2)
Modelling of $t\bar{t}$	± 1.1
Modelling of $W + 4$-jets background	± 1.0
Noise and multiple interactions	± 1.3
Jet energy scale	± 3.3
Parton distribution function	± 0.1
Acceptance correction	± 0.3
Bias correction	± 0.5
Total	± 3.9

This re-analysis of the lepton + jets events from Run-I is consistent with the previous DØ Run-I measurement in the same channel at the ≈ 1.4 standard deviation level. It now yields a precision which is comparable to all previous Run-I measurements of DØ and CDF combined ($m_{\text{top}} = 174.3 \pm 5.1$ GeV, [266]). The improvement in the statistical uncertainty over the previous measurement is equivalent to a factor of 2.4 more data.

Using the procedure described in [473], the new measurement can be combined with the one obtained using the dilepton sample at DØ during Run-I [398], yielding the new Run-I DØ value for the mass of the top quark:

$$m_{\text{top}} = 179.0 \pm 3.5 \, (\text{stat.}) \pm 3.8 \, (\text{syst.}) \, \text{GeV}/c^2 \,. \tag{229}$$

This new method provides substantial improvement in both the statistical and systematic uncertainties. This is due to two main differences compared to the previous analysis: (i) each event now has an individual probability as a function of the mass parameter, and therefore well-measured events having a narrower likelihood contribute more to the extraction of the top quark mass than those that are poorly measured, and (ii) all possible jet and neutrino combinations are included, which guarantees that all signal events contribute to the measurement. This analysis

inspired the matrix element analyses and dynamic likelihood analyses by CDF and DØ in Run-II.

Matrix element method in Run-II

DØ has also measured the top quark mass in lepton + jets events in 320 pb^{-1} of Run-II data [177, 474, 475]. In Run-II, the method has been improved significantly, most notably by the addition of a second parameter to the likelihood function which addresses the jet energy scale as the dominant source of systematic uncertainty.

The event selection is essentially adopted from the corresponding topological/kinematic cross section analysis (Sect. 4.3.2), yielding 70 and 80 events in the e + jets and μ + jets channel, respectively. The matrix element likelihood is extended by an additional "JES" parameter, which is defined as a global scale factor for all jets in the event relative to the Monte Carlo reference scale. The signal probability $P_{\text{sgn}}(x; m_{\text{top}}, \text{JES})$ is sensitive to the jet energy scale parameter JES, because of the mass of the hadronically decaying W-boson is constrained in the $t\bar{t}$ matrix element. The total event probability is therefore given by the combination of signal and background probabilities according to:

$$P_{\text{evt}}(\mathbf{x}; m_{\text{top}}, \text{JES}) = f_{\text{top}} P_{\text{sgn}}(\mathbf{x}; m_{\text{top}}, \text{JES}) + (1 - f_{\text{top}}) P_{\text{bkg}}(\mathbf{x}; \text{JES}) \,, \tag{230}$$

where $\mathbf{x}$ denotes all kinematic variables of the reconstructed lepton and jets. The transverse momentum of the neutrino is obtained from the p_T imbalance of the five detected final state objects. f_{top} is the signal fraction in the sample under study. Technically, the integrations are now performed with the Monte Carlo integration algorithm VEGAS [476, 477], as provided by the GNU Scientific library [478]. In order to extract the top quark mass from a set of n events with measurements $x_1, \ldots, x_n$, a likelihood function is built from the event probabilities:

$$-\ln \mathcal{L}(\mathbf{x}_1, \ldots, \mathbf{x}_n; m_{\text{top}}, \text{JES}) = -\sum_{i=1}^{n} \ln P_{\text{evt}}(\mathbf{x}_i; m_{\text{top}}, \text{JES}) \,. \tag{231}$$

The top quark mass is determined by minimising $-\ln \mathcal{L}$ with respect to m_{top} and JES simultaneously, taking all correlations between both parameters into account. The signal fraction f_{top} is also fitted simultaneously.

The method is tested in large ensembles of Monte Carlo events with top quark pole masses of 160, 170, 175, 180 and 190 GeV/c^2. In addition, samples with $m_{\text{top}} = 175$ GeV/c^2 and all jets scaled by 0.92, 0.96, 1.04 and 1.08 are prepared in order to calibrate the JES fit. The resulting pull width for both, m_{top} and JES is found to be in good agreement with 1.0, indicating a reliable estimate of the statistical uncertainties by the likelihood procedure. Only small corrections of both parameters are applied to the result obtained in the data sample.

Figure 108 shows the 2-dimensional (m_{top}, JES) fit with σ contours (left) along with 1-dimensional projections of the negative log-likelihood onto the m_{top} axis (middle) and

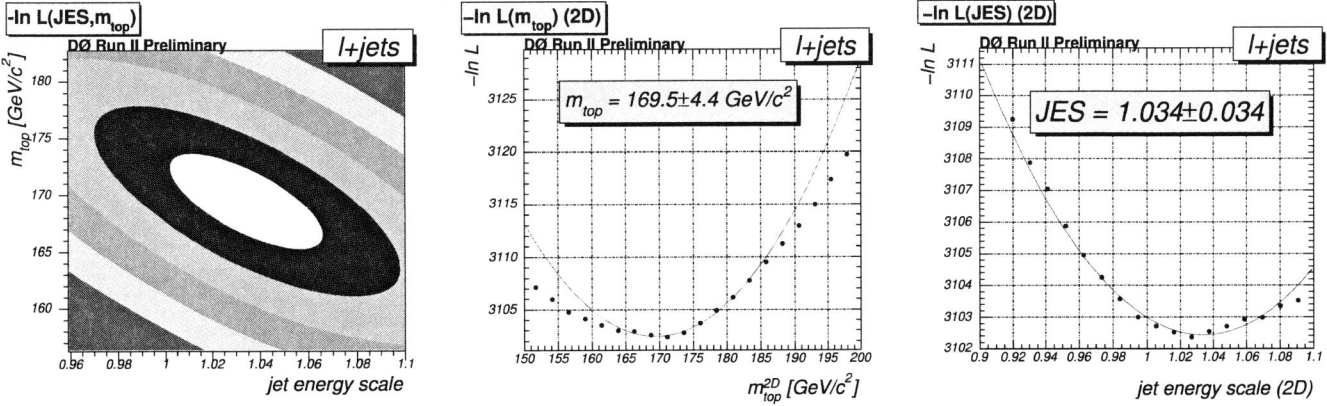

Fig. 108. *Left*: 2-dimensional $(m_{\mathrm{top}}, \mathrm{JES})$ fit with σ *contours*. The projection of the negative log-likelihood onto the m_{top} parameter is shown in the *middle*, taking correlations into account. The projection onto the JES axis is shown in the *right plot*, also taking correlations into account. The m_{top} and JES axes are corrected for the calibration result

Table 62. Summary of the systematic uncertainties for the matrix element method applied to DØ Run-II data in the lepton + jets channel

Source of systematics	Δm_{top} (GeV/c^2)
JES p_{T} dependence	± 0.70
b fragmentation	± 0.71
b response (h/e)	$^{+0.87}_{-0.75}$
Modelling of $t\bar{t}$	± 0.34
Modelling of $W + 4$-jets background	± 0.32
Signal fraction	$^{+0.50}_{-0.17}$
QCD contamination	± 0.67
MC calibration	± 0.38
Trigger	± 0.08
PDF uncertainty	± 0.07
Total	$^{+1.7}_{-1.6}$

onto the JES axis (right). The fit yields a signal fraction f_{top} of $0.316^{+0.049}_{-0.055}$ (stat.), in good agreement with the expectation from the corresponding cross section analysis. The fitted jet energy scale of 1.034 ± 0.034 indicates that the scale in the simulation is consistent with that in the data.

The jet energy scale uncertainty is already included in the error yielded by the likelihood. The other systematic uncertainties are determined in ensemble tests using Monte Carlo events and summarised in Table 62.

The application of the 2-dimensional matrix element method to DØ Run-II data yields a measurement of the top quark mass of:

$$m_{\mathrm{top}} = 169.5 \pm 4.4 \,(\mathrm{stat.} + \mathrm{JES})^{+1.7}_{-1.6} \,(\mathrm{syst.}) \,\mathrm{GeV}/c^2$$
$$= 169.5 \pm 3.0 \,(\mathrm{stat.}) \pm 3.2 \,(\mathrm{JES})^{+1.7}_{-1.6} \,(\mathrm{syst.}) \,\mathrm{GeV}/c^2\,. \tag{232}$$

Figure 109 shows the prospects for the expected development of the precision of the top quark mass measurement in the lepton + jets channel in DØ, using the *in situ* jet energy scale calibration from the $W \to q\bar{q}'$ invari-

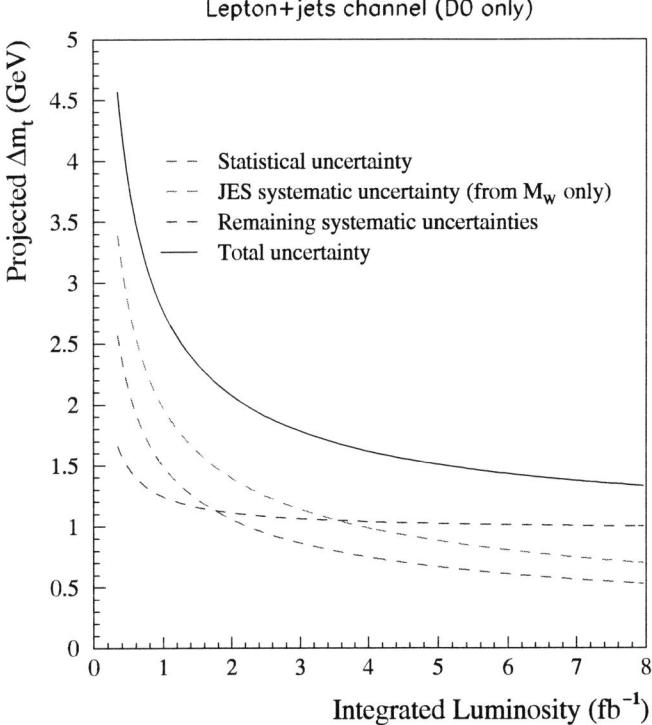

Fig. 109. Prospect for the precision on the top quark mass measurement in the lepton + jets channel in DØ

ant mass, as a function of the integrated luminosity. Already with an integrated luminosity of $2\,\mathrm{fb}^{-1}$, the statistical uncertainty will drop below $1\,\mathrm{GeV}/c^2$. Up to an integrated luminosity of 3–$4\,\mathrm{fb}^{-1}$, the jet energy scale uncertainty will be dominated by the *in situ* m_W calibration. Beyond this, the contributions from the external jet energy scale uncertainties, originating from Monte Carlo modelling of radiation and fragmentation effects and the difference between the b-jet and light quark jet energy scale, dominate. Improvements beyond those shown here might be possible in the future. For example, the b-jet energy scale might be included in the fit since a modification of this scale changes the helicity distribution of the W-boson decay products in the rest frame of the W-boson.

Dilepton channel. DØ has measured the top quark mass in the dilepton channel based on $230\,\mathrm{pb}^{-1}$, using the matrix element weighting technique, $\mathcal{MWT}$ [479]. This technique, which was one of the methods used for the DØ Run-I measurements of the top mass in the dilepton channel [370], solves the event kinematics for η of each neutrino requiring the sum of the neutrino $\mathbf{p}_\mathrm{T}$'s to equal the measured missing E_T vector. Then each event is given a weight according to the probability to observe the measured lepton p_T in $t\bar{t}$ events and the parton distributions functions. Since the event weight is calculated using the LO matrix element for the Standard Model $t\bar{t}$ production, this technique is called matrix element weighting method.

The event selection follows that of the corresponding cross section measurement (Sect. 4.3.1), yielding 8 events in the $e\mu$ channel, 5 events in the ee channel, and zero events in the $\mu\mu$ channel.

The basic idea for this method to reconstruct events from the decay of top-antitop quark pairs with two charged leptons (either electrons or muons) and two or more jets in the final state has originally been proposed by Dalitz and Goldstein [367, 399, 443]. Kondo has published similar ideas [400, 439–442]. Only the momenta of the two jets with the highest p_T are used in this analysis. These two jets are assigned to the b and $\bar{b}$ quark from the decay of the t and $\bar{t}$ quarks. Then, a likelihood is assigned to hypothesised values of the top quark mass between $80\,\mathrm{GeV}/c^2$ and $280\,\mathrm{GeV}/c^2$. For each event, the pairs of t and $\bar{t}$ momenta are found that are consistent with the observed lepton and jet momenta and the missing E_T vector. Such a consistent pair of top-antitop quark momenta is considered a solution. A weight is assigned to each solution, given by:

$$w = f(x)f(\bar{x})p\left(E_\ell^*|m_\mathrm{t}\right)p\left(E_{\bar{\ell}}^*|m_\mathrm{t}\right), \qquad (233)$$

where $f(x)$ is the parton distribution function for the initial quark to carry a momentum fraction x of the proton, and $f(\bar{x})$ is the corresponding value for the initial antiquark. The quantity $p(E_\ell^*|m_\mathrm{t})$ is the probability for the hypothesised top quark mass m_t that the lepton ℓ has the observed energy in the top rest frame [367, 399, 443].

There are two ways to assign the two jets to the b and $\bar{b}$ quarks. For each assignment of observed momenta to the final state particles, there may be up to four solutions for each hypothesised value of the top quark mass. The likelihood for each value of the top quark mass m_t is then given by the sum of the weights over all the possible solutions:

$$W_0(m_\mathrm{t}) = \sum_{\text{solutions}} \sum_{\text{jets}} w_{ij}. \qquad (234)$$

In this procedure the implicit assumption is made that all momenta are measured perfectly well. The weight $W_0(m_\mathrm{t})$ is therefore zero if no exact solution is found. However, the probability to observe a given event if the top mass has the value m_t does not have to be zero if no exact solution

is found, because of the finite resolution of the momentum measurements. This is accounted for by repeating the weight calculation with input values for the particle momenta that are drawn from normal distributions centred on the measured value with widths equal to the resolution of the momentum measurements. The $\not\!E_\mathrm{T}$ is corrected by the vector sum of the differences in the particle momenta from the measured values and the added random noise vector with x- and y-components drawn from a normal distribution with a mean of zero and an RMS of $8\,\mathrm{GeV}/c$. Finally, the weight curves for N such variations are averaged:

$$W(m_\mathrm{t}) = \frac{1}{N}\sum_{n=1}^{N} W_n(m_\mathrm{t}). \qquad (235)$$

Thus, effectively the weight $W(m_\mathrm{t})$ is integrated over the final state parton momenta, weighted by their experimental resolutions. The procedure is called resolution sampling. In $t\bar{t}$ Monte Carlo events, the resolution sampling reduces the number of events for which no solutions can be found from 10% to 1%.

For each event, the value of the hypothesised top quark mass at which $W(m_\mathrm{t})$ reaches its maximum is used as the estimator for the mass of the top quark. Using a large number of Monte Carlo events, the expected distributions of weight curve peaks are generated for top masses at 120, 140, 160, 175, 190, 210, and $230\,\mathrm{GeV}/c^2$ with the ALPGEN [224] plus PYTHIA [223] generators. The resulting distribution is called a template. Similarly, the background Monte Carlo events, as discussed in the dilepton cross section analyses, are used to construct a background template. The signal and background templates are added and normalised according to the expected signal-to-background ratio.

The peak mass distribution in data is compared to the Monte Carlo templates using a binned maximum likelihood fit. The likelihood is calculated as:

$$\mathcal{L}(m_\mathrm{t}) = \prod_{i=1}^{n_\text{bin}} \left[\frac{n_\mathrm{s}s_i(m_\mathrm{t}) + n_\mathrm{b}b_i}{n_\mathrm{s} + n_\mathrm{b}}\right]^{n_i}, \qquad (236)$$

where n_i is the number of data events observed in bin i, $s_i(m_\mathrm{t})$ is the normalised signal template content for bin i at top quark mass m_t, b_i is the normalised background template content for bin i. The product runs over all n_bins bins.

Figure 110 shows as an example the weight curves for the five data events in the ee channel with and without resolution sampling.

The method is tested in ensembles containing a large number of Monte Carlo events for the $e\mu$ and ee channels separately and combined. The results indicate that the calibration curve is perfectly consistent with unit slope and zero offset. The RMS of the pulls is close to one, so that the statistical uncertainties do not need any further rescaling. Systematic uncertainties are determined in ensemble testing using Monte Carlo events. A summary of the systematic uncertainties, including a comparison to the next-to-leading order event generator MC@NLO [215–217], is given in Table 63.

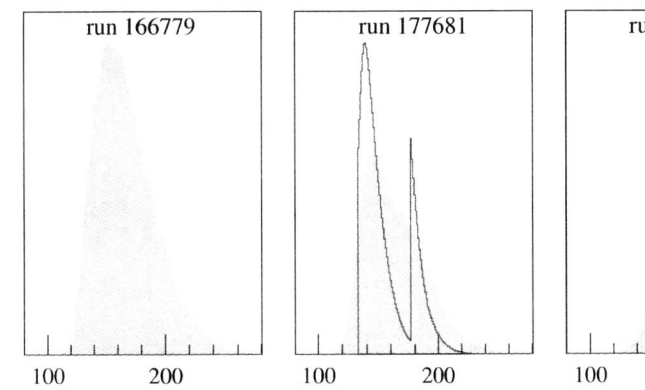
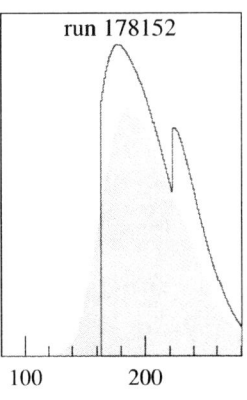
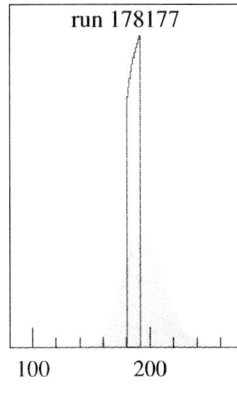
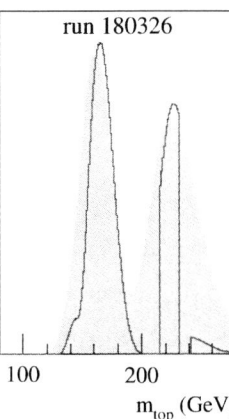

Fig. 110. Weight distributions from the five data events in the *ee* channel with (*solid/grey histograms*) and without resolution sampling (*open histograms*)

Table 63. Summary of the systematic uncertainties for the matrix element weighting method applied to DØ Run-II data in the dilepton channel

Source of systematics	Δm_{top} (GeV/c^2)
JES p_{T} dependence	± 5.6
Event generation (incl. MC@NLO)	± 3.0
PDF uncertainty	± 0.9
Underlying event simulation	± 1.0
Background	± 1.0
Calibration	± 1.1
Total	± 6.7

The application of the matrix element weighting method to DØ Run-II data yields a measurement of the top quark mass of:

$$m_{\text{top}} = 155^{+14}_{-13}\,(\text{stat.}) \pm 7\,(\text{syst.})\,\text{GeV/c}^2 . \quad (237)$$

All jets channel. The most recent DØ measurement of the top quark mass in the all-jets channel has been made in 110 pb^{-1} of Run-I data, recorded at $\sqrt{s} = 1.8$ TeV [480]. Since there are no neutrinos in this final state, the association of quarks and jets can be made unambiguously. Templates of the top quark mass, reconstructed in a kinematic fit, are compared using a binned maximum likelihood fit.

The event selection is very similar to that in the DØ all-jets cross section measurement in Run-II (Sect. 4.3.5). Events are required to contain at least six central and high-E_{T} jets. The signal-to-background ratio is improved by requiring that at least one jet has a soft-muon *b*-tag. The corresponding tagging efficiency in $t\bar{t}$ event is 15%–20%, the overall mistag rate on the dominant QCD multijet background only 2%.

Studies of $t\bar{t}$ Monte Carlo events, generated with HERWIG V5.7 [226], show that the mean of the invariant masses of two triplets of jets formed from the six highest-E_{T} jets, $M \equiv (\frac{m_{t_1}+m_{t_2}}{2})$, provide a satisfactory discriminant for distinguishing $t\bar{t}$ signal from background.

The two triplets are chosen to be those that minimise

$$\chi^2 = \left(\frac{m_{t_1} - m_{t_2}}{2\sigma_{m_t}}\right)^2 + \left(\frac{m_{W_1} + m_{W_0}}{\sigma_{m_W}}\right)^2$$
$$+ \left(\frac{m_{W_2} + m_{W_0}}{\sigma_{m_W}}\right)^2 , \quad (238)$$

where $M_{W_0} = 77.5$ GeV/c^2 is the mean value of the reconstructed W-boson mass in the all-jets $t\bar{t}$ Monte Carlo events, and m_{t_1}, m_{t_2} and M_{W_1}, M_{W_2}, are the calculated masses of the reconstructed jets that correspond to candidate top quarks and W-bosons, respectively, computed from the jet triplets and, within each triplet, the jet doublets. The standard deviations are determined in the $t\bar{t}$ Monte Carlo to be $\sigma_{m_t} \approx 31$ GeV/c^2 and $\sigma_{m_W} \approx 21$ GeV/c^2. Minimising the χ^2 provides the correct jet-quark combination in about 40% of the $t\bar{t}$ Monte Carlo events. The top quark mass is measured through the best fit of different admixtures of signal and background to the observed mass distribution. The posterior probability density $p(m_t, \sigma_{t\bar{t}}|\text{Data})$, is calculated for a set of mass values m_t. For each m_t value, the posterior probability density, numerically identical to the likelihood $\mathcal{L}$, is maximised by varying $\sigma_{t\bar{t}}$ to give the "maximal likelihood", $\mathcal{L}_{\max}(m_t)$ as a function of the hypothesised top quark mass, m_t. The "best fitted mass", m_{fit}, is taken to be the location of the minimum of the negative log-likelihood curve, $-\ln \mathcal{L}_{\max}(m_t)$.

The signal templates are generated using a Monte Carlo simulation of $t\bar{t}$ events for a discrete set of masses in the range 110 to 310 GeV/c^2 in 10 GeV/c^2 steps. The background is modelled using *untagged* events. These multijet events are weighted according to the tag-rate functions to tag each of their jets as a function of the jet η and E_{T}.

Since the jets in $t\bar{t}$ events tend to be more energetic, have a more isotropic momentum flow, and have larger transverse energies than those in light-quark jets, the event sample is enriched further by event discrimination based on a suitable set of kinematic variables. In this analysis, the following eight variables are used: $E_{\text{T5}} \times E_{\text{T6}}$, $|\eta_{W_1} \times \eta_{W_2}|$, $\sqrt{\hat{s}}$, $\mathcal{A}$, $\mathcal{S}$, $N_{\text{jet}}^{E_{\text{T}}}$, $H_{\text{T3}}/H_{\text{T}}$, and H_{T}/H, where E_{T1}

to E_{T6} and E_1 to E_6 are the transverse energies and energies, respectively, of the six jets, ordered in decreasing E_T; η_{W_1} and η_{W_2} are the pseudorapidities of the two hypothesised W-bosons; $\sqrt{\hat{s}}$ is the invariant mass of the N_{jets} system; $\mathcal{A}$ is the aplanarity and $\mathcal{S}$ the sphericity; $N_{\text{jet}}^{E_T}$ is the number of jets above a given E_T-threshold, over the range 10 GeV to 55 GeV, weighted by the threshold; $H_T = \sum_j E_{Tj}$; $H_{T3} = H_T - E_{T1} - E_{T2}$; and $H = \sum_j E_j$, where the sums are over all $\Delta R = 0.5$ cones jets with $|\eta| < 2.5$ and $E_T > 10$ GeV. The above variables are combined into a single discriminant, calculated using a neural network (NN) with eight inputs, a single hidden layer with three nodes, and a single output D_{NN}. The network is trained and tested on independent HERWIG $t\bar{t}$ Monte Carlo samples and untagged events for the background.

A cutoff $D_{\text{NN}} > 0.97$, optimised on Monte Carlo events to minimise the RMS of the fitted top mass distribution, yields 65 events in the final sample. Figure 111 shows a comparison between the observed mass distribution in the data and the sum of background and 175 GeV/c^2 top quark signal scaled to the observed number of top events. The fitting procedure, tested in Monte Carlo, is corrected for a small mass bias of 2.6 GeV/c^2, using the relation $m_{\text{fit}} = 0.712 m_{\text{t}} + 53.477$ GeV/c^2. The systematic uncertainties are studied in ensembles of Monte Carlo events.

This analysis of the DØ Run-I all-jets data yields a measured top quark mass of :

$$m_{\text{top}} = 178.5 \pm 13.7\,(\text{stat.}) \pm 7.7\,(\text{syst.})\ \text{GeV}/c^2 . \tag{239}$$

The corresponding $t\bar{t}$ production cross section in this fit is estimated to be $\sigma_{t\bar{t}} = 11 \pm 5$ pb, which is consistent with the DØ measurement from the dedicated cross section analysis ($\sigma_{t\bar{t}} = 5.6 \pm 1.4\,(\text{stat.}) \pm 1.2\,(\text{syst.})$ pb).

7.1.3 Combination of measurements

The TEVATRON Electroweak Working Group (TeVEWWG), responsible for the combined CDF/DØ average top mass, takes account of correlations between systematic uncertainties in the different measurements. They assume statistical uncertainties and fit uncertainties to be uncorrelated between all measurements, full correlation of uncertainties from uranium noise (for DØ only) and multiple interactions within each experiment, full correlation of the background uncertainty within each channel, and full correlation of signal and Monte Carlo uncertainty among all measurements of CDF and DØ. The dominant uncertainty from the jet energy scale is separated into various components derived from internal calibrations using the $W \to qq'$ mass scale, and from external calibration using calorimeter-track comparisons for isolated tracks in CDF and p_T balancing in photon+jet events in DØ. Jet energy scale uncertainties from the internal calibration are treated as uncorrelated between all measurements, the external calibration is split up further in contributions correlated among all measurements and in contributions correlated within each of the two experiments. The combinations are performed using a program implementing a numerical χ^2 minimisation as well as the

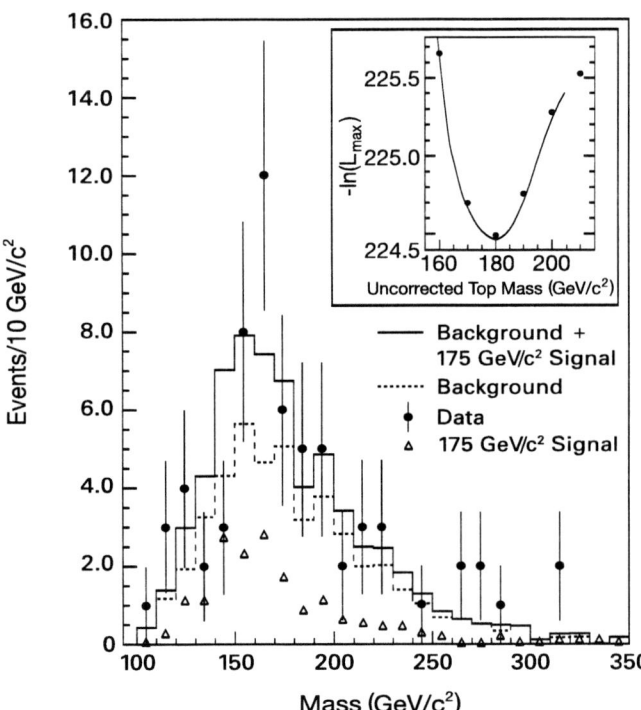

Fig. 111. Data and the sum of background and Monte Carlo signal as a function of the mean mass, M. The *insert* shows the $-\ln \mathcal{L}_{\text{max}}$ as a function of the top quark mass

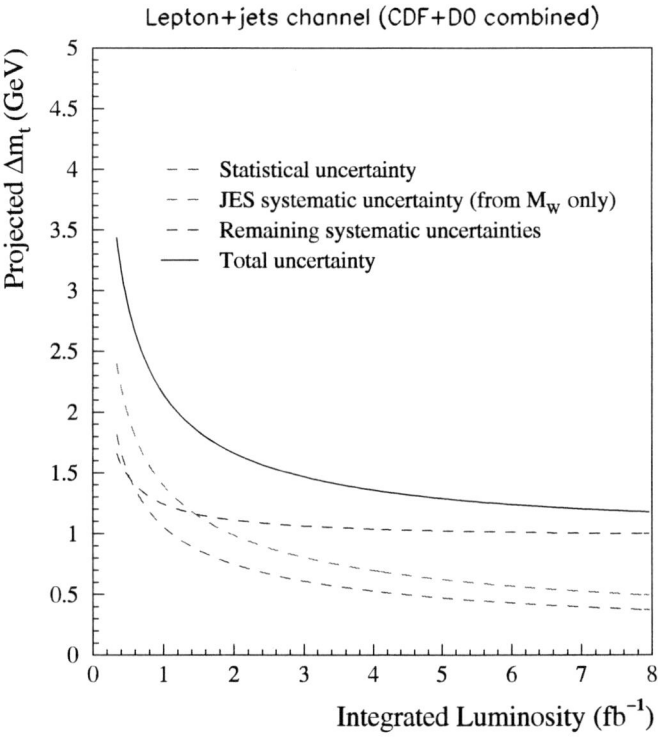

Fig. 112. Prospect for the precision on the top quark mass measurement in the lepton + jets channel in CDF + DØ

analytic BLUE method [315, 316, 481]. The two methods used are mathematically equivalent, and are also equivalent to the method used in an older combination [473], and give identical results for the combination. In addition, the BLUE method yields the decomposition of the error on the average in terms of error categories specified for the input measurements.

The combination of the five published Run-I measurements [396, 398, 446, 464, 482] yields a world-average on the top quark mass of [180]

$$m_{\rm top} = 178.0 \pm 2.7\,({\rm stat.}) \pm 3.3\,({\rm syst.})\,{\rm GeV}/c^2$$
$$= 178.0 \pm 4.3\,({\rm stat.} + {\rm syst.})\,{\rm GeV}/c^2\,. \qquad (240)$$

The TEVATRON Run-I measurements of the top quark mass yield a total precision of 2.4%.

Including, in addition, the most recent preliminary Run II measurements from CDF [176, 459] and DØ [177]

Table 64. Measurements of the top quark mass from DØ and CDF and their average

$m_{\rm t}$ (GeV/c^2)	Source	$\int \mathcal{L}\,{\rm dt}$	Ref.	Method
$173.3 \pm 5.6 \pm 5.5$	DØ Run I	125	[465, 484]	$\ell +$ jets, TM
$180.1 \pm 3.6 \pm 3.9$	DØ Run I	125	[396, 397] $\star$	$\ell +$ jets, ME
$168.4 \pm 12.3 \pm 3.6$	DØ Run I	125	[398] $\star$	$\ell\ell, \eta(\nu)/\mathcal{MWT}$
$178.5 \pm 13.7 \pm 7.7$	DØ Run I	110	[480]	all jets
$179.0 \pm 3.5 \pm 3.8$	DØ Run I	110-125	[396, 397]	DØ comb.
$170.0 \pm 6.5^{+10.5}_{-6.1}$	DØ Run II	160	[466] †	$\ell +$ jets/topo, TM
$169.9 \pm 5.8^{+7.8}_{-7.1}$	DØ Run II	230	[467] †	$\ell +$ jets/topo, TM (update)
$170.6 \pm 4.2 \pm 6.0$	DØ Run II	230	[467] †	$\ell +$ jets/b-tag, TM
$177.5 \pm 5.8 \pm 7.1$	DØ Run II	160	[466] †	$\ell +$ jets/topo, Ideogram
$169.5 \pm 3.0 \pm 3.6$	DØ Run II	320	[177] †$\star$	$\ell +$ jets/topo, ME with $W \to jj$
$155^{+14}_{-13} \pm 7$	DØ Run II	230	[479] †	$\ell\ell, \mathcal{MWT}$
$176.1 \pm 5.1 \pm 5.3$	CDF Run I	110	[446, 482, 485] $\star$	$\ell +$ jets
$167.4 \pm 10.3 \pm 4.8$	CDF Run I	110	[446] $\star$	$\ell\ell$
$186.0 \pm 10.0 \pm 5.7$	CDF Run I	110	[464] $\star$	all jets
176.1 ± 6.6	CDF Run I	110	[446]	CDFcomb.
$174.9^{+7.1}_{-7.7} \pm 6.5$	CDF Run II	162	[449] †	$\ell +$ jets, TM
$173.5^{+2.7}_{-2.6} \pm 3.0$	CDF Run II	318	[176] †$\star$	$\ell +$ jets, TM with $W \to jj$
$173.0^{+2.9}_{-2.8} \pm 3.3$	CDF Run II	318	[451] †	$\ell +$ jets, TM+Jet Prob.
$177.8^{+4.5}_{-5.0} \pm 6.2$	CDF Run II	162	[453] †	$\ell +$ jets, DLM
$173.2^{+2.6}_{-2.4} \pm 3.2$	CDF Run II	318	[452] †	$\ell +$ jets, DLM (update)
$172.0 \pm 2.6 \pm 3.3$	CDF Run II	318	[454] †	$\ell +$ jets, ME
$179.6^{+6.4}_{-6.3} \pm 6.8$	CDF Run II	162	[455] †	$\ell +$ jets, Multivar.
$207.8^{+27.6}_{-22.3} \pm 6.5$	CDF Run II	318	[457] †	$\ell +$ jets, Decay Length
$165.3 \pm 6.3 \pm 3.6$	CDF Run II	340	[459] †$\star$	$\ell\ell$, ME
$168.1^{+11.0}_{-9.8} \pm 8.6$	CDF Run II	197	[460] †	$\ell\ell, \nu(\eta)$
$170.6^{+7.1}_{-6.6} \pm 4.4$	CDF Run II	359	[461] †	$\ell\ell, \nu(\eta)$ (update)
$170.0 \pm 16.6 \pm 7.4$	CDF Run II	193	[462] †	$\ell\ell, \nu(\phi)$
$169.8^{+9.2}_{-9.3} \pm 3.8$	CDF Run II	340	[447] †	$\ell\ell, \nu(\phi)$ (update)
$176.5^{+17.2}_{-16.0} \pm 6.9$	CDF Run II	193	[463] †	$\ell\ell, p_z(t\bar{t})$
$170.2^{+7.8}_{-7.3} \pm 3.8$	CDF Run II	340	[448] †	$\ell\ell, p_z(t\bar{t})$ (update)
178.0 ± 4.3	CDF & DØ	110-125	[180] †	Run-I combination
172.7 ± 2.9 *	CDF & DØ	110-340	[46] †	world-average (2005)
171.4 ± 2.1 □	CDF & DØ	110-1030	[181] ‡	world-average (2006)

* World average for the Particle Data Group Review October 2005. It is a combination of Run I and Run II measurements (labelled with $\star$), yielding a χ^2 of 6.45 for 7 degrees of freedom.
□ World average for the Particle Data Group Review April 2006. It is a combination of Run I and Run II measurements, yielding a χ^2 of 8.1 for 8 degrees of freedom.
† Preliminary result, not yet submitted for publication as of October 2005.
‡ Preliminary result, not yet submitted for publication as of April 2006.

yields the new world-average on the top quark mass of [46, 483]

$$m_{\text{top}} = 172.7 \pm 1.7\,(\text{stat.}) \pm 2.4\,(\text{syst.})\,\text{GeV}/c^2$$
$$= 172.7 \pm 2.9\,(\text{stat.} + \text{syst.})\,\text{GeV}/c^2\,. \qquad (241)$$

The TEVATRON measurements of the top quark mass yield a total precision of 1.7%.

A recent update of the combined top quark mass measurement by the TEVATRON Electroweak/Top Working group yields $m_{\text{t}} = 171.4 \pm 1.2\,(\text{stat.}) \pm 1.8\,(\text{syst.})\,\text{GeV}/c^2 = 171.4 \pm 2.1\,(\text{stat.} + \text{syst.})\,\text{GeV}/c^2$ [181], i.e. a measurement with 1.2% precision. Since this update arrived after the ed-

itorial deadline for this review no further details could be included.

Figure 112 shows the prospects for the expected development of the precision of the top quark mass measurement in the lepton + jets channel in CDF+DØ, using the *in situ* jet energy scale calibration from the $W \to qq'$ invariant mass, as a function of the integrated luminosity. Already with an integrated luminosity of slightly more than $1\,\text{fb}^{-1}$, the statistical uncertainty will drop below $1\,\text{GeV}/c^2$. Up to an integrated luminosity of $1.4\,\text{fb}^{-1}$, the jet energy scale uncertainty will be dominated by the *in situ* m_W calibration. Beyond this, the contributions from the external jet energy scale uncertainties, originating from Monte Carlo modelling of radiation and fragmentation ef-

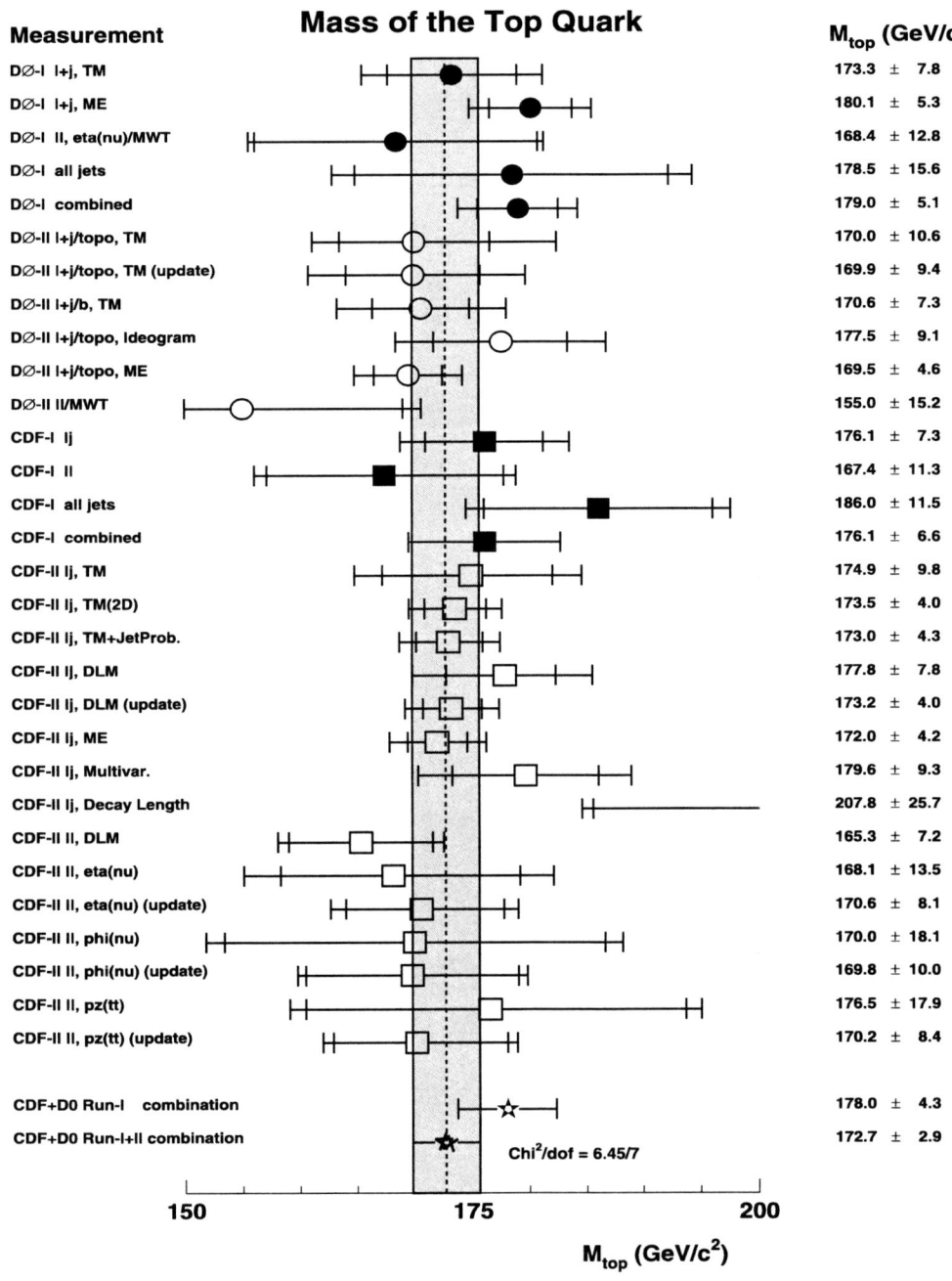

Fig. 113. Summary of the top-quark mass measurements by DØ (○) and CDF (□) in Run-I or Run-II. The *open symbols* indicate preliminary results, the *filled symbols* are published measurements. The top mass combinations [46, 180] by the TEVATRON Electroweak Working Group are shown as *stars*

fects and the difference between the b-jet and light quark jet energy scale, dominate. Depending on the total integrated luminosity recorded per experiment at the end of the TEVATRON Run-II, the total uncertainty of the top quark mass measurement from the lepton + jets channel alone could be as low as $1.3\,\mathrm{GeV/c^2}$. In combination with all the Run-I measurements and with the Run-II measurements in the dilepton and the all-jets channel, a total precision of $\Delta m_{\mathrm{top}} \approx 1\,\mathrm{GeV/c^2}$ appears to be achievable. This, however, also assumes the presently observed b-tagging performance and jet energy resolution. Both are expected to be degraded with increasing integrated and instantaneous luminosity.

The different measurements of the top quark mass, described in this chapter, are summarised in Table 64 and Fig. 113. Given the experimental technique used to extract the top mass, these mass values should be taken as representing the top *pole mass*. The top pole mass, like any quark mass, is defined up to an intrinsic ambiguity of order $\Lambda_{\mathrm{QCD}} \sim 200\,\mathrm{MeV}$ [178].

High energy physicists around the world have started planning for a future e^+e^- linear collider, which may become operational in the next decade. Such a machine will offer new means for precision studies of the top quark properties and dynamics. For example, the top quark mass could be measured with a precision of $\approx 20\,\mathrm{MeV/c^2}$ from a threshold scan [172, 173, 185]. A detailed discussion of the prospects for top physics at such a machine, however, is beyond the scope of this review.

7.2 Electric charge of the top quark

It is widely believed that the heavy particle discovered by the CDF and DØ Collaborations at the TEVATRON collider in 1995 [42, 43, 437] is the long-sought top quark. The currently measured properties are still only poorly known. In particular, its electric charge, one of the most fundamental quantities characterising a particle, has not been measured yet. In fact, the top quark is the only quark whose electric charge has not been measured. In general, quark charges are easily accessible in e^+e^- production by scanning the centre-of-mass energy up to above two times the quark mass and measuring the ratio $R = \frac{\mathrm{Rate}(e^+e^- \to hadrons)}{\mathrm{Rate}(e^+e^- \to \mu^+\mu^-)}$. The observed step, ΔR is related to the square of the quark produced at this energy threshold. However, the maximum centre-of-mass energy available at e^+e^- was $\sqrt{s} = 209\,\mathrm{GeV}$ at LEP 2, which was too low to produce $t\bar{t}$ pairs. Therefore, it still remains not only to be confirmed that the discovered quark has charge $2/3e$ and hence the Standard Model quantum numbers, but also to measure the strength of its electromagnetic (EM) coupling to rule out anomalous contributions to its EM interactions.

Furthermore, it is possible to interpret the discovered particle as either a charge $2/3e$ or $-4/3e$ quark. In the top quark analysis of the CDF and DØ Collaboration so far, the correlations of the b-quarks and the W-bosons in $p\bar{p} \to t\bar{t} \to W^+W^-b\bar{b}$ are not determined. As a result, there is a twofold ambiguity in the pairing of

W-bosons and b-quarks, and, consequently, in the electric charge assignment of the "top quark". In addition to the Standard Model assignment, $t \to W^+b$, also the decay $"t" \to W^-b$ is certainly conceivable, in which case the top quark would actually be an exotic quark with charge $q = -4/3e$. Current $Z \to \ell^+\ell^-$ and $Z \to b\bar{b}$ data can be fitted with a top quark of mass $m_t = 270\,\mathrm{GeV/c^2}$, provided that the right-handed b-quark mixes with the isospin $= 1/2$ component of an exotic doublet of charge $-1/3e$ and $-4/3e$ quarks, $(Q_1, Q_4)_R$ [300, 486, 487]. In this scenario, the $-4/3e$ charge quark is the particle discovered at the TEVATRON, and the top quark, with mass of $270\,\mathrm{GeV/c^2}$, would have so far escaped detection.

DØ has measured the top quark electric charge using $\approx 366\,\mathrm{pb^{-1}}$ of lepton + jets events with two or more identified b-jets and a jet-charge algorithm to discriminate between b- and $\bar{b}$-jets [488]. The experimental procedure to rule out one of the hypotheses comprises three steps.

The first step is to select a pure sample of $t\bar{t}$ events in data in the lepton + jets channel. So each selected $t\bar{t}$ events has two "legs", one with a leptonically decaying $W (t \to Wb \to \ell\nu b)$ and one with a hadronically decaying $W (t \to Wb \to q\bar{q}'b)$.

The second step of the analysis consists of assigning the correct jets and leptons to the correct "leg" of the event, uniquely specifying which b-jet comes from the same top (or anti-top) quark as the lepton. To make this assignment, a constrained fit is performed. In each $t\bar{t}$ event the observable $|Q|$ is computed, which is the sum of the lepton charge from the W-boson decay and the charge of the b-jet associated to the same leg as the lepton by the kinematic fit. The charge of the b-jet is computed using a jet charge algorithm calibrated from data. The goal of this analysis is to discriminate between the Standard Model hypothesis $Q_{\mathrm{top}} = +2/3e$ and the exotic hypothesis $Q_{\text{"top"}} = -4/3e$. Each event has a top and an antitop quark. Hence, assuming that charge is conserved, only the cases $|Q_{\text{"top"}}| = 4/3e$ and $|Q_{\mathrm{top}}| = 2/3e$ need to be considered without any loss of information. Consequently, the top quark charge can be measured twice in each event.

The third step is to use the shape of the jet charge for b-jets in data to derive the expected shape of $|Q|$ for the Standard Model and the exotic scenario. The top quark charge shapes are mixed with charge distributions expected for the small background contribution to the sample. The distribution of $|Q|$ is then compared with the data and a likelihood method is used to discriminate between the scenarios.

The lepton + jets events are selected as in the corresponding $t\bar{t}$ cross section measurement (Sect. 4.3.3). By requiring events with at least two secondary vertex (SVT) b-tagged jets, the $t\bar{t}$ to background ratio is significantly enhanced. The $Wb\bar{b}$ background is expected to represent only $\approx 5\%$ of the sample. The second largest source of background is single top production, which is expected to contribute around 1% of the selected events.

Discrimination between b- and $\bar{b}$-jets is achieved by using tracks of charged particles inside the SVT-tagged jets. The track momenta and charge are measured with the DØ central tracking system. All tracks within a cone of $\Delta R = 0.5$ from the SVT-tagged jet axis are used. The tracks must have $p_T > 0.5$ GeV and be within 0.1 cm of the primary vertex in the z-direction along the beam axis. The jet charge q_{jet} is defined as the p_T weighted average of the track momenta

$$q_{\mathrm{jet}} = \frac{\sum_i q_i p_{T_i}^{0.6}}{\sum_i p_{T_i}^{0.6}}, \qquad (242)$$

where the subscript i runs over the selected tracks. The exponent 0.6 and the cone size are the result of an optimisation using fully simulated Monte Carlo $t\bar{t}$ events. The discriminating power for b- versus $\bar{b}$-jets and the jet charge distributions are directly derived from $b\bar{b}$ data. The jet charge distributions are then applied to the $t\bar{t}$ Monte Carlo, in order to predict the expected distribution of the top quark charge. Important differences, in particular in the η and p_T distribution, between b-jets in $b\bar{b}$ data and b-jets in $t\bar{t}$ events are taken into account and result in the dominant systematic uncertainty for this measurement. The shape of the jet charge distribution for b- and $\bar{b}$-jets is obtained from a $b\bar{b}$ data sample, where both jets are SVT-tagged and one jet contains a muon in addition. The former jet is the 'probe jet', the latter one the 'tag jet'. The jet charge distribution of the 'probe jet' is corrected for the fraction of c-jets in the sample ($\approx 6 \pm 2\%$) using Monte Carlo. Similarly, a correction for the muons in the $b\bar{b}$ sample to come from a cascade decay of a b-meson rather than from its direct decay or for possible b-meson oscillation is determined from Monte Carlo. The resulting b- and $\bar{b}$-jets charge distributions, normalised to unity, can be interpreted as the probability density functions to measure a certain jet charge Q given the type of quark (b or $\bar{b}$) initiating the jet. These probability density functions are denoted $f_b(Q)$ and $f_{\bar{b}}(Q)$. Systematic uncertainties are assigned to the rate of swapped muon signs due to cascade decays, b-mixing, and charge misidentification of the measured tagging muon.

In order to measure the top quark charge and to determine a confidence level for the measurement, an observable and an expectation for the observable in the case of $|Q_{\mathrm{top}}| = 2/3e$ (Standard Model – SM scenario) and in the case $|Q_{\text{"top"}}| = 4/3e$ (exotic scenario) is needed. The expected shapes of jet charge for b-jet and $\bar{b}$-jet in data are used here to form distributions of the reconstructed charges for the SM top and exotic charges. The top quark charge is measured twice per event. One top quark charge is constructed as the sum of the charge of the lepton (e or μ) and the jet charge of the b-jet from the same top quark. The second top quark charge is constructed as the sum of the second b-jet charge *minus* the charge of the charged lepton. The two observables in each event are defined as:

$$Q_1 = |q_\ell + q_b|,$$
$$Q_2 = |-q_\ell + q_B|, \qquad (243)$$

where q_ℓ is the charge of the charged lepton, q_b is the charge of the b-jet on the leptonic leg of the event (j_b) and q_B is the charge of the b-jet on the hadronic leg of the event (j_B). The top quark charge observables Q_1 and Q_2 are produced using a constrained kinematic fit for the $t\bar{t}$ hypothesis with the top mass fixed at 175 GeV/c^2 to determine the association of the two b-jets to the W-bosons. The kinematic fit considers all jets in the event. The expected shapes of the top quark charge distributions Q_1 and Q_2 are derived from $t\bar{t}$ Monte Carlo using jet-parton matching for the flavour tagging. If the true flavour of j_b (j_B) is b then the jet charge q_b (q_B) is set to a randomly chosen value according to the probability density function $f_b(Q)$. If it is a $\bar{b}$ then the function $f_{\bar{b}}(Q)$ is used. The resulting distributions of the observables $Q_1 = |q_\ell + q_b|$ and $Q_2 = |-q_\ell + q_b|$ provide the Standard Model top charge templates. The shape of the exotic top quark charge templates are determined by exchanging the jet charge of the SVT-tagged jets on the leptonic and hadronic side of the event, yielding two observables $Q_1 = |q_\ell + q_B|$ and $Q_2 = |-q_\ell + q_b|$. The resulting exotic and Standard Model templates, normalised to unity, are used as probability density functions p^{ex} and p^{sm}, respectively, in the confidence level calculation. In that, systematic uncertainties from the jet energy calibration, the jet energy resolution, the jet reconstruction efficiency, the uncertainty in the background composition, and the uncertainty in the top quark mass are taken into account.

In 17 out of 21 selected double tag events, the kinematic fit converges, yielding a total of 34 measurements of the top quark charge. Figure 114 (left) shows the measured values in data overlaid with the Standard Model and exotic top quark charge distributions. Using these values, the ratio of the likelihood of the observation assuming the Standard Model case divided by the likelihood of the observation as-

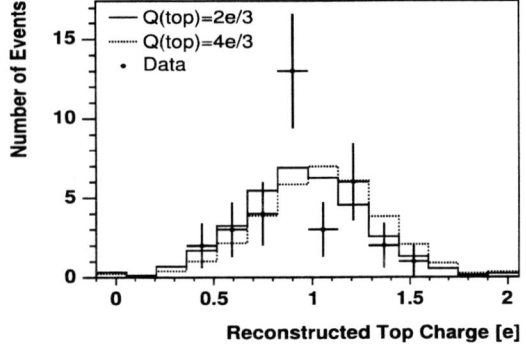

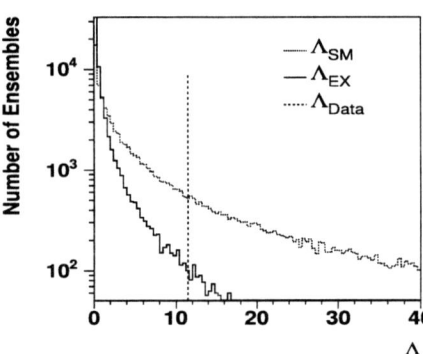

Fig. 114. *Left*: The 34 measured values of the top quark charge compared to the Standard Model and the exotic scenario templates. *Right*: The distribution of the likelihood ratio Λ^{sm} and Λ^{ex} for 100 000 ensembles. The measured value of Λ^{data} is shown as straight *(dashed)* line

suming the exotic scenario is formed:

$$\Lambda = \frac{\prod_i p^{\mathrm{sm}}(q_i)}{\prod_i p^{\mathrm{ex}}(q_i)}, \qquad (244)$$

where $p^{\mathrm{sm}}(q_i)$ is the probability to observe the top quark charge q_i in the Standard Model scenario, and $p^{\mathrm{ex}}(q_i)$ is the probability to observe the top quark charge q_i in the exotic scenario. The subscript i runs over all 34 available measurements of the top quark charge. The data yields: $\Lambda^{\mathrm{data}} = 11.5$.

The value of Λ^{data} is compared to the distribution of expected Λ^{sm} and Λ^{ex} obtained by generating pseudo-experiments to emulate the Standard Model and exotic cases, respectively. For both scenarios, the signal and background fractions are allowed to fluctuate according to their errors, where the systematic uncertainties are incorporated using the nuisance parameter method [332]. 100 000 pseudo-experiments with 34 pseudo-observations each yield the expected distributions of Λ^{ex} and Λ^{sm}, which are shown in Fig. 114 (right) together with the value observed in data. As can be seen, Λ^{sm} is more likely if the data contains Standard Model top quarks rather than exotic top quarks. The probability for the exotic case to give $\Lambda > \Lambda^{\mathrm{data}}$ is only 6.3%, giving an exclusion of the $4/3e$ scenario at the 93.7% CL. The expected confidence level for the exotic scenario to fluctuate above the median of the Standard Model value is 89.0%. The consistency of the observation with the Standard Model, computed as the probability for the Standard Model to give an outcome $\Lambda^{\mathrm{sm}} > \Lambda^{\mathrm{data}}$ is 34.0%. Therefore the observation is in excellent agreement with the Standard Model expectation for the top quark charge.

In summary, the scenario that the top quark charge is $4/3e$ is ruled out at the 94% confidence level. With more data being available for analysis at the TEVATRON, more stringent tests of the top quark charge scenarios ($4/3e$ and $2/3e$) will be possible in the near future.

8 Anomalous top quark production

8.1 Limits on $\ell\ell/\ell + $jets cross section ratio

The measurements of the $t\bar{t}$ production cross section in the different top quark decays channels, described in Sects. 4.2 and 4.3, clearly demonstrate that the top quark is produced and experimentally observed. However, it is a priori not obvious, that the 'top quark', observed in the dilepton decay mode is identical to the 'top quark' in the lepton + jets decay mode. If both decay modes result exclusively from the decay of the Standard Model top quark, they should have the same production cross section. If the production or the decay of the top quarks had non-Standard Model contributions, as discussed in Sect. 4.1, one mode might be enhanced with respect to the other.

Based on the $t\bar{t}$ cross section measurements, using $\sim 125\,\mathrm{pb}^{-1}$ of Run II data, CDF measures the cross section ratio $R_\sigma = \sigma_{\ell\ell}/\sigma_{\ell j}$ [489]. In taking the ratio, the small

event statistics are handled carefully, and, most importantly, the systematic uncertainties that are correlated between the two analyses cancel. In addition, the cross-section measurements assume that the top quark decays with Standard Model branching fractions, e.g. there is virtually always a W in the decay, which decays with its usual branching fractions. If this is not true, then the measured value for R_σ would not be consistent with unity. Thus, one can use R_σ to extract limits on non-standard branching fractions of the top quark. These limits are by construction model-dependent.

Here, a simple model is used which does not have any particular physics motivation, but should give a reasonable estimate of the efficiency to detect the non-standard decay. This allows to test if the top quark decays into something else than $t \to Wb$. For example, in 2-Higgs doublet models, the top quark could decay to a charged Higgs boson $t \to H^\pm b$ with $H^\pm \to cs$ or $H^\pm \to \tau\nu$. In general, the considered cases are a fully hadronic decay $t \to Xb$, where $B(X \to qq') = 100\%$ or a fully leptonic decay, i.e. $t \to Yb$, where $B(Y \to \ell\ell') = 100\%$. The decay branching ratios of the top quark in those decays are labelled $\beta = B(t \to Xb)$ and $\beta' = B(t \to Yb)$.

Figure 115 shows the probability distribution function for the cross section ratio $R_\sigma = \sigma_{\ell\ell}/\sigma_{\ell j}$, as observed in the data. Taking into account the correlation of systematic uncertainties, the cross section ratio R_σ is found to be:

$$R_\sigma = 1.45^{+0.83}_{-0.55}, \qquad (245)$$
$$R_\sigma > 0.46 \quad \text{at 95\% CL}, \qquad (246)$$
$$R_\sigma < 4.45 \quad \text{at 95\% CL}. \qquad (247)$$

Therefore, limits on the fully hadronic or the fully leptonic decay of the top quark are set at:

$$B(t \to Xb) = \beta < 0.46 \quad \text{at 95\% CL}, \qquad (248)$$
$$B(t \to Yb) = \beta' < 0.47 \quad \text{at 95\% CL}. \qquad (249)$$

With larger statistics, these results are expected to improve significantly. Figure 116 shows the expected development of the lower limit on the top quark decay branching

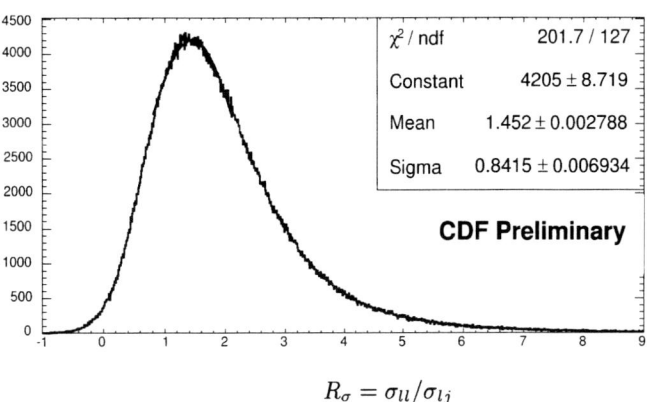

$$R_\sigma = \sigma_{ll}/\sigma_{lj}$$

Fig. 115. Probability distribution $\mathrm{d}N/\mathrm{d}R_\sigma$ function for the ratio $R_\sigma = \sigma_{\ell\ell}/\sigma_{\ell j}$, as observed in $\sim 125\,\mathrm{pb}^{-1}$ of CDF data

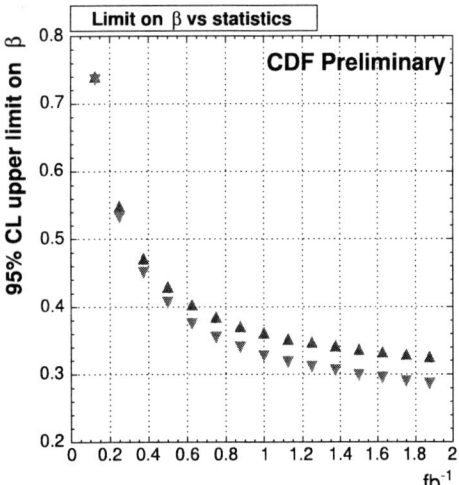 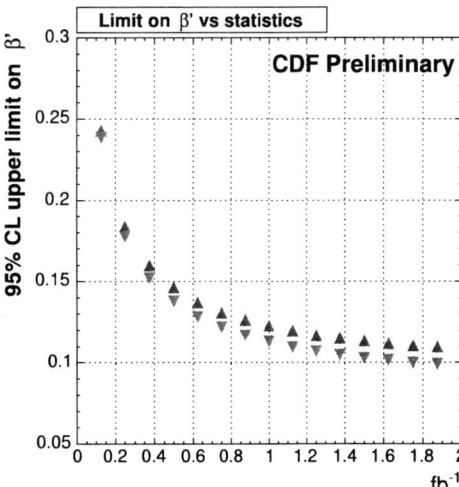

Fig. 116. Development of the expected lower limits on β (branching fraction to an all-hadronic decaying top) and β' (branching fraction to an all-leptonic decaying top) as a function of the integrated luminosity. The *red points* (▲) show the result without cancellation of correlated systematic uncertainties, the *green points* (▼) include such cancellations

fraction β to an all-hadronic final state (left) and β' to an all-leptonic final state (right). Taking correlations of systematic uncertainties into account, improves the result by relative $\sim 10\%$. With $1\,\mathrm{fb}^{-1}$ of data, CDF is expected to have a sensitivity to β values down to 35% and to β' values as low as 12%.

8.2 Studies of $t\bar{t}$ kinematics

The measurements of the $t\bar{t}$ production cross section in the dilepton channel by CDF and DØ Collaborations in Run-I [490, 491] showed a slight excess over Standard Model predictions. Several of the events observed in the Run-I data had missing transverse energy ($\not{E}_T$) and lepton p_T's large enough to call into question their compatibility with Standard Model top decay kinematics. It was suggested that the kinematics of these events could be better described by the cascade decays of heavy squarks [492].

CDF has searched for anomalous kinematics in $t\bar{t} \to$ dilepton events in $200\,\mathrm{pb}^{-1}$ of Run-II data [320]. In this study, a detailed analysis of the kinematics of the Run-II dilepton sample, which consists of 13 events (Sect. 4.2.1, [302]) is performed, searching for new physics based on the comparison of kinematic features of observed events with those expected from the Standard Model. An *a priori* choice of which kinematic quantities to investigate is made, in order to be sensitive to a wide range of new physics: (i) the event's $\not{E}_T$, (ii) the transverse momentum of the leading lepton p_T^ℓ, and (iii) the angle $\phi_{\ell m}$ between the leading lepton and the direction of the $\not{E}_T$ in the plane transverse to the beam, (iv) the fourth variable, T, is introduced which represents how well the kinematics of an event satisfy the $t\bar{t}$ dilepton decay hypothesis; a non-$t\bar{t}$ dilepton event has on average a small value of T compared to $t\bar{t}$ events. Imposing the mass constraints $m(\ell_1\nu_1) = m(\ell_2\nu_2) = m_W$ and $m(\ell_1\nu_1 b_1) = m(\ell_2\nu_2 b_2) = 175\,\mathrm{GeV/c}^2$ leaves two of the six neutrino momentum components unspecified when solving the kinematics of the system. A scan is performed over these two remaining degrees of freedom and the resulting summed transverse momentum of the neutrinos $\not{E}_T^{\mathrm{pred}}$ is compared to the measured one $\not{E}_T^{\mathrm{obs}}$ by computing

$$\mathcal{T}(\not{E}_T^{\mathrm{pred}}) = \exp\left\{ -\left|\not{E}_T^{\mathrm{pred}} - \not{E}_T^{\mathrm{obs}}\right|^2 / 2\sigma_{\not{E}_T}^2 \right\},\tag{250}$$

where $\sigma_{\not{E}_T}$ parameterises the uncertainty on $\not{E}_T$ due to the mismeasurement of the underlying event. The variable T is now defined as the square root of the integral $\mathcal{T}$ over the possible values of $\not{E}_T^{\mathrm{pred}}$, determined from the scan and summed over the two-fold ambiguity in the lepton-b-jet pairing.

This search is concentrated on events with large values of $\not{E}_T$, p_T^ℓ, and $\phi_{\ell m}$ and small values of T, as that is the region where new physics is expected. Therefore the following weight is assigned to each event:

$$W = \left(w_{\not{E}_T} w_{p_T^\ell} w_{\phi_{\ell m}} w_T \right)^{1/4},\tag{251}$$

where $w_{\not{E}_T}$, $w_{p_T^\ell}$, $w_{\phi_{\ell m}}$, and w_T represent probabilities (assuming the Standard Model) for an event to have a $\not{E}_T$, p_T^ℓ, $\phi_{\ell m}$ larger than that observed and a T smaller than that observed, respectively. Then, 13 subsets of the data are constructed ("K-subsets"); the first subset ($K = 1$) contains only the event with the lowest weight W, the second subset ($K = 2$) contains only the two events with the lowest weight, and so on.

To quantify the departure of the K-subsets from the Standard Model predictions, a shape comparison is made using the Kolmogorov–Smirnov statistics. For each of the four variables i, the KS deviation $\Delta_{K,i}$ between the Standard Model cumulative function and the cumulative function of the K-subset is computed. The probability of this deviation is assessed using a large number of $t\bar{t}$ Monte Carlo pseudo-experiments, where the number of events from the Standard Model processes are Poisson fluctuated. Only pseudo-experiments with a total of 13 events are accepted. In each pseudo-experiment, K-subsets are formed and the respective $\Delta_{K,i}$ for each subset are calculated. This way, probability distributions for $\Delta_{K,i}$ are built from which the KS probability $p_{K,i}$ can be computed. Next, the geometric mean Π_K of the four $p_{K,i}$'s is calculated for each pseudo-experiment and the probability distribution functions $\mathcal{F}_K$ are formed such that the quantity

$$P_K = \int_0^{\Pi_K^{\text{obs}}} \mathcal{F}_K(\Pi)\,\mathrm{d}\Pi \qquad (252)$$

determined how well each K-subset agrees with the Standard Model expectation based on the combined information from the four variables. Q is defined to be the value of the subset K with the smallest P_K. By isolating this "unlikely" subset Q, the dilution of a possible signal from the presence of Standard Model events is minimised.

The quantity P_Q is used as the test statistic to quantify the discrepancy of the data with the Standard Model. Calculating P_Q for a large number of pseudo-experiments, the probability distribution function $\mathcal{L}(P_Q)$ is built such that the significance of the departure of the Q-subset from Standard Model events is

$$\alpha = \int_0^{P_Q^{\text{data}}} \mathcal{L}(P_Q)\,\mathrm{d}P_Q . \qquad (253)$$

α is the p-value of the test, representing the probability to obtain a data sample less consistent with the Standard Model than what is actually observed. Sufficiently low values of α would indicate the presence of new physics in the data sample, and the Q events would represent the subsample of the data with the largest concentration of new physics.

The method is tested on Monte Carlo events with a $50\% : 50\%$ mixture of Standard Model $t\bar{t}$ events and SUSY events, a $W + \geq 3$ jet sample and a $W + 4$-jet sample. It was found to give stable results and a better performance than a simple Kolmogorov–Smirnov test.

The distributions of the four selected variables in the dilepton data set are shown in Fig. 117. The outlined technique is applied to this data sample. The most unlikely subset of events is found to be the entire data set (i.e. $Q = 13$), with a p-value of 1.6%. This result is entirely driven by the excess of leptons at low $p_T (< 40\,\text{GeV}/c)$ seen in Fig. 117b. Six of the nine low-p_T events contain at least one identified b-jet and more than half of the low-p_T events are consistent with the $t\bar{t}$ kinematic hypothesis with large values of T. It is therefore concluded that new physics scenarios invoked to describe the high-p_T^ℓ/high-$\not{E}_T$ events, observed in Run-I, are not favoured by the Run-II data. Including the effect of systematic uncertainties results in p-values ranging from 1.0% to 4.5%.

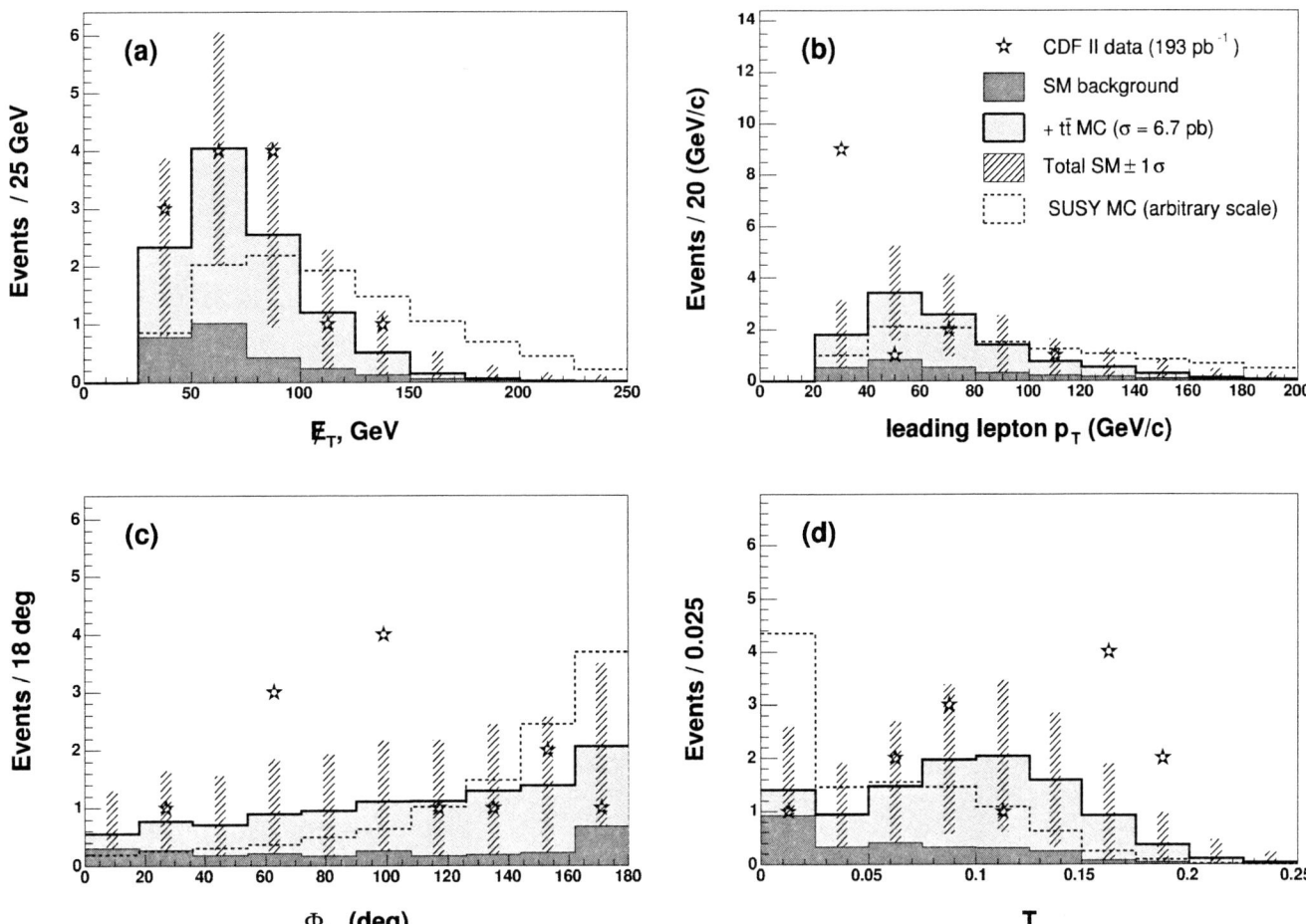

Fig. 117. leading lepton p_T, $\phi_{\ell,m}$ and T distributions for the top dilepton sample. The *hatched regions* represent the Poisson uncertainty on the expectation in a given bin. The *dashed histograms* are the expected distributions from a SUSY Monte Carlo

8.3 Measurement of the top quark p_T spectrum

The existence of the top quark is by now clearly established, as demonstrated by the top quark pair production cross section measurements in Sect. 4 and the top quark mass measurements in Sect. 7. In the Standard Model, the dominant top quark production mechanism at the Tevatron is quark–antiquark annihilation. However, a number of theoretical investigations [137, 438, 493, 494] have concluded that alternative production mechanisms may play an important role in top production at the Tevatron. In particular, many exotic models predict sizeable enhancements in the cross section for the production of top quarks having transverse momentum $p_T > 200\,\mathrm{GeV/c}$.

Using $106\,\mathrm{pb}^{-1}$ of Run-I data, CDF has measured the true p_T distribution of pair-produced top quarks [495]. A previous study by DØ [465] compares the top quark p_T distribution with the Standard Model prediction. The CDF analysis selects events in the lepton + jets channel (lepton = e or μ) according to the standard Run-I selection criteria. In order to increase the signal significance, either the lowest E_T jet is required to satisfy tight jet cuts (in terms of E_T and η) or that at least one jet be associated with a b-quark decay from a secondary vertex tag. The events are reconstructed by a kinematic fit similar to that used in the top quark mass analysis. As opposed to using this fit to measure the top quark mass, here the mass is constrained to $175\,\mathrm{GeV/c}^2$, a value close to the world average. Requiring events to have a $\chi^2 > 10$ from this three-constraint kinematic fit leaves 61 events in the data sample. Simulations using the HERWIG event generator show that there is only a weak correlation between the measured and the true top quark p_T in events for which the incorrect jet-parton assignment is made. There is a strong correlation between the measured p_T's for the top and antitop quarks in a given event. Therefore, the measurement is performed using only the fully reconstructed hadronic top quark decay candidates. The background contribution from W + jets events and QCD multijet events is estimated using a combination of Monte Carlo and data, as done in the corresponding cross section measurement [293, 490, 496].

The distribution of measured top quark p_T for the 61 selected events is shown in Fig. 118. The Kolmogorov–Smirnov probability to observe a difference between the Standard Model prediction and the reconstructed p_T distribution as large as the one measured, is calculated to be 5%, varying between 1% and 9.4% when the systematic uncertainties are taken into account. To correct for the p_T bias due to the reconstruction and resolution effects, an unsmearing procedure appropriate for small data samples is used. This procedure extracts the fraction of top quarks that are produced in each of four p_T bins of width $75\,\mathrm{GeV/c}$, spanning the range between 0 and $300\,\mathrm{GeV/c}$. An unbinned likelihood fit to the measured p_T distribution is performed, using a superposition of the response functions, determined from Monte Carlo, and the background template. The logarithm of the likelihood function to be

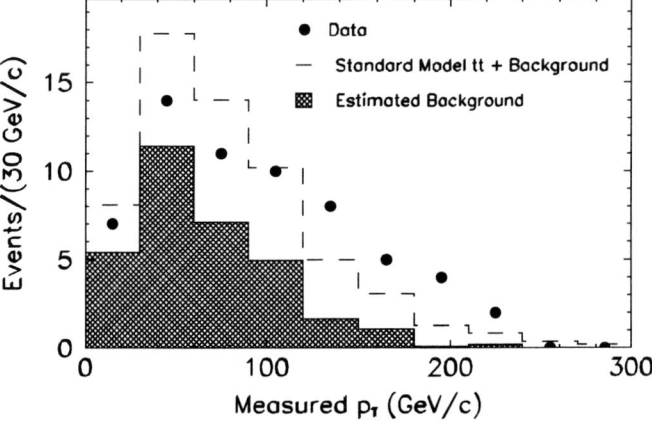

Fig. 118. The measured p_T distribution for the hadronically decaying top quarks in the 610 event sample. The *hatched distribution* is the estimated background distribution, normalised to the estimated number of background events. The *dashed distribution* is the Standard Model prediction, normalised to the observed number of candidate events

maximised is:

$$
\ln \mathcal{L}(B) = \\
\sum_{i=1}^{n_{\mathrm{data}}} \left(\ln \left[\sum_{j=1}^{n_{\mathrm{bin}}} \left[(1-B) R_j T_j \left(p_T^i \right) \right] + B V(p_T^i) \right] \right) \\
- \frac{(B - \mu_b)^2}{2\sigma^2(\mu_n)} . \tag{254}
$$

In this equation, R_j is the fitted fraction of top quarks produced in true bin j, while the $T_j(p_T)$ are the response functions for the $t\bar{t}$ signal and $V(p_T)$ is the background template. The fit parameter B is the fitted background fraction and $\mu_b \pm \sigma(\mu_b)$ is the estimated background fraction. The data is separated into two "tagging subsamples", one of which consists of the subset of events with one or more b-tags, the other consisting of those events with no b-tag. For both samples the respective, appropriate transfer functions are used. The response functions $T_j(p_T)$ depend on the form of the true p_T distribution within each p_T bin. Thus, an iterative technique is employed that interpolates the true p_T distribution across a given bin based upon the current R_i parameter values. The resulting R_i fit values are corrected for the fact that the $t\bar{t}$ acceptance is a function of the top quark p_T. Systematic uncertainties considered include the top quark mass, effects of the simulation of initial and final state radiation, the jet energy calibration, the background modelling, and the shape of the p_T spectrum within in each bin.

The resulting values for the four R_i are compared to the Standard Model prediction in Table 65. Also shown is the result for $R_1 + R_2$, the fraction of top quarks that are produced with $p_T < 150\,\mathrm{GeV/c}$ (due to a strong negative correlation between the fitted values of R_2 and R_2, the fractional uncertainty in this result is much smaller than it is for the individual estimates for R_1 and R_2). By combining the statistical and systematic uncertainties, using a convolution of the likelihood function for R_2 with a Gaussian

Table 65. Measurement of the top quark p_T distribution. The Standard Model (SM) expectation is generated using the HERWIG Monte Carlo program

p_T Bin	Parameter	Measurement	SM Expectation
$0 \leq p_T < 75\,\text{GeV}/c$	R_1	$0.21\,^{+0.22}_{-0.21}\,(\text{stat.})^{+0.10}_{-0.08}\,(\text{syst.})$	0.41
$75 \leq p_T < 150\,\text{GeV}/c$	R_2	$0.45\,^{+0.23}_{-0.23}\,(\text{stat.})^{+0.04}_{-0.07}\,(\text{syst.})$	0.43
$150 \leq p_T < 225\,\text{GeV}/c$	R_3	$0.34\,^{+0.14}_{-0.12}\,(\text{stat.})^{+0.07}_{-0.05}\,(\text{syst.})$	0.13
$225 \leq p_T < 300\,\text{GeV}/c$	R_4	$0.000\,^{+0.031}_{-0.000}\,(\text{stat.})^{+0.024}_{-0.000}\,(\text{syst.})$	0.025
$0 \leq p_T < 150\,\text{GeV}/c$	$R_1 + R_2$	$0.66\,^{+0.17}_{-0.17}\,(\text{stat.})^{+0.07}_{-0.07}\,(\text{syst.})$	0.84

distribution, G, representing the systematic uncertainties, yields an upper limit on R_4:

$$R_4 < 0.16 \quad \text{at } 95\% \text{ CL}. \quad (255)$$

In this analysis, CDF also searches for top quark production with true $p_T > 300\,\text{GeV}/c$ by modifying the final response function to incorporate a possible high-p_T component and subsequently recalculating the upper limit. Since the largest limit is obtained by assuming no high-p_T component, the above upper limit is extended into a conservative upper limit on the fraction of top quarks produced with p_T in the range 225–$425\,\text{GeV}/c$. Above this p_T value, the relative acceptance for top quarks begins to fall, reducing to 50% of the acceptance at $225\,\text{GeV}/c$ for top quarks produced with $p_T = 500\,\text{GeV}/c$.

At present, there is no equivalent analysis of the top quark p_T spectrum from the higher-statistics Run-II data. When such analyses are available, it will be possible to test details of the top quark production mechanism and the top quark kinematics with much improved precision.

8.4 Limits on $t\bar{t}$ resonance production and $t\bar{t}$ mass spectrum

Narrow resonances decaying to $t\bar{t}$ pairs are predicted in the Standard Model as a possible decay mode of the Higgs boson ($H \to t\bar{t}$) [139] or by several theories beyond the Standard Model. For instance, in the top-colour-assisted technicolour model [497] which combines top-colour [137, 138] and technicolour [498–501] models. The technicolour interactions at the electroweak scale are responsible for electroweak symmetry breaking, and extended technicolour generates the masses of all quarks and leptons except that of the top quark. The large top quark mass is accounted for by predicting the existence of a residual global symmetry $SU(3) \times U(1)$ at energies below 1 TeV, creating a new strong gauge force. The strong top-colour interactions, broken near 1 TeV, induce a massive dynamical $t\bar{t}$ condensate X and all but a few GeV of the top quark mass, and contribute little to electroweak symmetry breaking. The $SU(3)$ results in the generation of top gluons, which have been searched for by CDF in Run-I in the $b\bar{b}$ channel [502]. The $U(1)$ gives the Z'-boson. The $t\bar{t}$ condensate X, or the heavy Z' boson, couples preferentially to the third generation. In one of the scenarios of the top-colour-assisted technicolour model, the heavy Z'

boson couples weakly and symmetrically to the first and second generations and strongly to the third generation of quarks, and has no coupling to the leptons (leptophobic). The cross section for the Z' boson in this model is large enough for it to be observed over a wide range of masses and widths in data available at the TEVATRON. Both, CDF and DØ perform model-independent searches for a narrow-width resonance X decaying into a $t\bar{t}$ pair in the lepton + jets channel. This search is performed by examining the reconstructed $t\bar{t}$ invariant mass distribution resulting from a constrained fit to the $t\bar{t}$ hypothesis. In Run-I, CDF and DØ performed such searches at $\sqrt{s} = 1.8$ TeV, finding no evidence for a $t\bar{t}$ resonance. The resulting limits on $\sigma_X \times B(X \to t\bar{t})$, where σ_X is the resonance production cross section, are used to exclude a leptophobic Z' boson with width $\Gamma_{Z'} = 0.012 M_{Z'}$. The resulting Z' mass limits from CDF and DØ at 95% CL are $M_{Z'} > 480\,\text{GeV}/c^2$ [503] and $M_{Z'} > 560\,\text{GeV}/c^2$ [504], respectively.

In Run-II, DØ has searched for $t\bar{t}$ production via an intermediate, narrow-width, heavy resonance in the lepton + jets channel, using lifetime tagging in $370\,\text{pb}^{-1}$ of data [505]. The lepton + jets events with four or more jets are selected as described in Sect. 4.3.2, where at least one jet is required to be identified as a b-jet using the secondary vertex tagger, as described in Sect. 4.3.3. Furthermore, the constrained kinematic fit, described later in this section, must converge for the selected events.

Monte Carlo samples corresponding to resonant $t\bar{t}$ production are generated with PYTHIA 6.202 [223], using CTEQ 5L [94] as the set of parton distribution functions, for ten different choices of the resonant mass M_X (GeV/c^2): 350, 400, 450, 500, 550, 600, 650, 750, 850, and 1,000. In all cases, the width of the resonance is set to $\Gamma_X = 0.012 M_X$. This qualified X as a narrow resonance since its width is smaller than the expected mass resolution of the DØ detector (about $0.4 M_X$ in Run-I). Therefore the obtained upper limits on $\sigma_X \times B$ are valid for all choices of Γ_X that are reasonably small compared to the detector resolutions. The generated resonance is forced to decay into $t\bar{t}$ and only the events with one W-bosons decaying leptonically (including $W \to \tau\nu$) and the other W boson decaying hadronically are selected for further processing through the detector simulation and reconstruction chain.

The $t\bar{t}$ invariant mass is reconstructed using a constrained kinematic fit similar to the one used for the measurement of the top quark mass (see Sect. 7.1.2). The lepton and jet resolutions used in the fit have been updated to reflect those of the Run-II DØ detector. The following constraints are used in the fit:

- the two jets must form the invariant mass of the W-boson ($M_W = 80.4\,\mathrm{GeV/c^2}$),
- the lepton and the $\not{E}_T$, taking into account the longitudinal neutrino momentum, must form the invariant mass of the W-boson,
- the masses of the two reconstructed top quarks have to be equal, and are set to $175\,\mathrm{GeV/c^2}$.

Only the four highest p_T jets are considered in the kinematic fit. From the resulting twelve possible jet-parton assignments, the one with the lowest χ^2 is chosen. This is found to give the correct solution in about 65% of the $t\bar{t}$ events. In this version of the analysis, the b-tagging information is not used to reduce the number of possible permutations.

Background estimates are obtained in the same way as described for the corresponding cross section measurement in Sect. 4.3.3. After b-tagging, only $\sim 4\%$ of the $W +$ jets but $\sim 60\%$ of $t\bar{t}$ events remain, which makes Standard Model $t\bar{t}$ production become the dominant background in this analysis. After applying the tagging rate measured in data (tag rate functions), the expected $t\bar{t}$ yield is normalised to the theoretical Standard Model prediction for the $t\bar{t}$ production cross section: $\sigma_t = 6.77 \pm 0.42$ pb for $m_t = 175$ GeV [114]. The $W +$ jets background is estimated from a combination of data (tagging rates) and Monte Carlo information (jet flavour, p_T, and η). The shape of the reconstructed $t\bar{t}$ invariant mass distribution is obtained from the Monte Carlo simulation. The multijet background is completely determined from data. The total number of expected events is estimated by applying the matrix method on the tagged sample. This method allows to determine the total normalisation for this background source but, due to limited available statistics, not the shape of the reconstructed $t\bar{t}$ invariant mass distribution, which is derived from a larger sample of events with the lepton failing

the strict isolation requirements and without requiring b-tagging. The kinematic biases resulting from the b-tagging requirement are mimicked by folding per-jet tag rate functions measured in data.

This analysis relies on the prediction of the overall normalisation as well as the shape of the reconstructed $t\bar{t}$ invariant mass distribution for both signal and the different backgrounds. The systematic uncertainties can be classified as those affecting only normalisation and those affecting both normalisation and shape of the $t\bar{t}$ invariant mass distribution for one or more processes (signal or background). The systematic uncertainties affecting only the normalisation include e.g. the experimental uncertainties on the Monte Carlo-to-data correction factors, the theoretical uncertainty on the Standard Model prediction for $\sigma_{t\bar{t}}$ (6%) and $\sigma_{\mathrm{singletop}}$ (12%) and the uncertainty on the integrated luminosity (6.5%). The systematics affecting the shape of the $t\bar{t}$ invariant mass distribution in addition to the normalisation have been studied both on signal and background samples. These include e.g. uncertainties on the jet energy calibration, jet reconstruction efficiency, b-tagging parameterisations for b, c and light jets, and the present precision with which the top quark mass is known. The relative systematic uncertainties on the overall normalisation of the Standard Model background amount to $\approx \pm 14\%$, and the shape uncertainties of the $t\bar{t}$ invariant mass distributions are taken into account in addition.

After selection cuts, 57 events remain in the $e +$ jets channel and 51 events in the $\mu +$ jets channel. Figure 119 (left) shows the $t\bar{t}$ invariant mass distribution for the combined $\ell +$ jets channels for the selected events in data and for the Standard Model background prediction. Assessment of the probability for known sources to reproduce the data is still being worked on. Assuming there is no resonance signal, a Bayesian approach is used to calculate 95% CL upper limits on $\sigma_X \times B(X \to t\bar{t})$ for each hypothesised M_X. A Poisson distribution is assumed for the number of observed events in each bin,

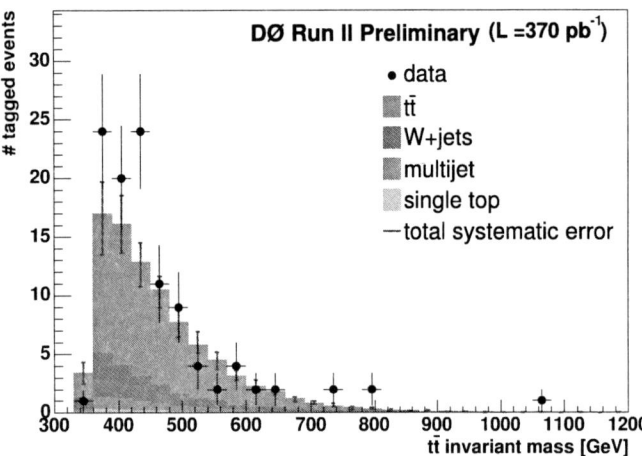
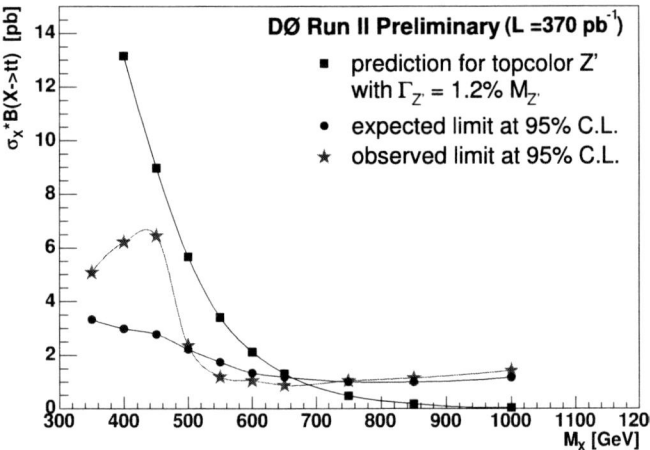

Fig. 119. *Left*: The resulting $t\bar{t}$ invariant mass distribution of the combined $\ell +$ jets channels. The *error bars* drawn on top of the Standard Model background indicate the total systematic uncertainty, which has significant bin-to-bin correlations. *Right*: Expected and observed 95% CL upper limit on $\sigma_X \times B(X \to t\bar{t})$ compared to the predicted topcolour-assisted technicolour cross section from a Z' boson with a width of $\Gamma_{Z'} = 0.012 M_{Z'}$ as a function of the resonance mass M_X

as well as flat prior probabilities for the signal cross section. Systematic uncertainties on the signal acceptance and background yields are implemented via a convolution procedure of a multivariate Gaussian distribution implementing a full covariance matrix including correlations. The expected and observed 95% CL upper limits on $\sigma_X \times B(X \to t\bar{t})$ as a function of M_X are displayed in Fig. 119 (right). This figure also includes the predicted $\sigma_X \times B(X \to t\bar{t})$ for a leptophobic Z' boson. By analysing the reconstructed $t\bar{t}$ invariant mass distribution and using a Bayeseian method, model-independent upper limits on $\sigma_X \times B(X \to t\bar{t})$ are obtained for different hypothesised masses of a narrow-width heavy resonance decaying into $t\bar{t}$. Within a topcolour-assisted technicolour model [497], the existence of a leptophobic Z' boson with $M_{Z'} < 680\,\mathrm{GeV}$ and width $\Gamma_{Z'} = 0.012 M_{Z'}$ is excluded at 95% CL.

9 Anomalous top quark decays

9.1 Top quark decays to charged Higgs

One of the open questions in the Standard Model of particle physics is the mechanism of electroweak symmetry breaking. Within the Standard Model, it is speculated that a single scalar field doublet breaks the symmetry, generating gauge boson masses for the $W^\pm$ and Z, while leaving the photon massless, and resulting in a single observable particle, the Higgs boson [20–22]. To date, current searches have resulted in the exclusion of the Standard Model Higgs boson with masses up to $114.4\,\mathrm{GeV/c^2}$ [47]. The simplest extension of the Standard Model Higgs sector is built by the introduction of another Higgs doublet, resulting in a 2 Higgs doublet model (2HDM) [297]. In these models, electroweak symmetry breaking results in five physical Higgs bosons, three of which are neutral (h^0, H^0, A^0), and two of which are charged ($H^\pm$). Supersymmetry theories attempt to solve a number of problems in the Standard Model, including divergent loop corrections to the Higgs mass. The minimal supersymmetric extension of the Standard Model (MSSM) consists of a 2HDM (type 2) sector, in which one doublet couples to the up-type quarks and leptons, and the other to the down-type quarks and neutrinos [297, 506].

At the TEVATRON, direct production of H^+H^- via the weak interaction is expected to have a relatively small cross section, of the order of 0.1 pb [507]. The $t\bar{t}$ production, with a Standard Model expected production cross section of $6.7^{+0.7}_{-0.9}$ pb [113, 114, 116, 117], may offer another source of charged Higgs production. If kinematically allowed, the top quark can decay to H^+b, competing with the Standard Model top decay $t \to W^+b$. This mechanism might provide a larger production of charged Higgs boson and offer a much cleaner signature than that of direct $H^\pm$ production.

Previous searches for the charged Higgs boson have been performed in the $\tau_h + \not{E}_T + jets + X$ channel, where $X = e, \mu$ in [382] and where $X = e, \mu$ or τ_h in [508], where τ_h

denotes the detection of a τ through its decay to hadrons. On the assumption that the charged Higgs decays exclusively to $\tau\bar{\nu}$, both these searches set limits directly on $B(t \to H^+b)$ based on the measured production rate, and interpret their results in terms of the ratio of the vacuum expectation value of the two Higgs doublets, $\tan\beta$. In most models, the decay $H^+ \to \tau\bar{\nu}$ is predominant at large values of $\tan\beta$.

Most recent searches in the $\not{E}_T + jets + X$ channel, where $X = e$ or μ [509] obtain limits in the $(m_{H^+}, \tan\beta)$ plane assuming that $H^\pm$ decays to $\tau\bar{\nu}$, $c\bar{s}$ and $t^*\bar{b}$, the latter resulting in a $Wb\bar{b}$ final state. Previous searches that predict the $t \to H^+b$ and charged Higgs branching fraction as a function of $\tan\beta$ do so at tree level in the context of the MSSM. It is now known that higher order radiative corrections significantly modify this prediction. The corrections strongly depend on the model parameters and are particularly large at high values of $\tan\beta$ [510].

CDF has searched for charged Higgs bosons in $t\bar{t}$ decay products using $192\,\mathrm{pb}^{-1}$ of Run-II data [511]. In this analysis, the assumption is made that the charged Higgs boson may decay either to $c\bar{s}$, $\tau\bar{\nu}$, $t^*\bar{b}$ or W^+h^0. In the latter case only the decay of h^0 to $b\bar{b}$ final states is considered. Thus, for a single top quark, five possible decay modes are considered:

- $t \to W^+b$
- $t \to H^+b$ with $H^+ \to \tau\bar{\nu}$
- $t \to H^+b$ with $H^+ \to c\bar{s}$
- $t \to H^+b$ with $H^+ \to t^*\bar{b}$
- $t \to H^+b$ with $H^+ \to W^+h^0$ and $h^0 \to b\bar{b}$.

This search is based on the observed number of events in the $e/\mu + \not{E}_T + jets + X$ channels, where $X = e$ or μ (dilepton), $X = \tau_h$ (lepton+tau), $X = 1$ or more jets with a displaced vertex (lepton + jets, ≥ 1 tags), and $X = 2$ or more jets with displaced vertex (lepton + jets, ≥ 2 tags). Depending on the top quark and Higgs boson branching ratios, the number of expected events in these decay channels can show an excess or deficit when compared to the Standard Model expectation.

The measurements of the $t\bar{t}$ production cross section under the assumption $B(t \to H^+b) = 0$ are reported in References [302] (dilepton), [380] (lepton+tau), [309] (lepton + jets, ≥ 1 tags), and [307] (lepton + jets, ≥ 2 tags). In this analysis, extra requirements are applied to each channel in order to force the association of every event to a single channel. In particular, the lepton + jets channels are separated into lepton + jets+exactly 1 tag, and lepton + jets + 2 or more tags. The background contribution to each of these "exclusive" channels was recalculated, and the changes from the original cross section analyses found to be negligible.

The acceptance of the detector for channel k is then:

$$\epsilon_k = \sum_{i,j=1}^{5} B_i B_j \epsilon_{ij,k}(\Gamma_t, \Gamma_{H^\pm}, m_{H^\pm}, m_{h^0}), \quad (256)$$

where B_i (B_j) represent the branching fractions of the top quark (anti-quark) to decay via mode i (j), and $\epsilon_{ij,k}$ is the efficiency to detect a $t\bar{t}$ event decay whose top quarks decay to i and j in the channel k.

The 5 branching ratios B_i can be written in terms of $B(t \to H^+b)$, $B(H^+ \to c\bar{s})$, $B(H^+ \to t^*\bar{b})$, $B(H^+ \to W^+h^0)$ and $B(h^0 \to b\bar{b})$. $B(H^+ \to \tau\bar{\nu})$ is then given by $B(H^+ \to \tau\bar{\nu}) = 1 - B(H^+ \to c\bar{s}) - B(H^+ \to t^*\bar{b}) - B(H^+ \to H^+h^0)$. The dependence of $\epsilon_{ij,k}$ on the width of the top (Γ_t), the width of the charged Higgs ($\Gamma_{H\pm}$), the mass of the charged Higgs ($m_{h\pm}$) and the mass of the h^0 (m_{h^0}) is explicitly calculated.

The efficiencies $\epsilon_{ij,k}$ are obtained from Monte Carlo simulation of $t\bar{t}$ using the PYTHIA [223] generator, modified to include the decay $H^+ \to t^*\bar{b}$. The expected number of events in channel k (μ_k) is:

$$\mu_k = \sigma_{t\bar{t}}^{\mathrm{prod}}\epsilon_k(\rho) + n_k^{\mathrm{bkg}}, \qquad (257)$$

where ρ represents a generic set of parameters from which the nine quantities (five branching ratios, Γ_t, $\Gamma_{H\pm}$, $m_{H\pm}$ and m_{h^0}) needed to calculate the acceptance can be derived, $\sigma_{t\bar{t}}^{\mathrm{prod}}$ is the production cross section, and N_k^{bkg} is the number of expected background events in the channel k. A likelihood is obtained by comparing the number of observed versus the number of expected events in all channels. The computation of the likelihood takes into account correlations between different channels.

In the MSSM the nine quantities can be predicted from a specific set of MSSM parameters, including $m_{H\pm}$ and $\tan\beta$. For this analysis, CPsuperH [512] is used to calculate all the Higgs masses and branching ratios. This program includes QCD, SUSY-QCD and SUSY-EW radiative corrections. In addition, corrections to the top and bottom Yukawa couplings are implemented in a consistent way.

When comparing the number of expected events to what is expected when $B(t \to H^+b) = 0$ for the four channels, a Bayesian approach with a flat prior on $\log_{10}\tan\beta$ is used. The probability is integrated over its maximum density region to obtain upper and lower limits in $\tan\beta$ at the 95% CL. The resulting exclusion region in the ($m_{H\pm}, \tan\beta$) plane is shown in Fig. 120. In all the used MSSM benchmark scenarios, the low $\tan\beta$ region is excluded in a similar region as shown in this figure. The large $\tan\beta$ exclusion re-

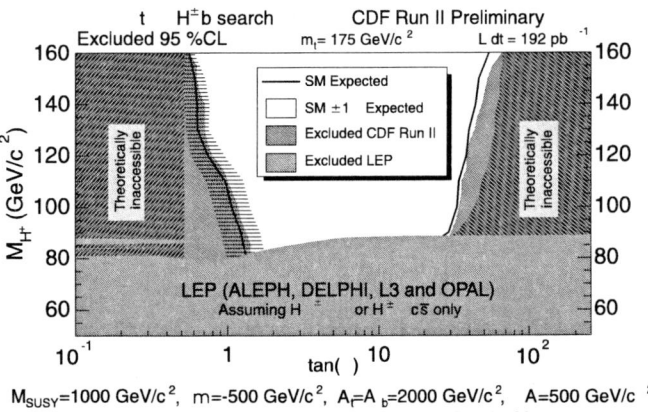

Fig. 120. MSSM exclusion region. The expected exclusion limits are indicated by a *black solid line* and the 1-sigma confidence band around it is obtained from pseudo-experiments. The contour of the *red solid region* indicates the observed limits at the 95% CL. The *lower green region* is the LEP combined results from direct searches

gion, however, can be significantly reduced, and even vanishes, depending on the parameters of the benchmark used.

Present bounds on the Standard Model Higgs boson mass, reinterpreted in the context of the MSSM, put strong constraints on values of $\tan\beta$ [513]. The region of large values of $\tan\beta$ is also theoretically appealing since it is consistent with the approximate high-energy unification of the top and bottom Yukawa couplings [514–516].

In this region, the decay $H^+ \to \tau\bar{\nu}$ is expected to dominate in a large fraction of the MSSM parameter space. Under this assumption the charged Higgs branching ratio of $H^+ \to \tau\bar{\nu}$ is explicitly set to unity (Tauonic model), and the posterior probability is evaluated as a function of $B(t \to H^+b)$. Furthermore, in this case the charged Higgs and top quark widths are set to $\Gamma_{H\pm} = 1.4 \text{ GeV}/c^2$ and $\Gamma_t = \frac{\Gamma_W}{1-B(t\to H^+b)}$. A posterior probability density of $B(t \to H^+b)$ is obtained using a flat prior that is constant between 0 and 0.9 and null elsewhere. The 95% CL is obtained by integrating the posterior probability density over the max-

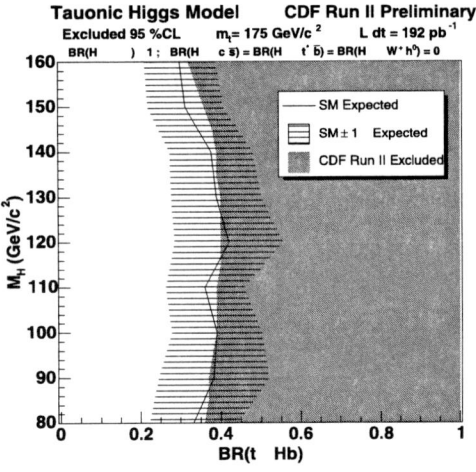

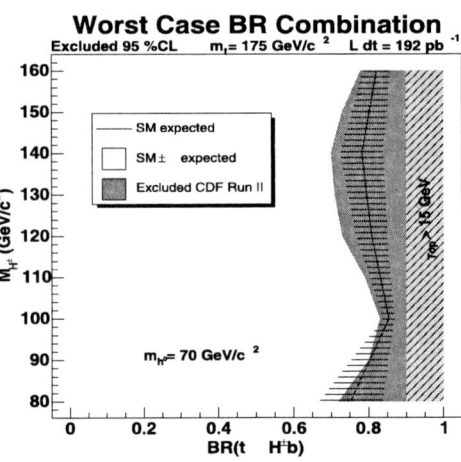

Fig. 121. The *red solid region* represents the CDF Run-II excluded region in the ($m_{H\pm}, B(t \to H^+b)$) plane, while the expected exclusion, assuming only the presence of Standard Model background, is shown as the *black solid line*. *Left*: The tauonic model with $B(H^+ \to \tau\bar{\nu}) \equiv 1$). *Right*: The worst case branching ratio combination

imum density region. This procedure is repeated for different charged Higgs boson masses and the resulting 95% CL excluded region is shown in Fig. 121 (left) as a function of $m_{H^\pm}$ yielding an overall limit of $B(t \to H^+ b) < 0.4$ at 95% CL.

The Tauonic Higgs model is a particular choice of the charged Higgs branching ratios with $B(H^+ \to \tau\bar{\nu}) = 1$. The analysis is repeated considering all possible branching ratio combinations, i.e. making no assumptions on the decay of the charged Higgs boson. For a specific charged Higgs mass, the charged Higgs branching ratio to each decay is divided in 21 bins. This results in 1771 possible combinations subjected to the relation $B(H^+ \to \tau\bar{\nu}) + B(H^+ \to c\bar{s}) + B(H^+ \to t^*\bar{b}) + B(H^+ \to W^+ h^0) \equiv 1$. Looping over all the 1771 possible combinations, a limit on $B(t \to H^+ b)$ is obtained in each bin. The worst limit is quoted for the used charged Higgs boson mass and the analysis is repeated for charged Higgs masses in the 80 to 160 GeV/c^2 range, as shown in Fig. 121 (right). These results are model-independent and therefore the most conservative limits on $B(t \to H^+ b)$, yielding an overall limit of $B(t \to H^+ b) < 0.85$ at 95% CL.

10 New physics in events with $t\bar{t}$ topology

10.1 Search for heavy fourth generation t' quarks

The cross section for $t\bar{t}$ production has been measured in Run-II in several different decay channels and methods (Sect. 4). The results are in good agreement with the Standard Model prediction. It is of interest to study whether the present data allow or preclude the production of hypothetical new quarks which decay to final states with a high p_T lepton, large $\not{E}_T$, and multiple hadronic jets, having large total transverse energy H_T. There are several possibilities arising from extensions to the Standard Model, in which this may be the case, and are not excluded by precision electroweak data or other direct searches.

CDF has performed a search for heavy fourth generation t' quarks in the lepton+jets channel using 200 pb^{-1} of Run-II data [517]. Here, the hypothetical new quark is referred to as t', but it need not be a standard fourth-generation up-type heavy quark. For the purpose of this analysis, a new quark is considered which:

– is pair-produced strongly,
– has mass greater than the top quark, and
– decays promptly to Wq final states.

In particular, it is not necessary to demand the charge of the quark to be $+2/3$, nor need it even be a fermion. In the case of a new scalar quark, however, the production rate will be reduced due to the β^3 factor in its production cross section.

A fourth generation of matter fermions with light neutrino ν_4 with mass $m(\nu_4) < m_Z/2$ is excluded by precision data from LEP 1. However, as pointed out in [298] and [299], a heavier fourth generation of fermions with $m_Z/2 < m_f < \mathcal{O}(\langle H \rangle)$ is consistent with existing precision electroweak data. The present bounds on the Higgs in such scenarios

are relaxed; The Higgs boson mass could be as large as 500 GeV/c^2. Additional fermion families can be accommodated in 2-Higgs doublet scenarios and $N = 2$ SUSY models, and possibly remove the requirement of the weak-mixing assumption. In that case, the decay $t' \to Wq$ may predominate, assuming $m(t') > m(b') + m(W)$, for example.

Other theoretical possibilities lead directly to the scenario of interest here. In one version of the "beautiful mirrors" model [300], there exists an up-type quark with the same quantum numbers as the top, which decays as $\chi \to Wb$. In this scenario, the slightly anomalous results from LEP on the b forward-backward asymmetry are accommodated naturally, and the electroweak fits are improved (with a relaxed upper limit on the Higgs boson mass).

Recent theoretical developments lead to the hypothesis of the existence of a heavy t'. Little Higgs models [518] evade the hierarchy problem by introducing a minimal set of gauge and fermion fields in the context of a large-extra-dimension framework. The minimal version of these models, however, results in new quarks which have mass of order 1 TeV/c^2, too heavy for Tevatron studies. Non-minimal Little Higgs scenarios, however, are of course possible.

So the basic conclusion is that there exist enough theoretical scenarios and ideas involving new heavy quarks, which need to be searched for in as many channels as possible. In order to allow the widest possible theoretical interpretation, the results are expressed in a model-independent way as limits on the t' pair production cross section times branching ratio $t' \to Wq$, leading to the high-H_T lepton + jets + $\not{E}_T$ signature with an acceptance determined from a generic fourth generation quark decaying to Wb.

This CDF t' analysis uses the same data set and event selection as the kinematic top cross section measurement (Sect. 4.2.2, [304]), using the requirement of ≥ 4 jets. The observed distribution of total transverse energy in the event, H_T, is used to distinguish the t' signal from the backgrounds by fitting it to a combination of t' signal, $t\bar{t}$, W+jets, and QCD background shapes. A likelihood method is used to extract the t' signal and/or to set an upper limit on its production rate. As in the top cross section measurement, a binned likelihood in H_T is calculated as a function of the t' cross section. Using Bayes' Theorem the likelihood is converted into a posterior density in $\sigma_{t'}$ from which the limit is derived. Unknown parameters, such as the production rate for W+jets, the $t\bar{t}$ production cross section, lepton ID data/MC scale factors true integrated luminosities etc. are considered to be systematic uncertainties and treated in the likelihood as nuisance parameters which float in the fit, but may be constrained within their expected distributions.

No evidence for a t' signal is observed. Figure 122 (left) shows the observed H_T distribution and the best fit to the distribution at the t' cross section point where the 95% CL upper limit is set. In this case the $t\bar{t}$ cross section has floated to 6.1 pb. Figure 122 (right) shows the final result of the 95% CL upper limit on the t' production rate as a function of the t' mass. The result includes three curves, that

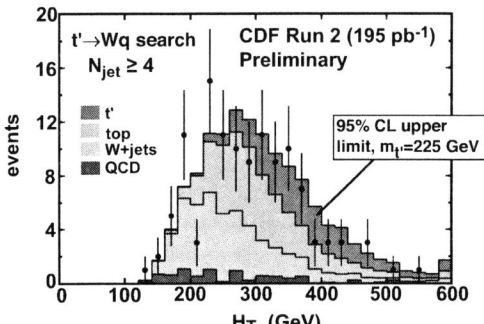 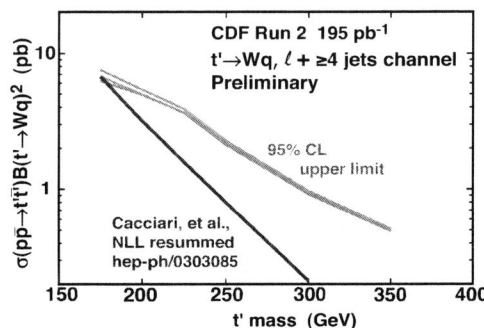

Fig. 122. *Left:* Distribution of H_T, showing the result of the fit for t' for $m(t') = 225\,\text{GeV}/\text{c}^2$. The normalisations of the various sources are those corresponding to the maximum likelihood when the cross section for t' is set to its 95% CL upper limit. *Right:* 95% CL upper limit on the production rate for t' as a function of the t' mass. The three *(blue) curves* correspond to top quark masses of 170, 175, and 180 GeV/c^2 (the *lowest curve* corresponds to the lowest mass). The *lowest (purple) curve* is the theoretical cross section calculated from PYTHIA and scaled to agree with NNLO calculation

in general overlap, for an assumed top quark mass of 170, 175, and 180 GeV/c^2. This plot also includes a curve with the theoretical prediction from PYTHIA [223], multiplied by a factor 1.055 to account for the fact that PYTHIA's prediction is smaller than the ones from the NNLO calculations [113, 114, 116, 117]. In conclusion, at 95% CL a t' mass larger than 175 GeV/c^2 is ruled out, if the true top mass is about the same value. For a smaller top mass, the excluded mass is lower, and vice versa for a higher mass.

11 Top quark physics at the LHC

11.1 LHC collider and experiments

In spite of the remarkable ability of the Standard Model of elementary particle physics to describe all existing accelerator data with high precision, the Standard Model fails to answer a number of fundamental questions, such as the mechanism of the electroweak symmetry breaking and the origin of fermion and boson mass, the relevance of the observed number of lepton and quark generations, the energy dependence of the running gauge couplings and a possible unification at a GUT scale, the hierarchy problem of fundamental energy scales to differ by many orders of magnitude etc. All these problems point towards new physics beyond the Standard Model, expected to show clear signatures at the TeV energy scale. To look for this new physics, CERN's Large Hadron Collider (LHC) is being constructed and expected to begin operation in 2007. At four interaction points the experiments ATLAS, CMS, LHC-B, and ALICE are being installed. ATLAS and CMS are omni-purpose experiments, which will investigate a multitude of research topics, such as Standard Model electroweak precision measurements, Higgs physics, top quark physics, the search for Supersymmetry, dark matter candidates, and other phenomena beyond the Standard Model, B-Physics ($\mathcal{CP}$ violation) and heavy ion physics. LHC-B is a collider experiment with single-spectrometer detector setup, optimised for measurements in the B sector and studies of $\mathcal{CP}$ violation, while ALICE is optimised for the investigation of heavy ion physics

and the quark-gluon plasma. In the following, the main focus of this review is placed on the top quark physics potential with the ATLAS and CMS experiments.

11.1.1 The LHC accelerator

The Large Hadron Collider [519–522] is a proton–proton collider, currently under construction at CERN. It is being built into the tunnel of the former LEP accelerator (Fig. 123), located about 100 m underground, so that most of the infrastructure could be recycled and only small civil engineering projects were necessary. The LHC will accelerate two beams of protons in opposite direction in a 27 km long ring up to a beam energy of 7 TeV. Operation with heavy ions is foreseen as well. The beams, each containing about 3×10^{14} protons, are brought to collisions at four interaction points. The design luminosity is $10^{34}\,\text{cm}^{-2}\,\text{s}^{-1}$. To keep the particles on track, the LHC will be equipped with high-field superconducting NbTi dipole magnets (up to 8.34 T), which are operated in superfluid helium. The main parameters of the LHC are listed in Table 8 in comparison to the $Sp\bar{p}S$ and the TEVATRON.

The LHC will benefit from existing accelerator facilities at CERN (Fig. 123), namely the Linac, the Booster, the proton synchrotron (PS), and the super proton synchrotron (SPS). The protons are obtained from a hydrogen source. They are pre-accelerated in the Linac to energies of 50 MeV. Then they enter the Booster, which increases the energy to 1.4 GeV. Successive acceleration of the protons takes place in the PS and SPS to energies of 25 GeV and 450 GeV, respectively, before they are injected into the LHC. The particles are accelerated by an RF system which operates at a temperature of 4.5 K and at 400.8 MHz, the second harmonic of the SPS frequency. Superconducting cavities are used which are sputtered with a thin film of Niobium. The design voltage of the RF system is 16 MV per beam, providing an average bunch length of 7.5 cm. The bunch spacing is 25 ns, i.e. ten RF periods.

Since two beams of particles with the same charge must be accelerated in opposite directions, two independent magnetic channels are needed. However, they will be housed in the same yoke and cryostat system. The magnet coils are

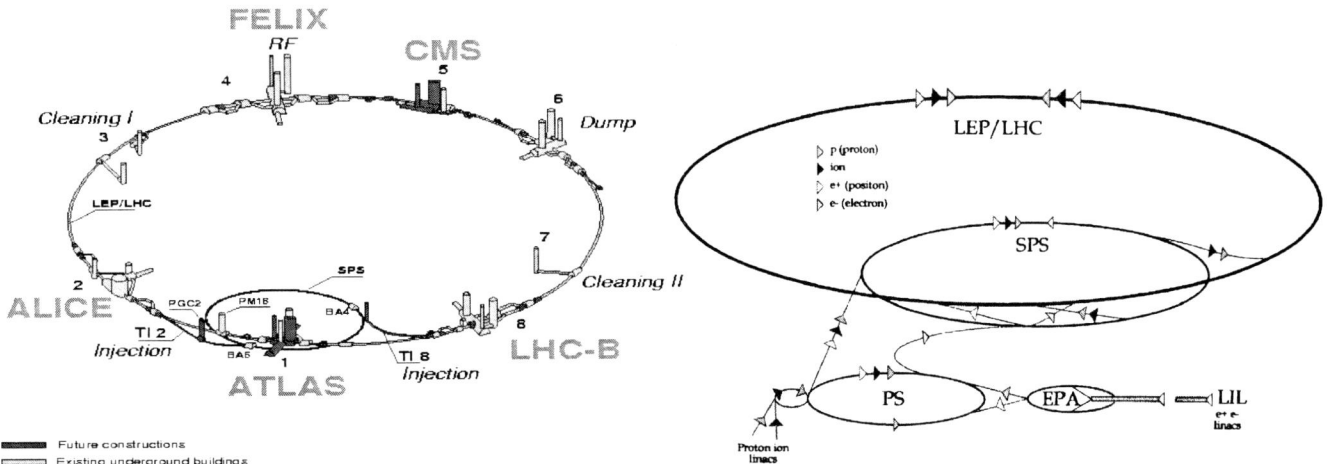

Fig. 123. *Left*: Layout of the LEP tunnel including the new LHC infrastructure (shown in *dark grey/red*) with the SPS pre-accelerator. *Right*: Chain of particle accelerators at CERN

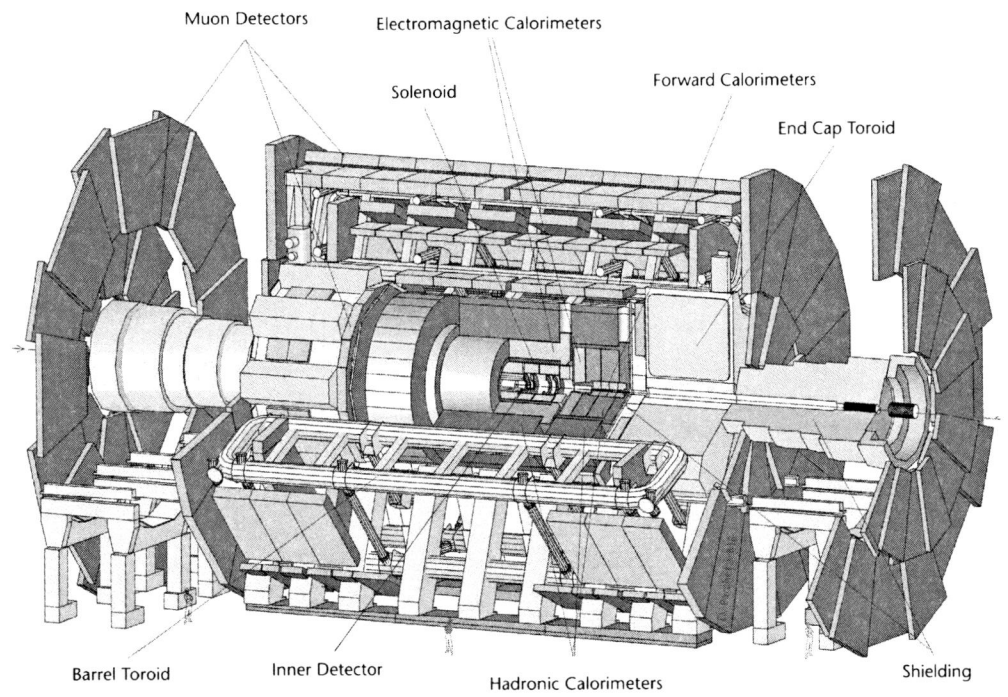

Fig. 124. A cutaway view of the ATLAS detector and its components

made of copper-clad Niobium-Titanium cables. They are operated at 1.9 K with a current of 15 000 A, and have to withstand forces of some hundred tons per meter during the ramping of the magnetic field. The LHC will consist of 1232 main dipoles and 392 main quadrupoles, the latter producing gradients of 233 T/m. In case of quenches, the stored energy must safely be released from the magnets. The energy is absorbed by resistors, which can be switched into the circuit, heating eight tons of steel to about 300°.

11.1.2 The ATLAS detector

The ATLAS[20] [523] detector is an omni purpose detector, designed to explore the full physics program of the LHC.

---

[20] The name is an acronym for 'A Toroidal LHC ApparatuS'.

Figure 124 shows a sketch of the ATLAS detector. The detector has cylindrical symmetry with a height (diameter) of 22 m, a length of 46 m, and a weight of almost 7000 tons. The parameters of the LHC and the challenging physics program put high demands on the components of ATLAS:

The inner detector. The inner detector (ID), embedded in a solenoidal magnetic field of 2 T, is contained inside a cylinder of 7 m length and radius 1.15 m, covering an acceptance of $|\eta| < 2.5$, and consists of three parts: A high resolution pixel detector with 123 million readout channels, a microstrip semi conductor tracker (SCT) with 6.2 million channels, and a transition radiation tracker (TRT) made of straw tubes with 420 000 channels. The silicon detectors provide a small number of high resolution space points (three in the Pixel, and four in the SCT), while the TRT provides on average 36 points for tracks in $|\eta| < 2.5$.

The Pixel detector, made of 1500 barrel and 700 disk modules of $61\,440$ pixels[21] each, consists of three radial layers. The layers are located at 5.05 cm, 8.85 cm, and 12.25 cm radius, each corresponding to $1.7\% X_0$ thickness. On each side of the barrel detector, three disks of pixel modules are placed at $z = 49.5$ cm, 58 cm, and 65.0 cm, in particular for the forward tracking and b-tagging.

The SCT is comprised of silicon microstrip modules, each containing four single-sided silicon detectors of 6.36×6.40 cm^2 area, segmented into 768 strips with a pitch of 80 μm. Two such modules are arranged back-to-back at a 40 mrad stereo angle to provide three-dimensional space point information. The barrel SCT contains four such double layers at radii between 40 cm and 52 cm, complemented by 9 disks on either side of the barrel for large-$|\eta|$ tracking.

The TRT is built from straw tube detectors with 4 mm diameter, filled with a Xe, CO_2, CF_4 mixture (70% : 20% : 10%) in a total volume of 3 m^3. Each straw contains a 30 μm thin goldplated W-Re wire at the centre. The barrel detector is 144 cm long and consists of 50 000 straws, arranged in three rings with a total of 73 layers at radii between 56 cm and 107 cm. The two end-caps each consist of 18 wheels. In the central region, the ATLAS tracker is expected to achieve a resolution of

$$\sigma_{1/p_{\mathrm{T}}} \approx 0.36 \oplus \frac{13}{p_{\mathrm{T}}\sqrt{\sin\theta}}\,(\mathrm{TeV}^{-1})\,. \tag{258}$$

The calorimeters. ATLAS has three types of calorimeters: An electromagnetic calorimeter covering the region $|\eta| < 3.2$, hadronic calorimeters in the barrel ($|\eta| < 1.7$), and in the endcaps ($1.5 < |\eta| < 3.2$), and forward calorimeters covering the region $3.1 < |\eta| < 4.9$. The electromagnetic calorimeter (EM) consists of liquid Argon (LAr) as active material with an accordion-like structure of Kapton electrodes and lead as absorber. The calorimeter is divided into several samples of $\Delta\eta \times \Delta\phi$ size varying from 0.003×0.1 to 0.1×0.1, depending on the pseudorapidity range. The EM has about 190 000 readout channels and is $24X_0$ thick. On the inner side of the EM, a presampler, consisting of a LAr layer of 1.1 cm (0.5 cm) thickness in the barrel (endcaps), is installed. The design goal for the EM energy resolution is

$$\frac{\sigma_E}{E} = \frac{0.1}{\sqrt{E}} \oplus \frac{0.3}{E} \oplus 0.01\,. \tag{259}$$

The hadronic calorimeter consists of a central barrel part ($|\eta| < 1.0$) and two identical extended barrels ($0.8 < |\eta| < 1.7$) based on an alternating structure of plastic scintillator plates (tiles) and iron absorbers. At larger pseudorapidity ($1.5 < |\eta| < 3.2$, endcap), the hadronic calorimeters are made from intrinsically radiation-hard LAr detectors. The very forward region ($3.1 < |\eta| < 4.9$) is covered by LAr detectors, placed at 4.7 m distance from the interaction point, and consisting of copper or tungsten as passive

material. The design goal for the energy resolution is

$$\frac{\sigma_E}{E} = \frac{0.5}{\sqrt{E}} \oplus 0.03\,. \tag{260}$$

The muon spectrometer. The muon system is an outstanding feature of the ATLAS detector. The muon spectrometer measures the deflection of muon tracks in superconducting air-core toroid magnets. They are equipped with trigger chambers and high-precision tracking chambers. The system covers the region $|\eta| < 1.0$, using magnetic bending by large barrel toroids, and the region $1.4 < |\eta| < 2.7$ using two smaller endcap magnets which are located on both ends of the barrel toroid. The precision measurement of muon trajectories is realised by two types of chambers, monitored drift tubes (MDT), and cathode strip chambers (CSC). There are three cylindric layers in the barrel, located at radii of 5, 7.5, and 10 m. The endcap chambers in the range $1 < |\eta| < 2.7$ are located at distances of 7, 10, 14, and 21–23 m from the interaction point. The MDTs consist of 3 cm-diameter aluminium tubes, filled with a 93% : 7% mixture of Argon and CO_2, and a 50 μm sense wire in the centre. The MDTs cover most of the η-range. The CSCs are multiwire proportional chambers of high granularity, filled with a 30% : 50% : 20% gas mixture of Argon, CO_2, and CF_4. The CSCs cover the region $2 < |\eta| < 2.7$. In addition, the muon system also contains resistive plate chambers (RPCs) in the barrel and thin gap chambers (TGCs) in the endcaps for triggering purposes. They have good time resolution of 2 ns and 5 ns, respectively.

The trigger and data acquisition system. The LHC proton bunches will cross at a rate of 40 MHz. At design luminosity of 10^{34} cm^{-2} s^{-1} there will be on average 23 inelastic pp collisions per bunch crossing, yielding an event rate of almost 1 GHz. A multiple-stage trigger system is required to reduce the data to a manageable amount and to filter for interesting events. In ATLAS, the Level-1 trigger performs the initial event selection and reduces the rate to less than about 75 kHz. It identifies regions in the detector with interesting features, so-called regions of interest (RoI). Information from all RoIs are combined and passed on to the Level-2 trigger, which applies a series of optimised selection algorithms to the event. The event filter (EF) processes the output of Level-2 with more sophisticated reconstruction and trigger algorithms using tools similar to the offline software. The EF then takes the final decision if the event is discarded or written to tape. Level-2 and the event filter form the high level trigger (HLT). The HLT output rate is about 100 Hz, and selected events are expected to have an average size of 1.5 MB. The mass storage therefore must be capable of recording a few hundred MB per second.

11.1.3 The CMS detector

The compact muon solenoid (CMS, [524]) is – similar to the ATLAS experiment – a multi-purpose detector

---

[21] A pixel covers the area $400 \times 50\,\mu\mathrm{m}^2$.

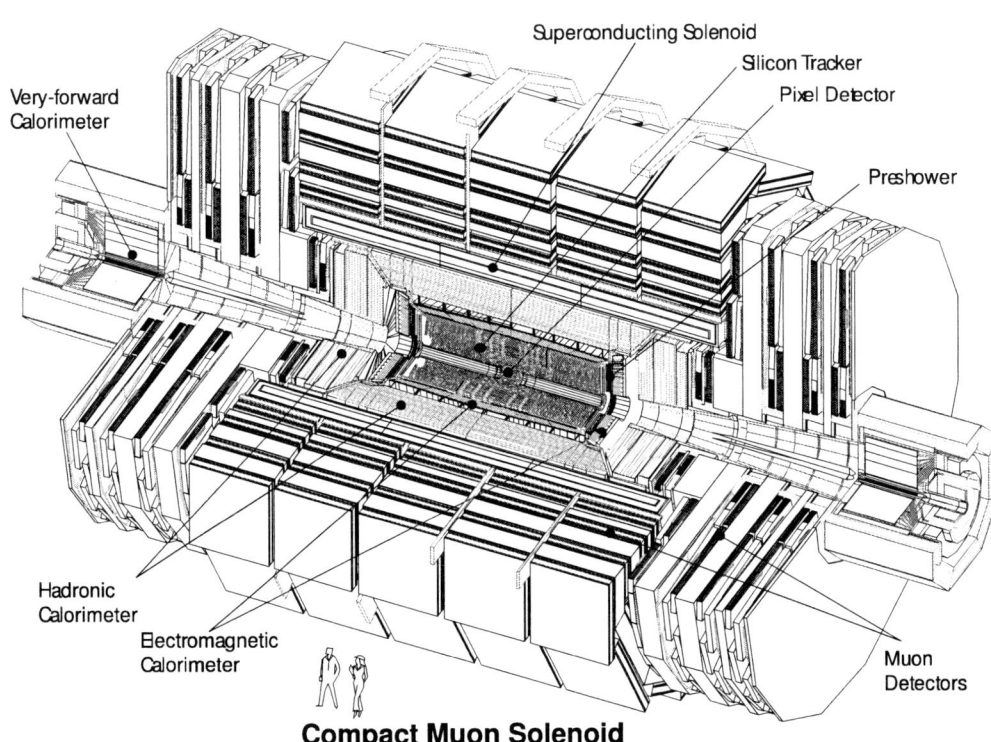

Compact Muon Solenoid

Fig. 125. A cutaway view of the CMS detector and its components

placed at a high luminosity intersection point of the LHC. It is designed to detect as many particle types as possible – leptons, photons, jets, and b-quarks – and isolate at each bunch crossing the events of interest for physics studies. A cutaway view of the CMS detector and its components is shown in Fig. 125. The CMS experiment is optimised for studies of the following physics processes: the search for Higgs bosons and SUSY particles, studies of the bottom and the top quark, precision tests of the Standard Model, QCD tests and examinations of the quark–gluon plasma. The resulting detector design consists of the following subsystems: the central tracker, the electromagnetic calorimeter (ECAL), the hadronic calorimeter (HCAL), the solenoid magnet, and the muon system. Except for the magnet, each subsystem is divided into a central 'barrel' part and two 'forward' parts. The main sub-detector systems are briefly described in the following.

The central tracker. The CMS tracker should provide robust and accurate tracking and vertexing information for charged particles. Due to the high radiation environment it has to be ensured that the detectors used are sufficiently radiation hard, to guarantee their functioning over the full operation period of the experiment. CMS has decided to built the entire tracking detector out of silicon pixel and silicon microstrip detectors, with a structural support made of carbon fibre composite. To reach sufficient radiation hardness and limit the leakage current of the devices, the whole tracker volume has to be cooled down to $-10\,^\circ$C. Due to the very high particle and track density near the interaction region, the innermost layers of the CMS tracker are made of silicon pixel de-

tectors, located at radii of 40 mm and 70 mm. Later on an additional layer at $r = 110$ mm will be added. In the forward regions, the pixel detector consists of two disks each. The pixel size was chosen to be $100 \times 150\,\mu\text{m}^2$. The expected 3-dimensional vertex resolution from the pixel detector is $15\,\mu\text{m}$, the impact parameter resolution transverse to the beam $35\,\mu\text{m}$, and $75\,\mu\text{m}$ along the beam line. In total, the pixel detector comprises about 4×10^7 readout channels with an expected occupancy of less than 10^{-4}.

The silicon microstrip tracker extends from a radius of 20 cm up to 110 cm. The expected particle rate allows the use of silicon microstrip detectors, which are arranged in four different sub-detectors: The tracker inner barrel (TIB) has four layers with ± 100 mrad stereo modules in layer one and two. There are two tracker inner disk modules (TID). One TID consists of three disks each equipped with three rings of sensors. Stereo modules are used in the inner two rings. The tracker outer barrel (TOB) has six layers with stereo modules and layers one and two. There are two tracker end caps (TEC), where each TEC is composed of nine disks, each with up to seven rings of modules of which those in rings one, two and five are stereo modules. Rectangular modules are used in the TOB and TIB, while wedge shaped sensors are used for TEC and TID. The sensors of all four sub-detectors have a strip pitch of $80\,\mu\text{m}$ in the inner detectors and up to $205\,\mu\text{m}$ in the endcap detectors with a varying sensor thickness of $320\,\mu\text{m}$ in the inner layers and up to $500\,\mu\text{m}$ in the endcaps. The silicon microstrip detector covers a total area of $224\,\text{m}^2$ and is therefore the largest semiconductor detector of that kind today. It ensures tracking coverage up to $|\eta| < 2.6$ with up to 12 measurement points. In total this detector comprises

10M channels and provides tracking at an occupancy of less than 1%. It will be aligned with a laser alignment system, aiming for spatial module alignment of better than 100 μm, which will be available from reasonable tracking at the trigger level. Higher precision will be achieved offline with track alignment. The tracker is embedded in the high-field superconducting solenoid of 5.9 m diameter, which achieves a uniform magnetic field of 4 T over a length of 12.48 m. The combination of this very strong magnet and the precision silicon tracker provide an expected momentum resolution of $\frac{\sigma_{p_T}}{p_T} \approx 0.1 \times p_T$ [TeV].

The calorimeters. The CMS calorimeter consists of two sub-detectors, the inner, electromagnetic calorimeter (ECAL), and the outer hadronic calorimeter (HCAL).

The ECAL aims at the measurement of the energy and direction of electrons and photons in the rapidity range up to $|\eta| < 2.6$. It is a crystal calorimeter made of $PbWO_4$ crystals, providing excellent energy resolution and radiation hardness at the same time. The scintillation speed is high, allowing easy bunch crossing identification and minimisation of pileup effects. The 23 cm long crystals of the barrel region correspond to 26 radiation lengths. In this region the crystals are read out via Avalanche Photo Diodes (APDs) which can be operated in the high magnetic field. In the forward region, the use of a preshower detector including 3 radiation lengths of lead makes it possible to use somewhat shorter crystals of 22 cm length. These are read out by vacuum photo triodes which are more radiation hard than APD's. The expected energy resolution ranges from $\frac{\sigma_E}{E} = \frac{2.7\%}{\sqrt{E}} x \oplus \frac{0.2}{E} \oplus 0.55\%$ in the barrel region to $\frac{\sigma_E}{E} = \frac{5.7\%}{\sqrt{E}} \oplus \frac{0.25}{E} \oplus 0.55\%$ in the endcap region (E in GeV).

The main task of the hadronic calorimeter is the measurement of hadronic jets and the so-called missing energy. The barrel and endcap parts of the HCAL are supplemented with two forward HCALs, increasing the acceptance to rapidities of $|\eta| < 5$. An outer tailcatcher calorimeter, covering $|\eta| < 1.3$ ensures total energy containment of showers in the barrel region. The HCAL is a sampling calorimeter, about 5–10 interaction lengths deep, depending on η. It consists of copper absorber plates interleaved with plastic scintillators. In the forward region the scintillators are replaced by quartz fibres, which are better suited for the high radiation dose expected there. The expected energy resolution ranges from $\frac{\sigma_E}{E} = \frac{70\%}{\sqrt{E}} \oplus$ 9.5% at $\eta = 0$ to $\frac{\sigma_E}{E} = \frac{172\%}{\sqrt{E}} \oplus 9\%$ in the forward region (E in GeV).

The muon system. The muon system will be used to identify muons and measure their momentum, offline as well as in the trigger. The muon detectors are placed in the return yoke of the magnet. In the barrel region, $|\eta| < 1.3$, drift tubes are installed, made of standard rectangular drift cells. In the endcaps, cathode strip chambers are used, due to their better performance in the varying magnetic field there. Also, being faster and finer segmented than the drift tubes, they are better suited for the higher particles rates in the forward direction. In addition, resistive plate chambers will be used in the barrel and the endcap region. They

will be used for trigger purposes, because of their fast response and good timing resolution. The expected momentum resolution is 6%–20% for $p_T < 100$ GeV and 15%–35% for $p_T = 1$ TeV. The global efficiency for detecting muons is larger than 90% for $p_T < 100$ GeV and remains about 70% up to the highest muon momenta expected at the LHC. Due to the strong magnetic field and the energy losses in the calorimeter, only muons with a transverse momentum of about 4 GeV (barrel) or 2 GeV (endcap) can reach the muon system.

The trigger system. At the nominal LHC luminosity of 10^{34} cm^{-2} s^{-1}, an average number of ~ 20 interactions per bunch crossing is expected every 25 ns. Considering the $\sim 10^8$ channels of the CMS detector, the data stream coming from the detector is estimated to be ~ 1 MB per event, resulting in 100 TB of data per second. At CMS, the challenge to reduce this enormous data stream from the 10^9 events per second (1 GHz) down to 100 Hz and a data rate of ~ 100 MB/s is met by employing a three-level trigger system:

The Level-1 trigger exploits only a small subset of data which can be collected very rapidly at each bunch crossing. Low resolution and coarse granularity information obtained from local pattern recognition in the muon system and macro-granular energy evaluation in the calorimeters is used to construct 'trigger candidates' for muons, electrons/photons, and jets (including τ candidates). For each trigger candidate, a position and transverse momentum measurement is provided, together with other properties such as quality indicators or bunch crossing identification. Threshold cuts are applied to the candidates so that the output rate after Level-1 is 50 (100) kHz in the low (high) luminosity operation mode of the LHC. The Level-1 decision is taken after 3.2 μs, of which ~ 1 μs is spent on the data processing, the rest on data transfer to Level-1. After a Level-1 decision is made, the precision data is transferred from pipeline memories to the Level-2.

Level-2 further reduces the event rate by a factor 10. It reconstructs the physics objects more accurately as it is provided with finer granularity and higher precision data. Also, first use is made of primary tracking information (Level-2.5). The algorithms employed are fairly sophisticated, yet fast, and they are executed on fully programmable commercial processors. On average, Level-2 processes a new event every 10 μs and yields an output rate of 10 kHz.

The Level-3 trigger exploits full event reconstruction, in particular the full tracker information, to perform an online analysis allowing to identify the physics process. The final output rate is 100 Hz, i.e. further reduction of a factor 100. Level-3 runs almost the complete offline algorithms on fully programmable commercial processors.

Level-2 and Level-3 together are called the high level trigger (HLT). All HLT algorithms are executed in a single processor farm with standard CPUs, allowing to profit maximally from the technological advances before 2007.

The design parameters of ATLAS and CMS are summarised in Table 66 in comparison to the TEVATRON experiments CDF and DØ.

Table 66. Design parameters of the two LHC experiments ATLAS and CMS, in comparison to the TEVATRON experiments CDF and DØ. For the energy resolutions, the energy E is given in GeV. For the track resolutions, the transverse momentum P_T is given in TeV for ATLAS and CMS, and in GeV for CDF and DØ

parameter		ATLAS	CMS	CDF	DØ
length		46 m	22 m	14 m	17 m
height		22 m	15 m	10 m	11 m
weight		7000 t	12 500 t	5000 t	4600 t
magnet	(solenoid)	2 T	4 T	1.4 T	2 T
	(toroid)	≈ 4 T	$--$		≈ 2 T
ECAL reso ($\frac{\sigma_E}{E}$)		$\frac{10\%}{\sqrt{E}} \oplus \frac{30\%}{E} \oplus 1\%$	$\frac{2.7\%}{\sqrt{E}} \oplus \frac{20\%}{E} \oplus 0.55\%$	$\frac{13.5\%}{\sqrt{E}} \oplus 2\%$	$\frac{15\%}{\sqrt{E}} \oplus 0.4\%$
HCAL reso ($\frac{\sigma_E}{E}$)		$\frac{50\%}{\sqrt{E}} \oplus 0.03$	$\frac{70\%}{\sqrt{E}} \oplus 9.5$ %	$\frac{75\%}{\sqrt{E}} \oplus 3\%$	$\frac{50\%}{\sqrt{E}}$
cal η		≤ 4.9	≤ 5.0	≤ 3.64	≤ 4.0
tracker η		≤ 2.5	≤ 2.6	≤ 2.0	≤ 3.0
nr. track points		$3+4+36$	12	$8+96$	$8+8$
track reso.$\sigma(1/p_T)^{\eta=0}$		$0.36 \oplus \frac{13}{p_T}$	≈ 0.1	0.0017	$0.0018 \oplus \frac{0.015}{p_T}$
muon η		≤ 2.7	≤ 2.4	≤ 1.5	≤ 2.0
muon reso.	($p_T < 0.1$ TeV)		9%–20%		10%–50%
	($p_T = 1.0$ TeV)	7%	15%–35%		

11.2 Brief summary of top quark physics at the LHC

Top quark physics will be one of the highlights of the LHC physics program. Although ATLAS and CMS have both extensively studied the sensitivity and expected performance of the respective detectors to the various top quark properties, most of the results shown here have been obtained by the ATLAS Collaboration. The reason is that the ATLAS Collaboration has published the *ATLAS Detector and Physics Performance Technical Design Report* [286], summarising the expected performance, in the year 1999, while the CMS Collaboration is planning to publish the corresponding document, *CMS Detector and physics performance technical design report* [287] at the end of 2005 or the beginning of 2006, i.e. too late to be included in this review. Despite the differences in the detector design, both experiments are expected to have comparable resolutions, signal efficiencies and background suppression potential. Therefore, all quoted results are meant to be representative for ATLAS as well as for CMS, as indicated in [288].

Top quarks are produced at the LHC by two types of processes: QCD and electroweak production. In QCD production (Sect. 2.1), top quarks are created in $t\bar{t}$ pairs via the processes $q\bar{q} \to t\bar{t}$ and $gg \to t\bar{t}$. While the centre-of-mass energy of the hadron collisions at the LHC is seven times higher than at the TEVATRON, the combined NLO cross section for these processes is a hundred times larger, $\approx 830 \pm 50$ pb^{-1} [113], where the uncertainty reflects the theoretical error obtained from varying the renormalisation scale by a factor of two. Due to the larger centre-of-mass energy available at the LHC, typical Bjorken-x values for the involved incoming partons are $x \approx 0.025$ or even lower. The large gluon density of the proton dominates at these values of x. Hence, $\approx 87\%$ of the $t\bar{t}$ contribution comes from the gluon–gluon fusion process, the remaining $\approx 13\%$ come from quark–antiquark annihilation, which is dominant at the TEVATRON. In contrast to $t\bar{t}$ production at threshold at the TEVATRON, at the LHC the relative

difference in the Bjorken-x values of the two incoming partons can be quite large, resulting in a strong forward boost of the $t\bar{t}$ system. The event topologies in top quark production at the LHC are less central and more boosted, i.e. more collimated. Both effects make the identification and reconstruction of top quark events more difficult at the LHC than at the TEVATRON. There will be 8 million $t\bar{t}$ pairs produced per year at a luminosity of 10^{33} cm^{-2} s^{-1}. Such large event samples will permit precision measurements of the top quark parameters. At "low" luminosity, $\mathcal{L} = 10^{33}$ cm^{-2} s^{-2}, there will be a $t\bar{t}$ pair produced every second, a $t\bar{t} \to \ell +$ jets event every four seconds. Thus, the LHC will be a "top factory".

In addition to the inclusive $t\bar{t}$ cross section, also differential cross sections will be of interest. For example, $d\sigma_{t\bar{t}}/d\eta$ gives access to parton distribution functions, or heavy particles decaying into a $t\bar{t}$ pair could show up as resonances in the $t\bar{t}$ mass spectrum, $d\sigma_{t\bar{t}}/dm_{t\bar{t}}$, and thus be an indication for new physics. Furthermore, the $t\bar{t}$ production cross section is in the Standard Model sensitive to the top quark mass, $\sigma_{t\bar{t}} \propto 1/m_t^2$.

The electroweak single-top quark production will be of keen interest at the LHC. It is composed of three channels: the s-channel, the t-channel (also referred to as W-gluon fusion) and the associated Wt production (see Sect. 2.2). The three processes have cross sections of 11 pb, 60 pb, and 247 pb, respectively. The cross section for the electroweak production is directly proportional to the CKM matrix element $|V_{tb}|^2$. A deviation of this quantity from the Standard Model prediction may indicate the existence of a fourth quark generation.

11.3 Measurement of the $t\bar{t}$ production cross section

The production cross section is so large that the top quark signal will be visible after the equivalent of one week of data taking at low luminosity in the lepton $+$ jet chan-

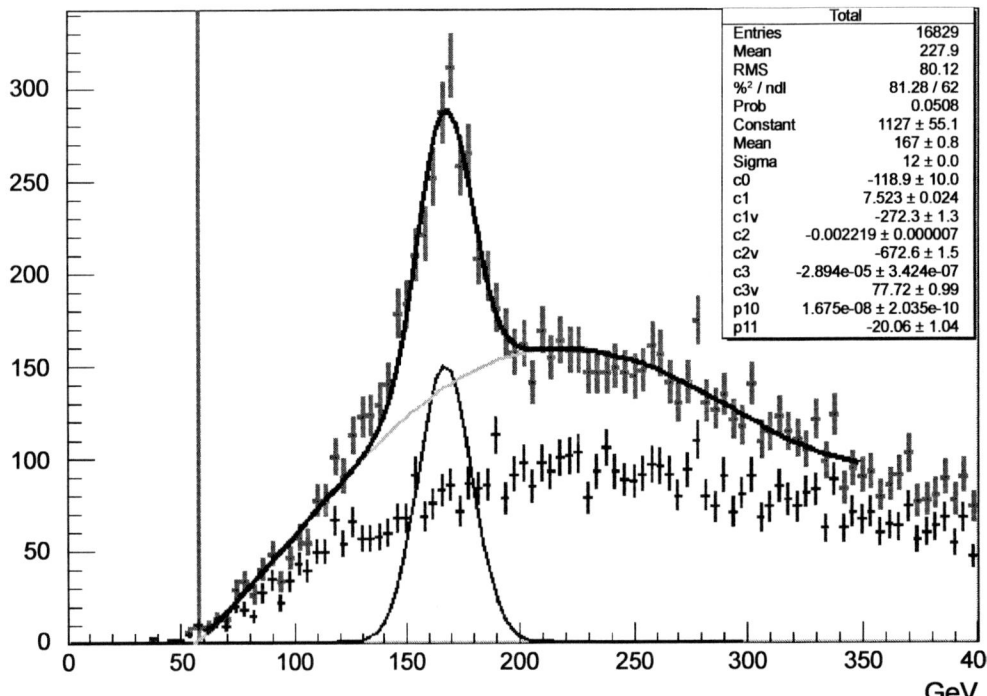

Fig. 126. Invariant mass distribution of the 3-jets combination with the highest vector p_T sum. The *full curve* fits the signal + background (W + jets and combinatorial), while the lower distribution (*points*) shows the contribution of W + jets only

nel [525]. By requiring one isolated lepton with transverse momentum $p_T > 20$ GeV/c, and exactly four jets with transverse energy ($E_T > 40$ GeV), the top signal is clearly visible above the $W + 4$-jets background in the invariant mass distribution of the most energetic jets (Fig. 126).

The cross section can thus be measured using this or similar approaches. With the large number of events collected, the statistical error will soon be negligible. After the equivalent of "one month" of data taking at low luminosity, it will be at the level of 0.4%. The overall error will be dominated by the systematic uncertainty related to the luminosity measurement. A 5% total uncertainty is achievable per experiment [286, 288]. Far more accurate measurements would be available from the ratio of the $t\bar{t}$ production to the inclusive W or Z production. A 1%–2% uncertainty on the extracted $t\bar{t}$ cross section would be far smaller than the uncertainty in the corresponding calculation. Because of its strong dependence on the top quark mass, a measurement of the production cross section together with a precise measurement of the top quark mass will provide a test of perturbative QCD. Alternatively, within the Standard Model, the cross section measurement provides a top quark mass estimate, with a potential accuracy of 3 GeV/c². A direct measurement, however, will be performed with significantly better precision.

11.4 Measurement of the single top production cross section

The top quark can also be produced in electroweak interactions. In this case, one single top quark is produced at a time. The total production cross section is expected to be ≈ 310 pb. The different production diagrams are shown in Fig. 18 in Sect. 2. The dominant process is the W-gluon fusion (t-channel) with a cross section of

≈ 250 pb [161, 162]. In the t and s channels at the LHC, the production rate of top quarks is $\approx 50\%$ higher than that of antitops [144, 161, 162], while at the TEVATRON they are identical[22]. The associate Wt production cross section is about 50–60 pb [145] and is one of the dominant backgrounds for the search for the Higgs boson in the $H \rightarrow WW^* \rightarrow \ell\nu\ell\nu$ channel [286]. The s-channel production with an expected cross section of ≈ 10 pb [144, 161] is expected to be the smallest contribution to single top production at the LHC.

It is interesting to study the three processes separately, since they have separate sets of backgrounds, their systematic errors for $|V_{tb}|$ are different, and they are sensitive to new physics in different ways. For example, the presence of a heavy W would result in an increase of the W^* signal. Instead, the existence of a FCNC process $gu \rightarrow t$ would be seen in the W-gluon fusion channel. Discriminants for the three signals are for example: the jet multiplicity (higher for Wt), the presence of more than one jet tagged as a b (this increases the W^* signal with respect to the W-gluon fusion contribution), the mass distribution of the 2-jet system (which has a peak near the W-boson mass for the Wt signal and not for the others).

The event preselection, as in the lepton + jets channels, requires a leptonically decaying W-boson. The different event topologies need dedicated final selection. The t channel is described here as an example. The full analysis is

---
[22] In pp collisions at the LHC the u-quark density is higher than the $\bar{u}$-, d- or $\bar{d}$-quark density so that more W^+ than W^- can be exchanged in the s- and t-channel, resulting in more t- than $\bar{t}$-quarks in the single-top production final state. In $p\bar{p}$ collisions at the Tevatron an equal number of u- and $\bar{u}$-quarks is present in the initial state, resulting in an equal number of t- and $\bar{t}$-quarks in the final state.

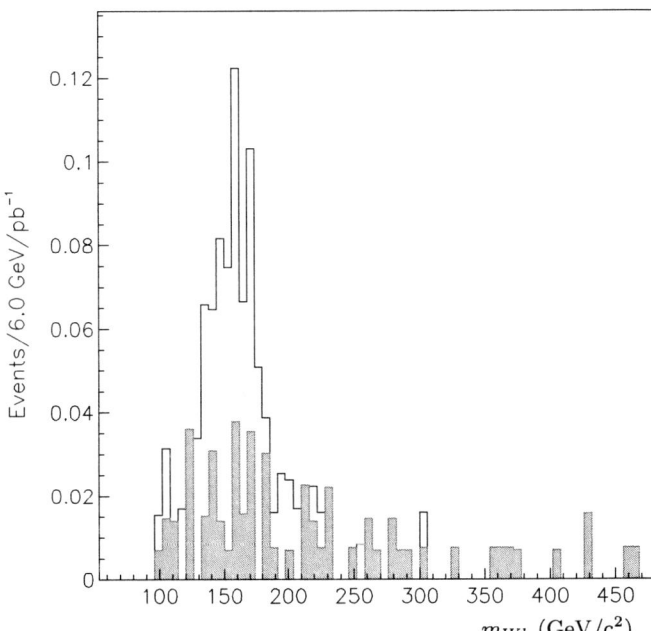

Fig. 127. Spectrum of the Wb invariant mass for the selected events obtained with a CMS full simulation. The *open histogram* represents the signal plus background expectations, the *shaded histogram* shows the background considered: $t\bar{t}$, $W + 2$-jets, and $W + 3$-jets

described in more detail in [286, 526]. The b-quark jet from the initial gluon splitting is lost in the beam-pipe. The events with a forward non b-tagged jet and one central b-tagged jet (coming from the top quark) are selected. The Wb invariant mass is then computed. The twofold ambiguity on the neutrino longitudinal momentum is resolved by choosing the one with smallest absolute value. This is true in only 55% of the cases. The result is shown in Fig. 127. The overall efficiency (including the $W \to \ell\nu$ branching ratio) is 0.3%. More than 6000 events are expected in $10\,\text{fb}^{-1}$ of integrated luminosity. The main backgrounds are $t\bar{t}$ and $W + \geq 2$ jet production. A signal-to-background ratio of 3.5 is obtained.

The single top production cross section is expected to be measured with a 10% precision. Since the electroweak production cross section is directly proportional to the square of the CKM matrix element $|V_{tb}|$, the precision on the cross section translates into an expected precision of 5% on the measurement of $|V_{tb}|$ from single top production. The statistical precision of the $|V_{tb}|$ determination from $30\,\text{fb}^{-1}$ has been estimated by ATLAS as: $\sigma(|V_{tb}|)_{(W-g)} \approx 0.4\%$, $\sigma(|V_{tb}|)_{(W-t)} \approx 1.4\%$, $\sigma(|V_{tb}|)_{(W*)} \approx 2.7\%$ [286, 527]. The dominating systematic uncertainties are expected to come from the theoretical uncertainties of the cross section calculations, in particular the scale dependence, the parton distribution functions, the experimental error on the mass of the top quark, and the luminosity measurement. The total systematic uncertainty for all three processes is estimated to be 6% [528]. A deviation of $|V_{tb}|$ from the Standard Model prediction, assuming unitarity and the existence of three quark generations, may indicate the ex-

istence of a fourth quark generation. The single top polarisation can also be measured in this channel with a 1.6% statistical precision with $10\,\text{fb}^{-1}$ [288].

11.5 Measurement of the top quark mass

At the LHC, direct measurements of the mass of the Higgs boson, the W-boson and the top quark will be carried out towards a consistency test of the Standard Model by testing the relation between m_t, m_W, and m_H.

Measurement in the lepton + jets channel. The lepton + jets $t\bar{t}$ channel is the most promising for the top quark mass measurement [529, 530]. Indeed, the leptonically decaying W-boson allows the top events to be efficiently triggered and selected. However, this channel is affected by combinatorial background. The assignment of the two b-jets to the hadronic and leptonic branches of the top quark decay is not unique, and can spoil the resolution of the top mass measurement. Studies are presently performed in ATLAS and CMS to evaluate the best strategy for the assignment of b-jets. ATLAS has found an assignment algorithm based on angular separation between the lepton and the b-jets, which results in an assignment purity of 70%–80% [531].

After the selection of the events with an energetic isolated lepton and $\not{E}_T$, the characteristics of the $t\bar{t}$ events are then used to improve the purity of the sample. The events must contain at least four energetic jets of which two must be identified as b-jets. The $b\bar{b}$ + jets, W + jets and Z + jets backgrounds are highly suppressed by this selection [529].

The top quark mass is reconstructed from the two light flavour jets from the W-boson decay and the b-jets from the top quark decay. For this reconstruction, the jet energy scale and angular resolutions are crucial. The non b-jet pairing minimising the $(M_{jj} - m_W)^2$ difference, where M_{jj} is the invariant mass of the two jets, is assumed to originate from the hadronically decaying W-boson. A difference smaller than $20\,\text{GeV}/c^2$ is required. It is finally combined with the b-jet giving the highest reconstructed top transverse momentum. The cone algorithm used to reconstruct the jets tends to underestimate the opening angle between the two jets from the W-boson. As detailed in Sect. 11.12.2, an *in situ* calibration can be applied to correct the energies and directions of the light flavour jets. However, of particular importance remains the knowledge of the b-jet energy scale, which has to be determined at the level of a percent to achieve the desired accuracy.

The distribution of the three jet invariant mass is displayed in Fig. 128 for ATLAS ([529], left) and CMS ([532], right) with an obtained signal-to-background ratio of $S/B \approx 65$. The reconstructed top quark mass is deduced from the fit value of the peak. The combinatorial background is dominant. With an integrated luminosity of $10\,\text{fb}^{-1}$, the statistical uncertainty on the top quark mass is at the level of $100\,\text{MeV}/c^2$, and hence negligible. The systematic uncertainties, summarised in [529], are dominated by two sources of systematic uncertainty, the final state radiation (FSR) and the b-jet energy scale. The FSR sys-

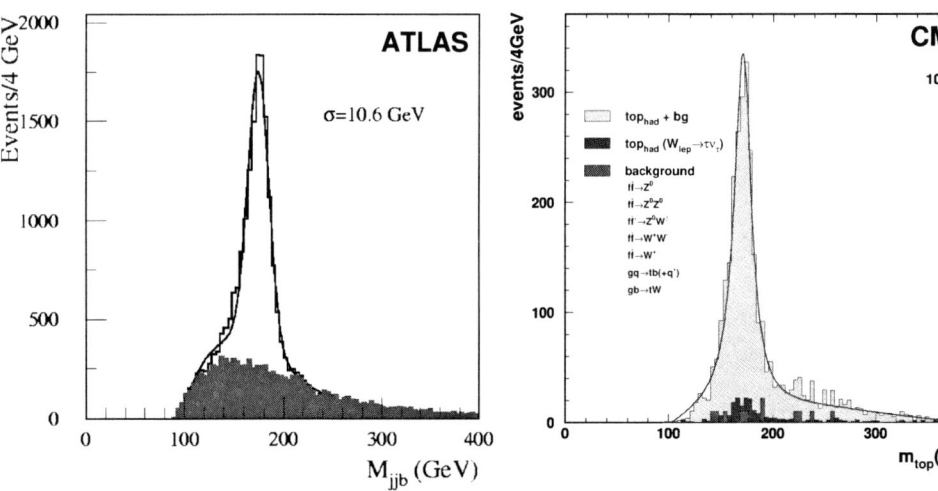

Fig. 128. Reconstructed top quark mass in the lepton +jets channel for ATLAS ([529], left) and CMS ([532], right), for $10\,\text{fb}^{-1}$ of integrated luminosity. For CMS, the combinatorial background within the event is included in the signal curve, i.e. only non-$t\bar{t}$ events are shown as background, whereas for ATLAS it is shown in the background curve

tematic error is conservatively evaluated to be 20% of the shift in the fitted top mass when disabling the FSR in the simulation and amounts to $1\,\text{GeV}/\text{c}^2$. At the LHC, the light flavour and b-jet energy scales are expected to be determined with a precision of 1% [286]. In this analysis, the b-jet energy scale systematic is $0.7\,\text{GeV}/\text{c}^2$, whereas the light flavour jet energy scale uncertainty is mostly cancelled by the *in situ* calibration and amounts to $0.2\,\text{GeV}/\text{c}^2$. Altogether, a $1.3\,\text{GeV}/\text{c}^2$ accuracy on the mass of the top quark is achievable. The effect of FSR can be lowered down to $0.5\,\text{GeV}/\text{c}^2$ if a kinematic fit is implemented. Indeed, the events with large FSR tend to have a high χ^2 and can be removed from the analysis. The systematic uncertainty thus becomes $0.9\,\text{GeV}/\text{c}^2$, dominated by the b-jet energy scale determination. An example of how to reduce the sensitivity to the heavy-flavour jet energy scale uncertainty is shown in the following.

Measurement in leptonic final states with J/ψ. A measurement of the top quark mass can be performed in lepton + jets events with an exclusively reconstructed J/ψ from a b-hadron decay in a b-jet, where the J/ψ

carries a large fraction of the b-hadron momentum because of its large mass (Fig. 129, left). The top quark is partially reconstructed from the isolated lepton coming from the W-boson and the J/ψ from the corresponding b-quark [533, 534].

To solve the twofold ambiguities on the b-quark origin, a flavour identification, requiring a muon of the same electric charge as the isolated lepton, is applied. The J/ψ can be precisely identified and reconstructed when it decays into a muon pair. As a result, one isolated lepton and three non-isolated muons are required, two of them being consistent with the J/ψ. This configuration is very rare. The invariant mass of the isolated lepton and the J/ψ is constructed (see Fig. 129, right) and the fit value of the peak turns out to depend linearly on the generated top quark mass [533, 534], as shown in Fig. 130.

The background is dominated by combinatorics, and its shape can be extracted from the data. The main systematic uncertainty comes from the b-quark fragmentation, in particular from the uncertainty in the b-hadron spectrum in top decays and that of the J/ψ spectrum in b-hadron

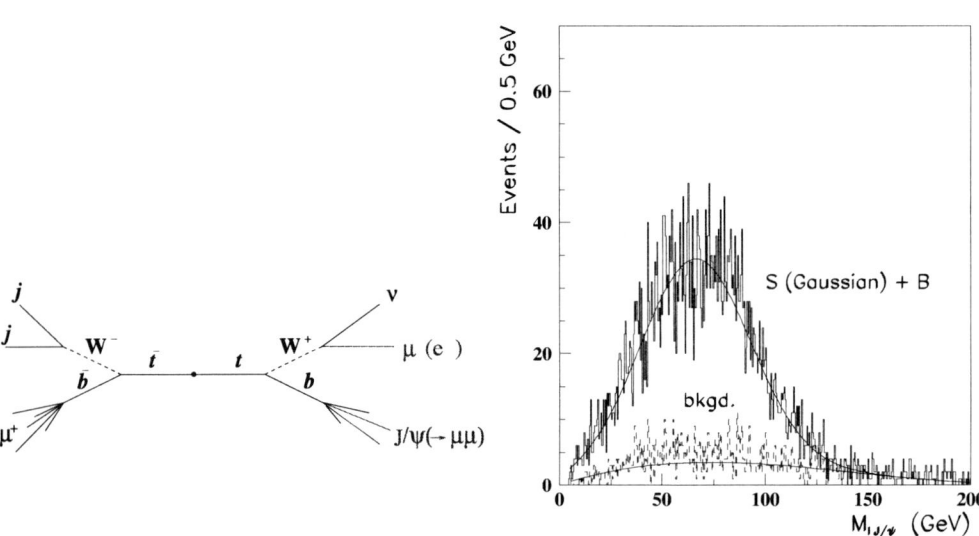

Fig. 129. *Left*: Diagram of the top decay to a leptonic final state with a J/ψ. *Right*: Example of lepton-J/ψ invariant mass in the four-lepton final state as obtained from a fast simulation of the CMS detector after four years at high luminosity running

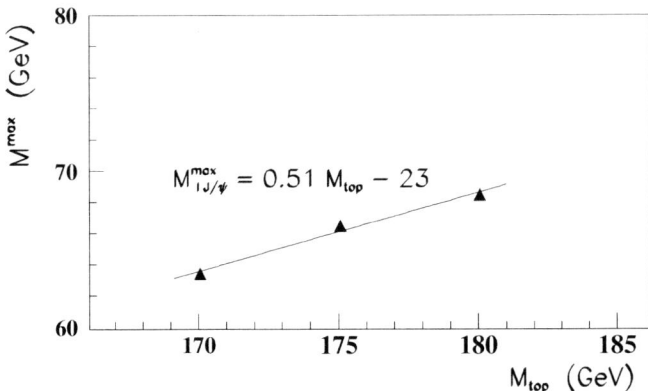

Fig. 130. Correlation between the fit value of the peak of the distribution of invariant mass between the isolated lepton and the J/ψ (M^{max}) and the generated top quark mass as obtained from a fast simulation of the CMS detector

decays. The B-factories can help in the determination of the latter. In general this method is rather insensitive to the jet energy scale and therefore complementary to the more traditional direct reconstruction of the top quark mass in lepton + jets events, adding additional information. Because of the tiny branching ratio of this final state, this analysis has to be carried out during the high luminosity phase of the LHC. One thousand events per year will be collected per experiment at high luminosity running ($\mathcal{L} = 10^{34}\,\mathrm{cm}^{-2}\,\mathrm{s}^{-1}$). Even after some $100\,\mathrm{fb}^{-1}$, the statistical error is expected to contribute significantly to the total error $\approx 1\,\mathrm{GeV}/c^2$ for a total error of $\approx 1.5\,\mathrm{GeV}/c^2$).

A more recent analysis using a similar approach considers the invariant mass spectrum of the high p_{T} lepton from the decay of the W-boson and the corresponding b-jet from the decay of the top quark, $M_{\ell b}$ [535]. In particular, the higher order moments of the invariant mass distributions up to $n = 15$ are considered, yielding a precision on the top quark mass of $500\,\mathrm{MeV}/c^2$.

Alternative measurements. A recently proposed different method for the experimental determination of the top quark mass in the lepton + jets channel is based on the mean distance of travel of b-hadrons in top quark events [458]. The dominant systematic uncertainties of this method do not depend on the b-jet energy scale and are therefore not correlated with those from the other methods, but a large number of events is required to achieve small statistical uncertainties. At the LHC, this method could be comparable to all other methods and yield a precision of $\approx 2.5\,\mathrm{GeV}/c^2$ for $10\,\mathrm{fb}^{-1}$ of integrated luminosity. Furthermore, with likely improvements in the understanding of b-hadron properties and $t\bar{t}$ events, the uncertainties associated with this method at the LHC could be substantially reduced, possibly yielding a precision on the top quark mass of ≈ 1.5–$2.0\,\mathrm{GeV}/c^2$ [458].

The top quark mass can also be measured in the fully leptonic and fully hadronic channels. The dilepton channel (ee, $e\mu$, $\mu\mu$) is very clean. However, the branching ratio is quite small and because of the presence of two neutrinos in the final state the kinematics is underconstrained.

The fully hadronic channel is experimentally very challenging. Efficient triggering is difficult and the QCD background is enormous. However, this channel has the largest branching ratio ($\approx 45\%$) and the kinematics is fully constrained. Accuracies of 2–3 GeV/c² seem feasible for these channels [529].

An ultimate overall error on the top quark mass of the order of 1 GeV/c² can be achieved when all effects are understood [529].

11.6 Electric charge of the top quark

The top quark charge has not yet been experimentally confirmed to be $+2/3$. Two approaches have been studied by ATLAS [536] to distinguish between $t(Q = 2/3) \to W^+ b$ and $t(Q = -4/3) \to W^- b$.

The first approach attempts to directly measure the top quark coupling through photon radiation in $pp \to t\bar{t}\gamma$ and $pp \to t\bar{t}$ with $t \to Wb\gamma$. Since the first process is at the LHC dominated by gluon–gluon fusion without any initial state photon radiation, the $t\bar{t}\gamma$ cross section is expected to be approximately proportional to Q_{t}^2. Using suitable selection criteria, in particular the transverse momentum of the radiated photons P_{T}^γ, the hard photon radiation off the top quarks can be enhanced.

The second approach measures the charges of all top quark decay products. Whereas for the leptonically decaying W-boson decay the charge is given by the charge of the lepton (e or μ), a jet-by-jet estimation of the b-quark charge is very difficult. Quantities correlated to the charge of the b-quark allow to distinguish between the cases on a statistical basis. As an example, the jet charge $Q_{\mathrm{jet}} = \frac{\sum_i q_i |P_{\mathrm{T}_i}|}{\sum_i |P_{\mathrm{T}_i}|}$, summing over all tracks in a jet, has been studied.

Both methods are expected to allow an unambiguous measurement of the top quark charge after one year of data taking at low luminosity ($10\,\mathrm{fb}^{-1}$) [536].

11.7 Helicity of the W-boson in top quark decay

The W polarisation (helicity) in the top quark decay is a sensitive probe of new physics in top production and decay. At the production level, a non-exhaustive list involves either anomalous $gt\bar{t}$ couplings, which naturally arise in dynamical electroweak symmetry breaking models such as technicolour or topcolour, or new interactions, as for example a strong coupling of the top quark with a heavy spin-0 resonance, such as a heavy (pseudo)scalar Higgs boson as predicted e.g. by SUSY models ($gg \to H \to t\bar{t}$), or the presence of extra dimensions. At the decay level, deviations from the Standard Model can for example arise from Wtb anomalous couplings, such as a $V + A$ contribution in the vertex structure [364, 366, 368], or from a decay to a charged Higgs boson.

Tests of the V-A nature of the tWb vertex through a measurement of the W helicity will be extended from the TEVATRON to the LHC. Current estimates are that the

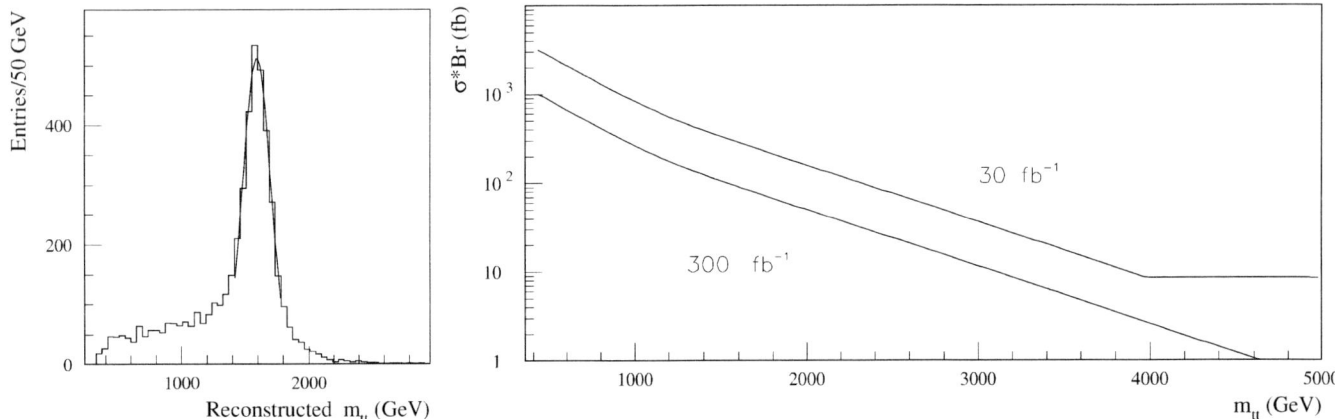

Fig. 131. *Left*: Measured $t\bar{t}$ invariant mass distribution for the reconstruction of a narrow resonance of mass $1600\,\mathrm{GeV/c^2}$ decaying to a $t\bar{t}$ pair. *Right*: Value of σB required for a 5σ discovery potential for a resonance decaying to a $t\bar{t}$ pair, as a function of its mass $m_{t\bar{t}}$

longitudinal fraction ($\mathcal{F}_0$) can be measured with a precision of about 5% [527, 537] with $10\,\mathrm{fb^{-1}}$ of data in CMS. ATLAS expects a precision of 2% on $\mathcal{F}_0$ and 1% on $\mathcal{F}_+$ with $10\,\mathrm{fb^{-1}}$ from a combination of the dilepton and the lepton + jets channel [538].

11.8 Top quarks in exotic models

Due to its large mass, the top quark could be part of the decay products of massive particles [528]. Its clear experimental signature makes it a very interesting tool to study the exotic decays. Some examples include: "Heavy top" in little Higgs models, signatures which include the top quark in models with extra-dimensions, search for $t\bar{t}$ production via intermediate resonances (as predicted in the Standard Model Higgs sector, MSSM Higgs, Technicolour models, strong electroweak symmetry breaking models, Topcolour, etc.). Physics beyond the Standard Model could affect cross section measurements for $t\bar{t}$ production in a variety of ways: Topcolour-Assisted Technicolour could result in the production of like-sign top quark pairs at the LHC, $pp \to tt + X$ [539]. In another scenario, a heavy resonance decaying to $t\bar{t}$ might enhance the cross section, and might produce a peak in the $t\bar{t}$ invariant mass spectrum. Because of the large variety of models and their parameters, a study of the sensitivity to a "generic" narrow resonance decaying to $t\bar{t}$ was performed for the ATLAS detector [286, 288]. Events of the single lepton topology were selected. In addition, between four and ten jets were required with $p_\mathrm{T} > 20\,\mathrm{GeV}$ and $|\eta| < 3.2$, with at least one of them tagged as a b-jet. After these cuts, the background is dominated by the $t\bar{t}$ continuum. The obtained mass resolution $\sigma(m_{t\bar{t}})/m_{t\bar{t}}$ is approximately 6.6%. As an example, Fig. 131 (left) shows the reconstructed $m_{t\bar{t}}$ distribution for a narrow resonance of mass $1,600\,\mathrm{GeV/c^2}$.

The reconstruction efficiency, not including branching ratios B, is about 20% for a resonance of mass $400\,\mathrm{GeV/c^2}$, decreasing gradually to about 15% for $m_{t\bar{t}} = 2\,\mathrm{TeV/c^2}$. For

a narrow resonance X, Fig. 131 (right) shows the required $\sigma B(X \to t\bar{t})$ for a discovery ($\geq 5\sigma$). Results are shown as a function of m_X for an integrated luminosity of $30\,\mathrm{fb^{-1}}$ and $300\,\mathrm{fb^{-1}}$. With $30\,\mathrm{fb^{-1}}$, it is estimated that a resonance can be discovered at $4\,\mathrm{TeV/c^2}$ for $\sigma B = 10\,\mathrm{fb}$, and at $1\,\mathrm{TeV/c^2}$ for $\sigma B = 1000\,\mathrm{fb}$.

11.9 Spin correlations in $t\bar{t}$ production

Top-antitop pairs, produced in QCD processes, are not polarised[23]. However, the top and antitop spins are correlated (Sects. 2.4.4 and 6.1). For QCD processes close to the production threshold, the $t\bar{t}$ system is produced in a ${}^3S_1$ state for $q\bar{q}$ annihilation, or in a ${}^1S_0$ state for gluon–gluon fusion. Hence, in the first case, the top and the antitop have parallel spin, while in the second case the spins are antiparallel. Since at LHC energies, the gluon–gluon process has a much larger overall cross section than the $q\bar{q}$ annihilation, NLO calculations predict an excess of top pairs with opposite spins, i.e. in most cases both top and antitop quark are either both left or both right handed [200, 201]. For this study, the spin of the top quarks is evaluated in the helicity basis, which corresponds to the top (antitop) direction of flight in the $t\bar{t}$ system. In this basis, the asymmetry parameter, which expresses the excess of same-helicity pairs, is given by:

$$\mathcal{A} = \frac{\sigma(t_L\bar{t}_L) + \sigma(t_R\bar{t}_R) - \sigma(t_L\bar{t}_R) - \sigma(t_R\bar{t}_L)}{\sigma(t_L\bar{t}_L) + \sigma(t_R\bar{t}_R) + \sigma(t_L\bar{t}_R) + \sigma(t_R\bar{t}_L)} . \quad (261)$$

At the LHC, this asymmetry is expected to be $\mathcal{A} = 0.327$ [200]. Simulations by ATLAS and CMS achieve with $\mathcal{A} = 0.31 \pm 0.03$ very consistent results [537, 540]. With

---

[23] Top and antitop quarks receive a small (2%) polarisation perpendicular to the scattering plane via QCD final state interactions [189, 190]. An additional, very small contribution of top/antitop quark polarisation is received from mixed QCD/weak interactions in the scattering plane [191].

an optimal choice of the quantisation axis, the maximum spin correlation asymmetry is 0.5 [205]. Deviations from the Standard Model expectations could be indications for alternative or additional $t\bar{t}$ production mechanisms, for example $gg \rightarrow H \rightarrow t\bar{t}$ [139].

For $t\bar{t}$ pairs produced with total invariant mass much larger than the production threshold, the asymmetry is diluted because of the presence of higher mass spin $t\bar{t}$ pairs, which are predominantly produced in unlike-spin configurations. Hence it is useful to introduce a cut on the total invariant mass $m_{t\bar{t}} < 550\,\mathrm{GeV/c^2}$ to maximise the power of the experimental analysis. A significant deviation of the measured asymmetry parameter from the theoretical value may indicate non-Standard Model physics, such as $t\bar{t}$ production via intermediate resonances or right-handed weak interactions.

The asymmetry is evaluated by studying the angular distributions of the top and antitop quark decay products, which are used as spin analysers [200]:

$$\frac{1}{N}\frac{\mathrm{d}^2 N}{\mathrm{d}\cos\theta_1 \cos\theta_2} = \frac{1}{4}\left(1 - \kappa \cos\theta_1 \cos\theta_2\right), \quad (262)$$

$$\frac{1}{N}\frac{\mathrm{d}N}{\mathrm{d}\cos\phi} = \frac{1}{2}\left(1 - \lambda \cos\phi\right), \quad (263)$$

where θ_1 (θ_2) are the angles between the decay products of the t $(\bar{t})$ quark in the t $(\bar{t})$ rest frame and the t $(\bar{t})$ direction in the $t\bar{t}$ frame, while ϕ is the angle between the decay products of the t and $\bar{t}$ quark in the respective rest frame [541]. The parameters κ and λ are the spin correlation variables. Note that the corresponding studies at the TEVATRON only use κ.

The measurement of the spin correlation parameters has been simulated in Monte Carlo events in the dilepton and the lepton + jets channels. For the dilepton channel, which has at LHC much higher statistics than at the TEVATRON and can be fully exploited, the natural choice for spin analysers are the two leptons from the decay of the W-bosons, since leptons are 100% polarised with respect to the top quark spin. In the lepton + jets channel, the best choice for spin analysers would be a lepton and a d-type quark, since the d-type quarks have 100% polarisation as well. However, since it is experimentally impossible to distinguish between light quark flavours, the second spin analyser is the least energetic jets in the t $(\bar{t})$ rest frame, which is polarised at 51%.

In the ATLAS study [541], the total generated Monte Carlo sample in this study corresponds to $10\,\mathrm{fb^{-1}}$ and generated with varying production parameters to study the systematic effects of: Q^2 factorisation scales, parton distribution functions, initial and final state radiation, b-fragmentation and hadronisation scheme, b-tagging and b-jet energy calibration, generated top quark mass. The results of the simulated spin correlation measurements in the two channels are summarised in Table 67. The spin correlation parameter λ appears more suitable for the measurement, since it is less sensitive to the systematic uncertainties.

Table 67. Comparison between the Standard Model predictions at Born-level (LO) and the simulated Monte Carlo measurement for the spin correlation parameters κ and λ in the dilepton and the lepton + jets samples with a cut on $M_{t\bar{t}} < 550\,\mathrm{GeV/c^2}$. The quoted precision combines statistical and systematic uncertainties for a measurement in $10\,\mathrm{fb^{-1}}$

Parameter	Dilepton	Lepton+jets
κ (SM)	0.42	0.21
$\Delta\kappa/\kappa$	8%	20%
λ (SM)	-0.29	-0.15
$\Delta\lambda/\lambda$	5%	13%

Using essentially the same technique, the CMS collaboration [527, 537] has estimated that the relative asymmetry can be measured to about 10% accuracy, shared almost equally between statistical and systematic uncertainties, with $30\,\mathrm{fb^{-1}}$ of data.

11.10 Measurements of top quark couplings

Measurement of the $tt\gamma$ and ttZ couplings. Currently, little is known about top quark couplings to the photon and Z-boson. There are no direct measurements of these couplings; indirect measurements, using LEP data, tightly constrain only the ttZ vector and axial vector couplings. All others are only very weakly constrained by LEP and/or $b \rightarrow s\gamma$ data. The ttV $(V = \gamma, Z)$ couplings can be measured directly in $e^+e^- \rightarrow t\bar{t}$ at a future e^+e^- linear collider. However, such a machine is at least a decade away. In addition, the process $e^+e^- \rightarrow t\bar{t}$ is simultaneously sensitive to $tt\gamma$ and ttZ couplings, and significant cancellations between various couplings may occur.

$t\bar{t}\gamma$ production and $t\bar{t}Z$ production at hadron colliders can be considered as tools to measure the ttV couplings [542, 543]. For $t\bar{t}\gamma$ production, the $\gamma\ell\nu b\bar{b}jj$ final state is considered, for $t\bar{t}Z$, the Z-boson is assumed to decay leptonically, yielding trilepton $(\ell'^+\ell'^-\ell\nu b\bar{b}jj)$ and dilepton $(\ell'^+\ell'^- b\bar{b} + 4j)$ final states. All relevant background processes are included. Once $t\bar{t}\gamma$ or $t\bar{t}Z$ selection cuts are imposed, the total background is substantially smaller than the signal. In all calculations, both b-quarks are assumed to be tagged.

At the TEVATRON, the $t\bar{t}Z$ cross section is too small to be observable. The $t\bar{t}\gamma$ cross section is large enough to allow a first, albeit not very precise, test of the $tt\gamma$ vector and axial vector couplings, provided that an integrated luminosity of more than $5\,\mathrm{fb^{-1}}$ can be accumulated. No useful limits on the dipole form factors $F_{2V,A}^\gamma$ can be obtained. Since $q\bar{q}$ annihilation dominates at TEVATRON energies, initial state photon radiation severely limits the sensitivity of $t\bar{t}\gamma$ production to anomalous top quark couplings.

This is not the case at the LHC, where gluon fusion is the dominant production mechanism. Combined with an integrated luminosity of $30\,\mathrm{fb^{-1}}$, one can probe the $tt\gamma$ couplings with a precision of $\approx 10\%\text{--}25\%$ per experiment. With $300\,\mathrm{fb^{-1}}$, a 4%–7% measurement of the

$t\bar{t}\gamma$ vector and axial vector couplings can be expected, while the dipole form factors $F^\gamma_{2V,A}$ can be measured with 20% accuracy. Finally, if the luminosity of the LHC can be upgraded by a factor of 10 without significant losses of particle detection efficiency for photons, leptons and b-quarks, these limits can be improved by another factor 2–3.

The $t\bar{t}Z$ cross section with leptonic Z decays is approximately a factor 20 smaller than the $t\bar{t}\gamma$ rate. It is therefore not surprising that the sensitivity limits on the $t\bar{t}Z$ couplings are significantly weaker than those which are expected for the $t\bar{t}\gamma$ couplings. For $300\,\text{fb}^{-1}$, the $t\bar{t}Z$ vector (axial vector) couplings can be measured with a precision of 45%–85% (15%–20%), and $F^Z_{2V,A}$ with a precision of 50%–55%. At an upgraded LHC, these bounds can be improved by factors of 1.4–2 (≈ 3) and 1.6, respectively.

Measurement of the top quark Yukawa coupling. For small Higgs boson masses ($\leq 130\,\text{GeV/c}^2$), the $H \to b\bar{b}$ decay channel is dominant. Unfortunately, it is impossible to efficiently trigger the acquisition of these events due to the huge dijet $b\bar{b}$ background present at the LHC. To observe the $b\bar{b}$ decay of the Higgs boson, an associate production mode, together with a W- or Z-boson or with a $t\bar{t}$ pair, has to be considered. The $t\bar{t}H$ production Feynman diagrams are shown in Fig. 132. These channels allow the measurement of the top quark Yukawa coupling. The expected corresponding cross section is small: $\sigma(m_H = 120\,\text{GeV/c}^2) = 0.8\,\text{pb}$, while the $t\bar{t}b\bar{b}$ background has a cross section of 3 pb.

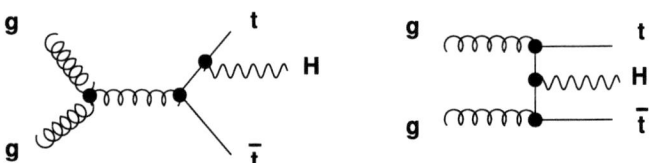

Fig. 132. Feynman diagrams for the associate $t\bar{t}H$ production at the LHC

The lepton + jets events are first selected as described before. This final state is intricate, since in addition to the "usual" lepton + jet event, two additional b-jets from the Higgs boson decay are present. As a result, the event selection requires at least six jets in the final state of which exactly four are b-jets.

Both W-bosons are fully reconstructed. The two b-jets from the top quark decay have to be identified and the pair giving the "best" reconstructed top quark pair is chosen. The remaining two b-jets are combined to reconstruct the Higgs boson. The resulting invariant mass distributions are shown in Fig. 133, demonstrating good agreement between the ATLAS and CMS analyses and a peak due to the presence of the Higgs boson.

The shape of the background can be extracted from $t\bar{t}jj$ data. With $30\,\text{fb}^{-1}$, 40 signal events are expected [288, 544–549], with a significance of 3.6σ. A 16% precision on the Yukawa coupling is expected to be achievable. The combination of the low and high luminosity runs given an integrated luminosity of $100\,\text{fb}^{-1}$ will allow a 4.8σ significance observation and a 12% measurement precision on the top quark Yukawa coupling to be reached. All these numbers are for a Higgs boson mass of $120\,\text{GeV/c}^2$.

A measurement of the process $pp \to t\bar{t}H + X$ at the LHC and a measurement of the Higgs boson branching ratios $B(H \to b\bar{b})$ and $B(H \to W^+W^-)$ at a future linear e^+e^- collider can be combined to determine the top quark Yukawa coupling [550]. For $300\,\text{fb}^{-1}$ at the LHC and $500\,\text{fb}^{-1}$ at a linear collider, the obtainable relative uncertainty is $\approx 15\%$ for a Higgs boson between 120 and 200 GeV/c^2. The purely statistical uncertainty ranges from 7% to 11%. The size of the expected total precision is comparable to those expected for the LHC alone [551, 552]. However, in contrast to the latter no model-dependent assumptions need to be made.

11.11 Rare decays

With its large mass, the top quark will couple strongly to the electroweak symmetry-breaking sector. Many models

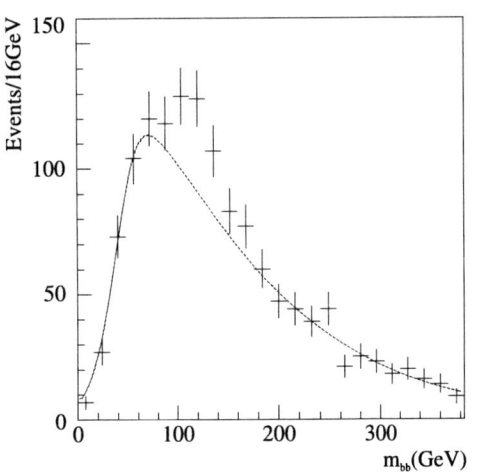

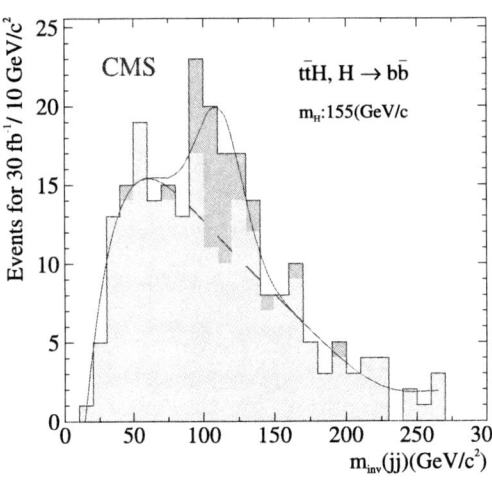

Fig. 133. Signal plus background expectation for the reconstructed Higgs boson mass in the $t\bar{t}H$ channel with $H \to b\bar{b}$ in the ATLAS (*left*) and CMS (*right*) detectors for $m_H = 120\,\text{GeV/c}^2$ and $m_H = 115\,\text{GeV/c}^2$ with $100\,\text{fb}^{-1}$ and $30\,\text{fb}^{-1}$, respectively. In both cases, fast simulations of the detectors have been used

of physics beyond the Standard Model include a more com-plicated electroweak symmetry breaking sector, with im-plications for the top quark decay. Examples include the possible existence of charged Higgs boson, or possibly large flavour changing neutral currents (FCNC) in the top de-cay. In the Standard Model, FCNC decays of the top quark are highly suppressed ($B < 10^{-13}$–10^{-10}). However, sev-eral extensions of the Standard Model can lead to very sig-nificant enhancements of the branching ratios (10^{-3}–10^{-2} or even higher). The sensitivity to some of these scenar-ios [288] has been investigated by both ATLAS [286] and CMS [528].

In particular the processes $t \to Zq$, $t \to \gamma q$, $t \to gq$, $t \to WbZ$, $t \to WbH$, and $t \to Hq$ have been studied. It has been shown that the limit of the branching ratios of these processes can be improved by orders of magnitude with re-spect to the current limits, and will range from 10^{-7} for $t \to WbZ$ to 10^{-3} for $t \to gq$ using $100\,\mathrm{fb}^{-1}$ of data.

The Wtb vertex structure can be probed using either top quark pair production or single top quark production processes. The $t\bar{t}$ cross section depends weakly on this ver-tex, but there are several sensitive observables, like C and P asymmetries, top quark polarisation, and spin corre-lations, which can provide interesting information [553]. The single top production rate is instead directly pro-portional to the square of the Wtb coupling. Defining the anomalous couplings for left- and right-handed top quarks:

$$F_{2L} = \frac{2M_W}{\Lambda}\eta^W(-f^W - ih^W),\qquad(264)$$

$$F_{2R} = \frac{2M_W}{\Lambda}\eta^W(-f^W + ih^W),\qquad(265)$$

where Λ is the energy scale associated with the new physics, f^W and h^W are the magnetic and electric dipole moments, respectively, and $\eta^W = \pm 1$ indicates if the new interaction interferes with the Standard Model con-structively ($\eta = +1$) or destructively ($\eta = -1$). With 10% systematic uncertainty on the single top produc-tion cross section, the TEVATRON will have sensitivity to $F_{2L}{}^{>+0.55}{}_{<-0.18}$ and to $F_{2R}{}^{>+0.25}{}_{<-0.24}$. With an ex-pected precision of 5% on the single top production cross section, the LHC will have sensitivity to $F_{2L}{}^{>+0.097}{}_{<-0.052}$ and to $F_{2R}{}^{>+0.13}{}_{<-0.12}$ [553].

11.12 Detector commissioning studies

The top pair production process is valuable for the *in situ* calibration of the LHC experiments in the detector commissioning stage: the large cross section and the large signal-to-background ratio for the lepton + jets $t\bar{t}$ channel allow the selection of high purity samples with large statis-tics in a short time. The experimental signature of $t\bar{t}$ events – high energy leptons and jets, b-jets, missing E_T – involves most parts of the detectors. Therefore top quark samples will play an important role in the calibration of the detec-tors in the initial phase.

11.12.1 Top quark analyses without b-tagging

The experimental signature for top quark events include one or more b-jets, arising from top quark decay. Thus, the b-tagging performance of the ATLAS and CMS ex-periments will play a crucial role for top quark analyses. However, efficient b-tagging relies on precise alignment of the tracking detectors, which will be reached in a first it-eration only after a few months of data taking. Feasibility studies [531] are undertaken in order to assess whether it will already during the commissioning stage be possible to reconstruct $t\bar{t}$ events without the use of any b-tagging.

The kinematic cuts applied to select the top quark en-riched sample require exactly one isolated lepton with $p_T > 20\,\mathrm{GeV}$ and exactly four jets, reconstructed with a cone algorithm of size $\Delta R = 0.4$, with $p_T > 40\,\mathrm{GeV}$. The analy-sis reconstructs exclusively the hadronic branch of the $t\bar{t}$ event: the combination of 3 out of 4 jets with the highest vector p_T sum is assumed to originate from the same top (antitop) branch. The invariant mass of the 3-jets combina-tion is an estimate of the top mass.

The analysis is performed on a Monte Carlo sample with $m_t = 175\,\mathrm{GeV/c^2}$, including $t\bar{t}$ and W + 4-jets back-ground. The generated sample corresponds to $150\,\mathrm{pb}^{-1}$, i.e. a few days of data taking in the initial luminosity phase. The invariant mass distribution, shown in Fig. 126, is fitted with a Gaussian curve for the top mass peak plus a poly-nomial curve which accounts for the W + jets signal and the combinatorial background. The fitted Gaussian curve peaks at $167.0\,\mathrm{GeV/c^2}$, with an RMS of $12\,\mathrm{GeV/c^2}$. This study shows that it will be possible to reconstruct $t\bar{t}$ events in the absence of b-tagging.

11.12.2 Energy scale calibration from w-bosons in top decay

In order to obtain a precise measurement of the mass of the top quark, precise knowledge of the absolute energy scale of hadronic jets is of paramount importance. Miscalibra-tions can arise from detector effects (dead channels, im-precise cell weighing), physics effects (final state radiation, pile-up), and cone algorithm effects (out-of-cone energy, jet overlap). The best method to achieve a reliable calibration is to use the physics data to reconstruct particle of known properties. One such technique reconstructs the hadronic W-boson from the lepton + jets $t\bar{t}$ sample. The kinematic cuts used to select the events are the same as the "commis-sioning top" cuts described above. In addition, a quality cut is applied: the 4-vectors of the fourth jet, the lepton and the $\not{E}_T$ are summed, and the invariant mass of the re-sult is formed. This mass is required to be in the range $140\,\mathrm{GeV/c^2} < M_{j_4\ell\nu} < 200\,\mathrm{GeV/c^2}$. With this set of cuts, the purity of the W-boson in the selected sample is $\approx 85\%$.

Since the mass of the W-boson is known to a precision of a few tens of MeV/c^2, M_W can be used to obtain calibra-tion factors for the energies of the jets originating from the decay of the W-boson. For the jets j_1, j_2 from the W-boson decay, momentum conservation implies

$$M_W^2 = 2E_{j_1}E_{j_2}(1 - \cos\theta_{j_1 j_2}),\qquad(266)$$

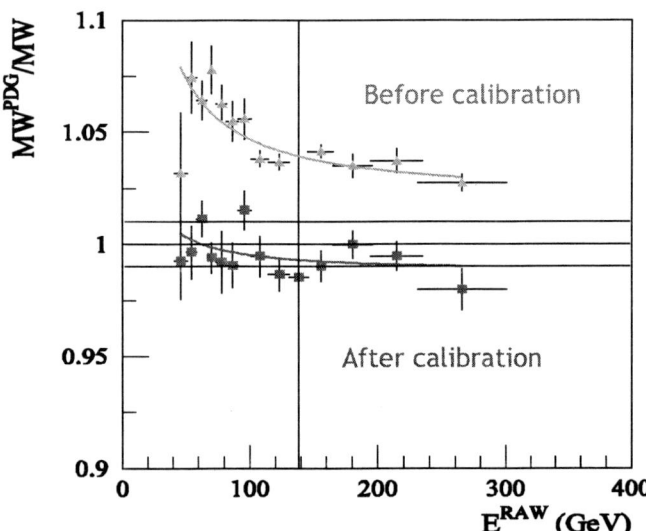

Fig. 134. Ratio between the nominal and reconstructed mass of the W-boson, as a function of the uncalibrated W energy, before and after the jet energy scale calibration

where E_{j_i} indicates the energy of jet i and $\theta_{j_1 j_2}$ indicates the angle between the two jets. Thus, the mass measurement is influenced both by the resolution of the jet energy scale and the angular measurement. To disentangle angular and energy contributions, the invariant mass of the two-jet system, the nominal mass of the W-boson and the kinematic properties of the jets are used as input in a constrained χ^2 fit:

$$\chi^2 = \left(\frac{m_{jj} - m_W}{\sigma_{m_W}}\right)^2 + \sum_{i,X} \left(\frac{X_i - \alpha_E^i X_i}{\sigma_X}\right)^2 , \quad (267)$$

where $X = E, \eta, \phi$. The result of the fitting procedure is the correction factor α_E, which is assumed to depend on the jet energy E, yielding a correction function $\alpha_E = f(E)$. The application of the correction function allows the reconstruction of the mass of the W-boson with a 1% precision, shown in Fig. 134 (right). To obtain such a level of precision, only $10\,000$ $t\bar{t}$ events are necessary. This amount of data will be available within one or two months of data taking at the LHC. The calibration factors have been applied

Table 68. Summary of observed and expected precision and sensitivity to top quark properties in and beyond the Standard Model at the TEVATRON and the LHC. Quoted uncertainties are the quadratic sum of statistical and systematic uncertainties per experiment unless specified otherwise

quantity	CDF/DØ		ATLAS/CMS	
$\Delta\sigma_{t\bar{t}}/\sigma_{t\bar{t}}$	11% with 1 fb^{-1}	[554]	5%–10% luminosity systematics dominated	[286, 288]
$\Delta\sigma_{\text{single-top}}/\sigma_{\text{single-top}}$	26% with 1 fb^{-1}	[554]	10% (< 2% stat. error with 10 fb^{-1}	[288])
$B(t \to Wb)$	3.3% with 1 fb^{-1}	[554]		
V_{tb} from $\sigma_{\text{single-top}}$	14% with 1 fb^{-1}	[554]	6.5%	[528]
V_{tb} from $B(t \to Wb)$	> 0.22 with 1 fb^{-1}	[554]	0.2% (stat. only)	[286]
single-top polarisation	–		1.6% with 10 fb^{-1}	[288]
$\Delta m_{\text{top}}/m_{\text{top}}$	$\leq 2\,\text{GeV}/c^2$	Sect. 7	$\approx 1\,\text{GeV}/c^2$	[286, 288]
spin correlation θ	40% (2 fb^{-1})	[538]	7% ($\ell\ell \oplus \ell+$jets) for 10 fb^{-1}	[538]
spin correlation ϕ	–		4% ($\ell\ell \oplus \ell+$jets) for 10 fb^{-1}	[538]
W-helicity $\mathcal{F}_0$	6.5% with 1 fb^{-1}	[554]	2%–5% with 10 fb^{-1}	[527, 537, 538]
W-helicity $\mathcal{F}_+$	2.6% with 1 fb^{-1}	[554]	1% with 10 fb^{-1}	[538]
electric charge q_t	distinguish $\frac{2}{3}$ and $\frac{4}{3}$ cases with 1 fb^{-1}	Sect. 7.2	distinguish $\frac{2}{3}$ and $\frac{4}{3}$ cases with 10 fb^{-1}	[536]
Yukawa coupling y_t	–		4.8σ, 16% (12%) with 30(100) fb^{-1}	[548, 549]
FCNC $B(t \to gq)$	$< 1.9 \times 10^{-2}$ with 2 fb^{-1}	[288, 555]	$< 1 \times 10^{-5}$ - $< 1.4 \times 10^{-3}$ (10 fb^{-1})	[288, 556]
FCNC $B(t \to Zq)$	$< 1.5 \times 10^{-2}$ with 1 fb^{-1}	[554]	$< 6.5 \times 10^{-4}$ - 1.3×10^{-3} with 10 fb^{-1}	[286, 288, 556]
FCNC $B(t \to \gamma q)$	$< 3.0 \times 10^{-3}$ with 1 fb^{-1}	[554]	$< 8.6 \times 10^{-5}$ - 1.9×10^{-4} with 10 fb^{-1}	[286, 288, 556]
FCNC $B(t \to WbZ)$	–		$< 10^{-7}$ with 100 fb^{-1}	[553]
$\Delta\sigma^{M_{Z'}=1\,\text{TeV}/c^2}$ $B(Z' \to t\bar{t})$	100 fb with 1 fb^{-1}	[554]	700 fb with 30 fb^{-1}	[286, 288]
anom. coupling	$F_{2L}{}^{>+0.55}_{<-0.18}$	[553]	$F_{2L}{}^{>+0.097}_{<-0.052}$	[553]
	$F_{2R}{}^{>+0.25}_{<-0.24}$	[553]	$F_{2R}{}^{>+0.13}_{<-0.12}$	[553]
$\Delta F_{1V,A}^Z$	–	[542]	15%–85% (300 fb^{-1})	[542]
$\Delta F_{1V,A}^\gamma$	$^{<+1.03...+2.60}_{>-1.17...-1.88}$ (8 fb^{-1})	[542]	15%–50% (30 fb^{-1}), 4%–7% (300 fb^{-1})	[542]
$\Delta F_{2V,A}^\gamma$	–	[542]	35% (30 fb^{-1}), 20% (300 fb^{-1})	[542]
$\Delta F_{2V,A}^Z$	–	[542]	55% (300 fb^{-1})	[542]

to a Z + jets sample, providing the same level of precision and demonstrating that the result of the calibration can be used for other physics signatures involving light quark jets.

11.13 Conclusion on top quark physics at the LHC

The physics of the top quark will be one of the main physics topics at the LHC. Many exciting analyses will be carried out. Only a few of them have been summarised in this section. Most of the analysis can be done with the first $10\,fb^{-1}$. Due to the large production cross section, the statistical uncertainty will be, in most of the cases, negligible. Table 68 summarises the expected precision and sensitivity to top quark properties in and beyond the Standard Model at the TEVATRON and the LHC.

The top quark mass measurement will be a key issue of top quark physics at the LHC. A $1\,GeV/c^2$ precision is expected to be reached, provided that an excellent understanding of the detectors to control the systematic uncertainties can be achieved. This will be a real challenge for the ATLAS and CMS experiments. The study of the top quark sector highlights several theoretical areas, where further improvements are expected such as the higher order QCD calculations and the b-fragmentation. Finally, most of the analyses carried out so far make use of fast simulation of the ATLAS and CMS detectors. At present, many of those studies are repeated with full simulation and a serious focus on the systematic uncertainties. First attempts are also started to reduce the systematic uncertainties and make measurements of selection efficiencies, resolutions and on the level of background contamination from data-type analyses in control samples rather than from the Monte Carlo simulations. These are crucial steps in the preparation for the upcoming LHC era ahead of us in the near future.

12 Summary

The discovery of the top quark ten years ago has opened a new and rich field of physics that is attracting much attention. In particular, the analyses of the TEVATRON Run II data in the last five years have allowed not only the measurement of the strong $t\bar{t}$ pair production cross section of about 7 pb, close to the Standard Model expectation of 6.7 pb, with a precision of $\approx 15\%$ but also the measurement of the top quark mass with a relative precision of 1.7%[24].

Since there is no significant discrepancy between the measurement of the total and the differential $t\bar{t}$ production cross section and the Standard Model predictions, extensions of the Standard Model in which $t\bar{t}$ production is enhanced ($t\bar{t}$ production via intermediate resonances, the Standard Model Higgs boson decay $H^0 \to t\bar{t}$, etc.)

or the top quark can decay to non-Standard Model particles (to a charged Higgs boson, via FCNC couplings, etc.) have been constrained. Further detailed comparisons of the top quark's production and decay characteristics from different channels are needed to test the Standard Model.

Many novel techniques have been developed over the recent years by the CDF and DØ experiments, increasing the precision of the measurements of top quark properties and the sensitivity for new physics in the top quark sector beyond any expectation.

The TEVATRON is running extremely well with further improvements on the delivered luminosity expected for the next two years, yielding 4–$8\,fb^{-1}$ of integrated luminosity by 2009. Also the Run II detector upgrades, to be completed by the spring of 2006, turned out to be very successful. Under those excellent conditions, CDF and DØ will focus on and enthusiastically pursue the top quark analyses in order to fully exploit the top physics potential of the TEVATRON and pave the way for the LHC. The present status and prospects are as follows:

– The $t\bar{t}$ pair production cross section via the strong interaction has been observed and measured in all decay channels with varying levels of assumptions and using different techniques. Analyses with τ-leptons in the final state are being pursued and to be released as first preliminary results soon. All measurements are so far consistent between the different channels, methods used and assumptions made, and between the two experiments. A total $t\bar{t}$ production cross section of about 7 pb has been measured with a precision of about 15% for individual measurements and close to 14% for the combination per experiment. Already now, the systematic and statistical uncertainties are comparable, in some analyses of larger data sets the systematic uncertainty is already dominant. The statistical uncertainty will be significantly reduced in analyses of $1\,fb^{-1}$ or more. All $t\bar{t}$ pair production cross section measurements are already now sufficient to allow a meaningful comparison of the $t\bar{t}$ production cross section with QCD predictions, which presently have 15% uncertainty.
– One of the most important top quark properties is its mass. The different measurements of the top quark mass by CDF and DØ are consistent with each other and have been combined by the TEVATRON Electroweak Working Group, yielding the large and precise value of $m_{top} = 172.7 \pm 2.9\,GeV/c^2$. In this forum, work is ongoing to prepare complete and comparable measurements of the top quark mass, which are already now dominated by systematic uncertainties partially correlated between the experiments, so that the expected ultimate precision of $\Delta m_{top} \leq 1.5\,GeV/c^2$ in combination can be achieved and the top quark physics potential at the TEVATRON be fully exploited. These prospects in particular result from the development of methods to reduce the dominant uncertainty from the jet energy scale by *in situ* calibration using the hadronic decay of the W-boson in top quark decays. Further studies are ongoing to also include the b-jet energy scale in such

---

[24] The latest update of the top quark mass measurement combination by the Tevatron Electroweak Working Group yields a precision of 1.2% [181]. However, this update arrived too late to be included in this review in more detail.

calibration fits, for example via its influence on the helicity distribution of the W-boson decays products in the W-boson rest frame. Therefore, the ultimate precision of the TEVATRON measurements of the top quark mass might improve even further. It will not be trivial for the LHC to reach similar or better precision.

– In the very near future, CDF and DØ should each have enough data to determine the electric charge of the top quark from jet pairing using jet charge techniques. DØ has already excluded the exotic scenario of a charge 4/3 top quark at the 94% CL.

– The observation of single top production via the electroweak interaction and the measurement of its cross section are one of the highlights of the top physics program at the TEVATRON. So far the s- and t-channel single-top quark production cross section has been excluded to be larger than 6.4 and 5.8 pb, respectively, in only 230 pb^{-1}. Analyses of larger data sets and the development of improved reconstruction of physics objects in the detectors or better signal-background separation using more sophisticated multivariate techniques will improve the sensitivity of the TEVATRON experiments dramatically. The observation of single top production is expected to be possible with a few fb^{-1}, provided a Standard Model cross section.

– Once the electroweak single-top quark production is observed and the single-top cross section measured, a direct measurement of the CKM matrix element $|V_{tb}|$ will be possible. At present, indirect determinations from the measurement of the ratio of top quark decay branching ratios $B(t \to Wb)/B(t \to Wq)$ yield limits of $|V_{tb}| > 0.80$ at 95% CL.

– First studies of hadronically decaying τ-leptons allow to set limits on the ratio of the decay branching ratio $B(t \to b\tau\nu)$ to its Standard Model expectation to be < 5.0 at 95% CL.

– Studies of the spin correlation in $t\bar{t}$ production, a consequence of the very short lifetime of the top quark compared to typical hadronisation times and of the $t\bar{t}$ production at the TEVATRON being dominated by $q\bar{q}$ annihilation, have so far only been performed in Run I $t\bar{t}$ to dilepton events and yielded a limit on the spin correlation coefficient of $\kappa > -0.25$ at 95% CL, whereas the Standard Model expectation if $\kappa = 0.93$ at NLO in those events. The ongoing Run II measurements are expected to yield significant measurements of κ with 1–2 fb^{-1}.

– The study of top quark decay angular distributions allows measurements of the helicity of the W-boson in top decays and hence tests of the electroweak tWb vertex. The best present measurements show that the fraction of longitudinally polarised W-bosons in top quark decay is $\mathcal{F}_0 = 0.74^{+0.22}_{-0.34}$, consistent with the Standard Model expectation of 70%. Contributions of right-handed W-bosons, in the Standard Model suppressed by m_b^2/m_t^2 and therefore expected to be negligible, are presently limited to $\mathcal{F}_+ < 0.18$ at 95% CL. Those analyses will profit enormously from the increase in statistics, eventually allowing simultaneous studies of $\mathcal{F}_0$ and $\mathcal{F}_+$.

– The studies of flavour-changing-neutral-current (FCNC) decays of top quarks at the TEVATRON run

Run I indicate that CDF and DØ have sensitivity to the anomalous couplings $\kappa_{tuZ} < 0.2$ and $\kappa_{tu\gamma} < 0.13$ with 2 fb^{-1} at 95% CL, i.e. potentially better sensitivity than LEP or HERA. The Run II studies are ongoing. First results are expected very soon.

– Generic tests of top quark kinematic distributions in $t\bar{t}$ and single top production will be continued and refined. Studies of the top quark p_T spectrum or general event kinematics in $t\bar{t}$ do not show any indication for deviations from the Standard Model expectation. The ratio of the measured $t\bar{t}$ production cross section in the dilepton and the lepton + jets channel has been measured to be $R_\sigma = 1.45^{+0.83}_{-0.55}$, consistent with the Standard Model expectation of 1.0. This result allows to set 95% CL limits on the top quark decay branching ratios for a fully hadronic decay $B(t \to Xb) < 0.46$ with $B(X \to qq') = 100\%$ and a fully leptonic decay $B(t \to Yb) < 0.47$ with $B(Y \to \ell\ell') = 100\%$.

– Searches for rare top quark decays such as $t \to Ws$ and $t \to Wd$ will be started in analyses of a few fb^{-1}, which allow a first direct determination of the corresponding CKM matrix elements $|V_{ts}|$ and $|V_{td}|$.

– Searches for new physics in the top sector such as the anomalous top quark production via resonances which decay to $t\bar{t}$, or via modified top quark couplings, etc. have been performed and will be pursued. $t\bar{t}$ production via a narrow-width heavy resonance, in particular via a leptophobic Z' is excluded at 95% CL for masses up to $M_{Z'} \approx 560$ GeV/c^2.

– Anomalous top quark decays such as $t \to H^{\pm}b$ have been studied. Resulting exclusion limits are set in the $M_{H\pm}$, $\tan\beta$ plane in the MSSM. Also decay branching ratios are limited to $B(t \to H^+b) < 0.4$ at 95% CL in the Tauonic Higgs model with $B(H^+ \to \tau\bar{\nu}) = 1$ and $B(t \to H^+b) < 0.85$ at 95% CL in the worst case scenario without any assumptions on the decay of the charged Higgs boson, yielding a model-independent limit.

The measurements of top quark properties and production rates presented in this review are used in global electroweak fits to test the Standard Model and to constrain the mass of the elusive Higgs boson. The top quark may lead to the discovery of new physics: its large mass may well indicate a special role in electroweak- and flavour-symmetry breaking, and particles yet unobserved may show up in its production or decay. It is also important to understand top quark events as fully as possible because they will constitute a strong background to many potential new-physics signals in other searches.

Starting physics data taking in 2008, the LHC will dominate the arena of top quark physics in most aspects as a real "top factory" with millions of top quark events produced every year. Amongst many other studies, the top quark mass is expected to be measured at the LHC with an ultimate precision of ≈ 1 GeV/c^2. Better understanding of QCD dynamics will be required to make full use of the rich statistics of top quark events at the LHC.

We are presently at the dawn of the top-physics era, going from first measurements to precision tests. The future promises a wealth of top physics at the TEVATRON and the LHC, and perhaps a high-energy e^+e^- linear collider. The large top-quark mass allows for accurate perturbative calculations of electroweak and strong top-quark processes. The experimental challenge is to match and surpass the accuracy of these calculations, in order to test the properties of the top quark[25] with the best possible sensitivity. Since the top quark is by far the heaviest fermion, it cannot be assumed that its properties are simply those predicted by the Standard Model. Perhaps the top quark is exotic in some way, and will give us our first glimpse of physics beyond the Standard Model or provide us with a hint to the solution of the puzzle of electroweak symmetry breaking.

Acknowledgements. It is thanks to the help of many people and organisations that I could do research in top quark physics and write this review: First of all, I thank my family for their wide spectrum of continuous support over the last years. In particular, Lisa and Ariane had to suffer a lot. I truly admire how they mastered those difficult years and I am most grateful for their love and help! I am grateful to Norbert Wermes for giving me the responsibility to lead the DØ-Bonn group as junior faculty. His faith and support were the basis for the success of this project. It was a thrilling experience to work with many such talented and enthusiastic members of this group, in particular Oleg Brandt, Tobias Golling, Markus Klute, Kevin Kröninger, Jörg Meyer, Su-Jung Park, Christian Schwanenberger, Markus Warsinsky and later also Marc-André Pleier and Eckhard von Törne. I also gratefully acknowledge the financial support of some group members by the German Academic Exchange Service (DAAD), the 'Studienstiftung des deutschen Volkes', the 'Heinrich-Herz Stiftung', and the 'Graduiertenförderung des Landes Nordrhein-Westfalen'. My own work was supported through a Feodor-Lynen Summer Research Fellowship by the Alexander-von-Humboldt Foundation, the University of Bonn and the University of Rochester (New York). Without this help and support neither the research work nor this review article would have been possible. Tom Ferbel as my host at the University of Rochester played a key role in this endeavour. It is to a large extent thanks to Tom that I had a very productive and enjoyable time at FERMILAB. Without his continuous support, encouragement, and wisdom things would have taken a different turn. In addition to his physics insights, I particularly appreciate his understanding for issues outside physics, his human side, his ability to listen, to understand, to sympathise, and to forgive. A truly great man! I am grateful to Erwin Hilger for his lasting support and help over many years. I enjoyed very much the combination of his humour and his lessons in what makes a great character. I also appreci-ate the physics and non-physics conversations with J. Kotcher and H. Montgomery. I very much enjoyed the research work with our experiment leaders and members of the CDF and DØ top physics working groups in my capacity as top physics convener, deputy physics coordinator or fellow physicist, in particular J. Blazey, V. Büscher, J. Hobbs, G. Landsberg, J. Qian, J. Womersley, T. Wyatt, E. Barberis, C. Gerber, G. Gutierrez, A. Juste, C. Tully, C. Clement, L. Chabalina, R. Demina, I. Fleck, A. Heinson, R. Schwienhorst, I. Iashvili, B. Kehoe, F. Rizatdinova, B. Vachon, E. Varnes, D. Wicke, M. Zielinski, U. Bassler, G. Bernardi, C. Biscarat, J. Cammin, M. Verzocchi, B. Heinemann, H. Höth, P. Schieferdecker, C. Schmitt, J. Starck, J. Thom, M. Vaupel, R. Wallny, all my DØ-German colleagues and friends, and the Tevatron Electroweak Working Group, in particular F. Canelli, D. Glenzinski, M. Grünewald, E. Halkiadakis, J. Konigsberg, T. Maruyama, and E. Thomson. I particularly enjoyed writing the Particle Data Group review on top quark physics together with Tony Liss. I am grateful for valuable discussion and/or proofreading of this document to W. Bernreuther, M. Carena, R.K. Ellis, T. Golling, S. Heinemeyer, A. Hoang, W. Hollik, M. Klute, K. Kröninger, E. Laenen, M.L. Mangano, C. Quigg, D. Rainwater, E. Rauter, C. Schwanenberger, P. Uwer, G. Weiglein, and S. Willenbrock. Thank you!

---

[25] In the Standard Model the quantum numbers of the top quark such as spin, weak isospin, electric charge and parity are uniquely defined while the mass of the top quark can only be inferred from global Standard Model fits to electroweak precision data, relying on the sensitivity of electroweak radiative corrections to the top quark mass. Therefore, only the consistency of the top quark mass with radiative corrections in the Standard Model or with the top quark Yukawa coupling to the Higgs boson and hence of the Higgs to $t\bar{t}$ decay branching ratio, also predicted in the Standard Model, can be tested.

References

1. S.J. Wimpenny, B.L. Winer, Ann. Rev. Nucl. Part. Sci. **46**, 149 (1995)
2. C. Campagnari, M. Franklin, Rev. Mod. Phys. **69**, 137 (1997) [hep-ex/9608003]
3. P.C. Bhat, H.B. Prosper, S.S. Snyder, Int. J. Mod. Phys. A **13**, 5113 (1998) [hep-ex/9809011]
4. K. Tollefson, E.W. Varnes, Ann. Rev. Nucl. Part. Sci. **49**, 435 (1999)
5. D. Chakraborty, J. Konigsberg, D. Rainwater, Ann. Rev. Nucl. Part. Sci. **53**, 301 (2003)
6. W. Wagner, Rep. Prog. Phys. **68**, 2409 (2005) [hep-ex/0507207]
7. G. Altarelli, L. Di Lella, Adv. Ser. Direct. High Energ. Phys. **4**, 177 (1989)
8. S.L. Glashow, Nucl. Phys. **22**, 579 (1961)
9. S. Weinberg, Phys. Rev. Lett. **19**, 1264 (1967)
10. A. Salam, Weak and Electromagnetic Interactions, ed. Nobel Symposium No. 8 (Almqvist and Wiksell, Stockholm, 1968)
11. S.L. Glashow, J. Iliopoulos, L. Maiani, Phys. Rev. D **2**, 1285 (1970)
12. H. Georgi, S.L. Glashow, Phys. Rev. Lett. **28**, 1494 (1972)
13. H.D. Politzer, Phys. Rev. Lett. **30**, 1346 (1973)
14. H.D. Politzer, Phys. Rep. **14**, 129 (1974)
15. D.J. Gross, F. Wilczek, Phys. Rev. D **8**, 3633 (1973)
16. S. Weinberg, Eur. Phys. J. C **34**, 5 (2004) [hep-ph/0401010]
17. G. 't Hooft, Nucl. Phys. B **35**, 167 (1971)
18. G. 't Hooft, M. Veltmann, Nucl. Phys. B **44**, 189 (1972)
19. G. 't Hooft, M. Veltmann, Nucl. Phys. B **50**, 318 (1972)
20. P.W. Higgs, Phys. Lett. **12**, 132 (1964)
21. F. Englert, R. Brout, Phys. Rev. Lett. **13**, 321 (1964)
22. G.S. Guralnik, C.R. Hagen, T.W.B. Kibble, Phys. Rev. Lett. **13**, 585 (1964)
23. N. Cabbibo, Phys. Rev. Lett. **10**, 531 (1963)

24. M. Kobayashi, T. Maskawa, Prog. Theor. Phys. **49**, 652 (1973)
25. G.L. Kane, M.E. Peskin, Nucl. Phys. B **195**, 29 (1982)
26. The CLEO Collaboration, A. Bean et al., Phys. Rev. D **35**, 3533 (1987)
27. D.P. Roy, S.U. Sankar, Phys. Lett. B **243**, 296 (1990)
28. The ARGUS Collaboration, H. Albrecht et al., Phys. Lett. B **192**, 245 (1987)
29. The ARGUS Collaboration, H. Albrecht et al., Phys. Lett. B **324**, 249 (1994)
30. The CLEO Collaboration, J. Bartelt et al., Phys. Rev. Lett. **71**, 1680 (1993)
31. The ALEPH, DELPHI, L3, OPAL, SLD Collaborations, the LEP Electroweak Working Group, the SLD Electroweak and Heavy Flavour Groups, Phys. Rep. **427**, 257 (2006) [hep-ex/0509008]
32. The PLUTO Collaboration, C. Berger et al., Phys. Lett. B **76**, 243 (1978)
33. The DESY-Heidelberg Collaboration, J.K. Bienlein et al., Phys. Lett. B **78**, 360 (1978)
34. The DASP Collaboration, W.E. Darden et al., Phys. Lett. B **76**, 246 (1978)
35. The TOPAZ Collaboration, A. Shimonaka et al., Phys. Lett. B **268**, 457 (1991)
36. The JADE Collaboration, E. Elsen et al., Z. Phys. C **46**, 349 (1990)
37. The CELLO Collaboration, H.J. Behrend et al., Z. Phys. C **47**, 333 (1990)
38. D. Schaile, P.M. Zerwas, Phys. Rev. D **45**, 3262 (1992)
39. The BES Collaboration, J.Z. Bai et al., Phys. Rev. Lett. **88**, 101 802 (2002) [hep-ex/0102003]
40. T. van Ritbergen, R.G. Stuart, Phys. Rev. Lett. **82**, 488 (1999) [hep-ex/9808283]
41. LEP-Electroweak Working Group and the LEP Collaborations: ALEPH, DELPHI, L3 and OPAL, Phys. Lett. B **276**, 247 (1992)
42. The CDF Collaboration, F. Abe et al., Phys. Rev. Lett. **74**, 2626 (1995) [hep-ex/9503002]
43. The DØ Collaboration, S. Abachi et al., Phys. Rev. Lett. **74**, 2632 (1995) [hep-ex/9503003]
44. The ALEPH, DELPHI, L3, OPAL, SLD Collaborations, the LEP Electroweak Working Group, the SLD Electroweak and Heavy Flavour Groups, A Combination of Preliminary Electroweak Measurements and Constraints on the Standard Model, CERN-PH-EP/2004-069 (2004) [hep-ex/0412015]
45. The ALEPH, DELPHI, L3, OPAL, SLD Collaborations, the LEP Electroweak Working Group, the SLD Electroweak and Heavy Flavour Groups, A Combination of Preliminary Electroweak Measurements and Constraints on the Standard Model, Updated for 2005 summer conferences: http://www.cern.ch/LEPEWWG, 2005
46. The Tevatron Electroweak Working Group, J.F. Arguin, et al., Combination of CDF and DØ Results on the Top-Quark Mass, FERMILAB-TM-2323-E, 2005 (unpublished) [hep-ex/0507091]
47. The LEP Working Group for Higgs Boson Searches, R. Barate et al., Phys. Lett. B **565**, 61 (2003) [hep-ex/0306033]
48. G.P. Zeller et al. (NuTeV), Phys. Rev. Lett. **88**, 091 802 (2002) [hep-ex/0110059]
49. S.W. Herb et al., Phys. Rev. Lett. **39**, 252 (1977)
50. The CELLO Collaboration, H.J. Behrend et al., Phys. Lett. B **144**, 297 (1984)
51. The JADE Collaboration, W. Bartel et al., Phys. Lett. B **88**, 171 (1979)
52. The JADE Collaboration, W. Bartel et al., Phys. Lett. B **89**, 136 (1979)
53. The JADE Collaboration, W. Bartel et al., Phys. Lett. B **99**, 277 (1981)
54. The MARK-J Collaboration, D.P. Barber et al., Phys. Lett. B **85**, 463 (1979)
55. The MARK-J Collaboration, D.P. Barber et al., Phys. Rev. Lett. **44**, 1722 (1980)
56. The MARK-J Collaboration, B. Adeva et al., Phys. Rev. Lett. **50**, 799 (1983)
57. The MARK-J Collaboration, B. Adeva et al., Phys. Rev. Lett. **51**, 443 (1983)
58. The MARK-J Collaboration, B. Adeva et al., Phys. Lett. B **152**, 439 (1985)
59. The MARK-J Collaboration, B. Adeva et al., Phys. Rev. D **34**, 681 (1986)
60. The PLUTO Collaboration, C. Berger et al., Phys. Lett. B **86**, 413 (1979)
61. The TASSO Collaboration, R. Brandelik et al., Phys. Lett. B **113**, 499 (1982)
62. The TASSO Collaboration, M. Althoff et al., Z. Phys. C **22**, 307 (1984)
63. The TASSO Collaboration, M. Althoff et al., Phys. Lett. B **138**, 441 (1984)
64. The AMY Collaboration, H. Sagawa et al., Phys. Rev. Lett. **60**, 93 (1988)
65. The AMY Collaboration, S. Igarashi et al., Phys. Rev. Lett. **60**, 2359 (1988)
66. The TOPAZ Collaboration, I. Adachi et al., Phys. Rev. Lett. **60**, 97 (1988)
67. The VENUS Collaboration, H. Yoshida et al., Phys. Lett. B **198**, 570 (1987)
68. The VENUS Collaboration, K. Abe et al., Phys. Lett. B **234**, 382 (1990)
69. The MARK II Collaboration, G.S. Abrams et al., Phys. Rev. Lett. **63**, 2447 (1989)
70. The ALEPH Collaboration, D. Decamp et al., Phys. Lett. B **236**, 511 (1990)
71. The DELPHI Collaboration, P. Abreu et al., Phys. Lett. B **242**, 536 (1990)
72. The OPAL Collaboration, M.Z. Akrawy et al., Phys. Lett. B **236**, 364 (1990)
73. CERN ISR Division, Annual Progress Report from the ISR Division for the Year 1981, CERN-ISR-DI/82-02, 1982 (unpublished)
74. The UA1 Collaboration, G. Arnison et al., Phys. Lett. B **147**, 493 (1984)
75. The UA1 Collaboration, C. Albajar et al., Z. Phys. C **37**, 505 (1988)
76. The UA1 Collaboration, C. Albajar et al., Z. Phys. C **48**, 1 (1990)
77. The UA2 Collaboration, T. Åkesson et al., Z. Phys. C **46**, 179 (1990)
78. The CDF Collaboration, F. Abe et al., Phys. Rev. Lett. **64**, 142 (1990)
79. The CDF Collaboration, F. Abe et al., Phys. Rev. Lett. **64**, 147 (1990)
80. The CDF Collaboration, F. Abe et al., Phys. Rev. D **43**, 664 (1991)

81. The CDF Collaboration, F. Abe et al., Phys. Rev. Lett. **68**, 447 (1992)
82. The CDF Collaboration, F. Abe et al., Phys. Rev. D **45**, 3921 (1992)
83. P. Grannis, Search for the Top Quark: Results from the DØ Experiment, Proc. 27th Int. Conf. High Energy Physics (ICHEP94), Glasgow, Scotland, 1994, ed. by P.J. Bussey, I.G. Knowles (IOP, 1995) [hep-ex/9409006]
84. The DØ Collaboration, S. Abachi et al., Phys. Rev. D **52**, 4877 (1995)
85. The DØ Collaboration, S. Abachi et al., Phys. Rev. Lett. **72**, 2138 (1994)
86. The CDF Collaboration, F. Abe et al., Phys. Rev. Lett. **73**, 225 (1994) [hep-ex/9405005]
87. The CDF Collaboration, F. Abe et al., Phys. Rev. D **50**, 2966 (1994)
88. K.W. Staley, The Evidence for the Top Quark: Objectivity and Bias in Collaborative Experimentation, (Cambridge Univ. Press, Cambridge, 2004)
89. C. Quigg, Phys. Today **50N5**, 20 (1997)
90. J.C. Collins, D.E. Soper, Ann. Rev. Nucl. Part. Sci. **37**, 383 (1987)
91. A.M. Cooper-Sarkar, R.C.E. Devenish, A. De Roeck, Int. J. Mod. Phys. A **13**, 3385 (1998) [hep-ph/9712301]
92. J. Pumplin, D.R. Stump, J. Huston, H.L. Lai, P. Nadolsky, W.K. Tung, J. High Energy Phys. **0207**, 012 (2002) [hep-ph/0201195]
93. A.D. Martin, R.G. Roberts, W.J. Stirling, R.S. Thorne, Phys. Lett. B **604**, 61 (2004) [hep-ph/0410230]
94. The CTEQ Collaboration, H.L. Lai et al., Eur. Phys. J. C **12**, 375 (2000) [hep-ph/9903282]
95. J.C. Collins, D.E. Soper, G. Sterman, Nucl. Phys. B **263**, 37 (1986)
96. R.K. Ellis, W.J. Stirling, B.R. Webber, QCD and Collider Physics, Cambridge Monographs on Particle Physics, Nuclear Physics and Cosmology (Cambridge Univ. Press, Cambridge, 1996)
97. P. Nason, S. Dawson, R.K. Ellis, Nucl. Phys. B **303**, 607 (1988)
98. W. Beenakker, H. Kuijf, W.L. van Neerven, J. Smith, Phys. Rev. D **40**, 54 (1989)
99. W. Beenakker, W.L. van Neerven, R. Meng, G.A. Schuler, J. Smith, Nucl. Phys. B **351**, 507 (1991)
100. P. Nason, S. Dawson, R.K. Ellis, Nucl. Phys. B **327**, 49 (1989)
101. G. Sterman, Nucl. Phys. B **281**, 310 (1987)
102. S. Catani, L. Trentadue, Nucl. Phys. B **327**, 323 (1989)
103. E. Laenen, J. Smith, W.L. van Neerven, Phys. Lett. B **321**, 254 (1994) [hep-ph/9310233]
104. E. Laenen, J. Smith, W.L. van Neerven, Nucl. Phys. B **369**, 543 (1992)
105. E.L. Berger, H. Contopanagos, Phys. Lett. B **361**, 115 (1995) [hep-ph/9507363]
106. S. Catani, M.L. Mangano, P. Nason, L. Trentadue, Phys. Lett. B **378**, 329 (1996) [hep-ph/9602208]
107. S. Catani, M.L. Mangano, P. Nason, L. Trentadue, Nucl. Phys. B **478**, 273 (1996) [hep-ph/9604351]
108. M. Cacciari, G. Corcella, A.D. Mitov, J. High Energ. Phys. **0212**, 015 (2002) [hep-ph/0209204]
109. N. Kidonakis, G. Sterman, Phys. Lett. B **387**, 867 (1996)
110. N. Kidonakis, G. Sterman, Nucl. Phys. B **505**, 321 (1997) [hep-ex/9705234]
111. N. Kidonakis, J. Smith, R. Vogt, Phys. Rev. D **56**, 1553 (1997) [hep-ph/9608343]
112. N. Kidonakis, Nucl. Phys. Proc. Suppl. **64**, 402 (1998) [hep-ph/9708439]
113. R. Bonciani, S. Catani, M.L. Mangano, P. Nason, Nucl. Phys. B **529**, 424 (1998) [hep-ph/9801375]
114. M. Cacciari, S. Frixione, M.L. Mangano, P. Nason, G. Ridolfi, J. High Energ. Phys. **0404**, 068 (2004) [hep-ph/0303085]
115. E. Laenen, G. Oderda, G. Sterman, Phys. Lett. B **438**, 173 (1998) [hep-ph/9806467]
116. N. Kidonakis, R. Vogt, Phys. Rev. D **68**, 114014 (2003) [hep-ph/0308222]
117. N. Kidonakis, Phys. Rev. D **64**, 014009 (2001) [hep-ph/0010002]
118. N. Kidonakis, R. Vogt, Eur. Phys. J. C **33**, 466 (2004) [hep-ph/0309045]
119. N. Kidonakis, Total uncertainty on the $t\bar{t}$ Cross Section Calculation (private communication, 2005)
120. A.D. Martin, R.G. Roberts, W.J. Stirling, R.S. Thorne, Eur. Phys. J. C **28**, 455 (2003) [hep-ph/0211080]
121. The H1 Collaboration, C. Adloff et al., Eur. Phys. J. C **21**, 33 (2001) [hep-ex/0012053]
122. The H1 Collaboration, C. Adloff et al., Eur. Phys. J. C **30**, 1 (2003) [hep-ex/0304003]
123. The ZEUS Collaboration, S. Chekanov et al., Eur. Phys. J. C **21**, 443 (2001) [hep-ex/0105090]
124. The ZEUS Collaboration, S. Chekanov et al., Phys. Rev. D **70**, 052001 (2004) [hep-ex/0401003]
125. C. Pascaud, F. Zomer, QCD Analysis from the Proton Structure Function F_2 Measurement: Issues on Fitting, Statistical and Systematic Errors, LAL-95-05, 1995 (unpublished)
126. The ZEUS Collaboration, S. Chekanov et al., Eur. Phys. J. C **42**, 1 (2005) [hep-ex/0503274]
127. A. Quadt, Measurement and QCD Analysis of the Proton Structure Function F_2 from the 1994 HERA Data using the ZEUS Detector, Ph.D. Thesis, Oxford University, RAL-TH-97-004, DESY-THESIS-1998-007 (1996)
128. W.T. Giele, S.A. Keller, D.A. Kosower, Parton Distribution Function Uncertainties (2001) [hep-ph/0104052]
129. W.T. Giele, S.A. Keller, Phys. Rev. D **58**, 094023 (1998) [hep-ph/9803393]
130. D. Stump, J. Pumplin, R. Brock, D. Casey, J. Huston, J. Kalk, H.L. Lai, W.K. Tung, Phys. Rev. D **65**, 014012 (2002) [hep-ph/0101051]
131. D. Stump, J. Pumplin, R. Brock, D. Casey, J. Huston, J. Kalk, H.L. Lai, W.K. Tung, Phys. Rev. D **65**, 014013 (2002) [hep-ph/0101032]
132. J. Pumplin, D.R. Stump, H.L. Lai, P. Nadolsky, W.K. Tung, J. High Energ. Phys. **0207**, 012 (2002) [hep-ph/0201195]
133. M. Botje, Eur. Phys. J. C **14**, 285 (2000) [hep-ph/9912439]
134. S.I. Alekhin, Statistical Properties of the Estimator Using Covariance Matrix (2000) [hep-ex/0005042]
135. S.I. Alekhin, Phys. Rev. D **68**, 014002 (2003) [hep-ph/0211096]
136. C.T. Hill, Phys. Lett. B **266**, 419 (1991)
137. C.T. Hill, S.J. Parke, Phys. Rev. D **49**, 4454 (1994) [hep-ph/9312324]
138. C.T. Hill, Phys. Lett. B **345**, 483 (1995) [hep-ph/9411426]
139. W. Bernreuther, M. Flesch, P. Haberl, Phys. Rev. D **58**, 114031 (1998) [hep-ph/9709284]

140. E. Eichten, K.D. Lane, Phys. Lett. B **327**, 129 (1994) [hep-ph/9401236]
141. S. Catani, Aspects of QCD, From the Tevatron to the LHC, Proc. Workshop on Physics at TeV Colliders, Les Houches, France, 1999, edited by Frontières (Gif-sur-Yvette, 1999) [hep-ph/0005233]
142. S.D. Willenbrock, D.A. Dicus, Phys. Rev. D **34**, 155 (1986)
143. C.-P. Yuan, Phys. Rev. D **41**, 42 (1990)
144. M.C. Smith, S. Willenbrock, Phys. Rev. D **54**, 6696 (1996) [hep-ph/9604223]
145. A. Belyaev, E. Boos, Phys. Rev. D **63**, 034012 (2001) [hep-ph/0003260]
146. S. Cortese, R. Petronzio, Phys. Lett. B **253**, 494 (1991)
147. T. Stelzer, S. Willenbrock, Phys. Lett. B **357**, 125 (1995) [hep-ph/9505433]
148. B. Harris, E. Laenen, E. Phaf, Z. Sullivan, S. Weinzierl, Phys. Rev. D **66**, 054024 (2002) [hep-ph/0207055]
149. Z. Sullivan, Phys. Rev. D **70**, 114012 (2004) [hep-ph/0408049]
150. Z. Sullivan, Phys. Rev. D **72**, 094034 (2005) [hep-ph/0510224]
151. S. Zhu, Phys. Lett. B **524**, 283 (2002)
152. J. Campbell, R.K. Ellis, F. Tramontano, Phys. Rev. D **70**, 094012 (2004) [hep-ph/0408158]
153. Q.H. Cao, C.-P. Yuan, Phys. Rev. D **71**, 054022 (2005) [hep-ph/0408180]
154. Q.H. Cao, R. Schwienhorst, C.-P. Yuan, Phys. Rev. D **71**, 054023 (2005) [hep-ph/0409040]
155. Q.H. Cao, R. Schwienhorst, J.A. Benitez, R. Brock, C.-P. Yuan, Phys. Rev. D **72**, 094027 (2005) [hep-ph/0504230]
156. J. Campbell, F. Tramontano, Nucl. Phys. B **726**, 109 (2005) [hep-ph/0506289]
157. R. Hamberg, W.L. van Neerven, T. Matsuura, Nucl. Phys. B **359**, 343 (1991)
158. K.G. Chetyrkin, J.H. Kühn, M. Steinhauser, Nucl. Phys. B **482**, 213 (1996) [hep-ph/9606230]
159. E. Laenen, S. Riemersma, J. Smith, W. van Neerven, Phys. Rev. D **49**, 5753 (1994) [hep-ph/9308295]
160. W.T. Giele, S. Keller, E. Laenen, Phys. Lett. B **372**, 141 (1996) [hep-ph/9511449]
161. B.W. Harris, E. Laenen, L. Phaf, Z. Sullivan, S. Weinzierl, Phys. Rev. D **66**, 054024 (2002) [hep-ph/0207055]
162. T. Stelzer, Z. Sullivan, S. Willenbrock, Phys. Rev. D **56**, 5919 (1997) [hep-ph/9705398]
163. M. Jeżabek, J.H. Kühn, Nucl. Phys. B **314**, 1 (1989)
164. I.I.Y. Bigi, Y.L. Dokshitzer, V.A. Khoze, J.H. Kühn, P.M. Zerwas, Phys. Lett. B **181**, 157 (1986)
165. A. Czarnecki, K. Melnikov, Nucl. Phys. B **544**, 520 (1999) [hep-ph/9806244]
166. K.G. Chetyrkin, R. Harlander, T. Seidensticker, M. Steinhauser, Phys. Rev. D **60**, 114015 (1999) [hep-ph/9906273]
167. The Particle Data Group, S. Eidelmann et al., Phys. Lett. B **592**, 1 (2004)
168. The DØ Collaboration, V.M. Abazov et al., Phys. Rev. Lett. **97**, 021802 (2006) [hep-ex/0603029]
169. The CKMfitter Group, J. Charles et al., Eur. Phys. J. C **41**, 1 (2005) [hep-ph/0406184]
170. T. Stelzer, Z. Sullivan, S. Willenbrock, Phys. Rev. D **58**, 094021 (1998) [hep-ph/9807340]
171. M. Martinez, R. Miquel, Eur. Phys. J. C **27**, 49 (2003) [hep-ph/0207315]
172. ECFA/DESY LC Physics Working Group, J.A. Aguilar-Saavedra et al., TESLA Technical Design Report, Part III: Physics at an e^+e^- Linear Collider, 2001, ed. by R.-D. Heuer, D. Miller, F. Richard, P.M. Zerwas, SLAC-REPRINT-2001-002, DESY-2001-011, ECFA-2001-209, 2001 (unpublished) [hep-ph/0106315]
173. NLC ZDR Design Group and the NLC Physics Working Group, S. Kuhlman et al., Physics and Technology of the Next Linear Collider: A Report Submitted to Snowmass '96, FERMILAB-PUB-96-112, 1996 (unpublished) [hep-ex/9605011]
174. R.B. Palmer, J.C. Gallardo, A. Tollestrup, A. Sessler, Sci. Cult. **13**, 39 (1998)
175. The CDF Collaboration, F. Abe et al., Phys. Rev. Lett. **80**, 5720 (1998) [hep-ex/9711004]
176. The CDF Collaboration, D. Acosta et al., Measurement of the Top Quark Mass using the Template Method in the Lepton plus jets Channel with *In Situ* $W \to jj$ Calibration at CDF-II, CDF note 7680, 2005 (unpublished)
177. The DØ Collaboration, V.M. Abazov et al., Top Quark Mass Measurement with the Matrix Element Method in the Lepton+Jets Final State at DØ Run II, DØ note 4874, 2005 (unpublished)
178. M. Smith, S. Willenbrock, Phys. Rev. Lett. **79**, 3825 (1997) [hep-ph/9612329]
179. S. Willenbrock, Rev. Mod. Phys. **72**, 1141 (2000) [hep-ph/0008189]
180. The Tevatron Electroweak Working Group, P. Azzi et al., Combination of CDF and DØ Results on the Top-Quark Mass, TEVEWWG-top 2004-01, 2004 (unpublished) [hep-ex/0404010]
181. The Tevatron Electroweak Working Group, J.F. Arguin, et al., Combination of CDF and DØ Results on the Top-Quark Mass, FERMILAB-TM-2355-E, 2006 (unpublished) [hep-ex/0608032]
182. A.H. Hoang, M. Beneke, K. Melnikov, T. Nagano, A. Ota, A.A. Penin, A.A. Pivovarov, A. Signer, V.A. Smirnov, Y. Sumino, T. Teubner, O. Yakovlev, A. Yelkhovsky, Eur. Phys. J. C direct **2**, 1 (2000) [hep-ph/0001286]
183. A.H. Hoang, A.V. Manohar, S. I. W., T. Teubner, Phys. Rev. Lett. **86**, 1951 (2001) [hep-ph/0011254]
184. U. Baur, M. Buice, L.H. Orr, Phys. Rev. D **64**, 094019 (2001) [hep-ph/0106341]
185. The American Linear Collider Working Group, T. Abe, et al., Resource Book for Snowmass 2001, SLAC-570, 2001 (unpublished) [hep-ex/0106055, hep-ex/0106056, hep-ex/0106057, hep-ex/0106058]
186. R. Harlander, M. Jeżabek, J.H. Kühn, T. Teubner, Phys. Lett. B **346**, 137 (1995) [hep-ph/9411395]
187. A. Brandenburg, M. Flesch, P. Uwer, Phys. Rev. D **59**, 014001 (1999) [hep-ph/9806306]
188. W. Bernreuther, A. Brandenburg, Phys. Rev. D **49**, 4481 (1994) [hep-ph/9312210]
189. W. Bernreuther, A. Brandenburg, P. Uwer, Phys. Lett. B **368**, 153 (1996) [hep-ph/9510300]
190. W.G.D. Dharmaratna, G.R. Goldstein, Phys. Rev. D **53**, 1073 (1996)
191. W. Bernreuther, M. Fücker, Z.G. Si, Phys. Lett. B **633**, 54 (2006) [hep-ph/0508091]
192. J.H. Kühn, Nucl. Phys. B **237**, 77 (1984)
193. G. Mahlon, S.J. Parke, Phys. Lett. B **411**, 173 (1997) [hep-ph/9706304]

194. G. Mahlon, S.J. Parke, Phys. Rev. D **55**, 7249 (1997) [hep-ph/9611367]
195. G. Mahlon, S.J. Parke, Phys. Rev. D **53**, 4886 (1996) [hep-ph/9512264]
196. T. Stelzer, S. Willenbrock, Phys. Lett. B **374**, 169 (1996) [hep-ph/9512292]
197. A. Brandenburg, Phys. Lett. B **388**, 626 (1996) [hep-ph/9603333]
198. S.J. Parke, Y. Shadmi, Phys. Lett. B **387**, 199 (1996) [hep-ph/9606419]
199. T. Arens, L.M. Sehgal, Phys. Lett. B **302**, 501 (1993)
200. W. Bernreuther, A. Brandenburg, Z.G. Si, P. Uwer, Nucl. Phys. B **690**, 81 (2004) [hep-ph/0403035]
201. W. Bernreuther, A. Brandenburg, Z.G. Si, P. Uwer, Phys. Rev. Lett. **87**, 242002 (2001) [hep-ph/0107086]
202. G.R. Goldstein, Spin Correlations in Top Quark Production and the Top Quark Mass, Spin 96: Proc. 12th Int. Symposium on High Energy Spin Physics, Amsterdam, 1996, ed. by C.W. deJager (World Scientific, Singapore, 1997) [hep-ph/9611314]
203. A. Czarnecki, M. Jeżabek, J.H. Kühn, Nucl. Phys. B **351**, 70 (1991)
204. A. Brandenburg, Z.G. Si, P. Uwer, Phys. Lett. B **539**, 235 (2002) [hep-ph/0205023]
205. P. Uwer, Phys. Lett. B **609**, 271 (2005) [hep-ph/0412097]
206. J.H. Kühn, G. Rodrigo, Phys. Rev. Lett. **81**, 49 (1998) [hep-ph/9802268]
207. J.H. Kühn, G. Rodrigo, Phys. Rev. D **59**, 054017 (1999) [hep-ph/9807420]
208. M.T. Bowen, S.D. Ellis, D. Rainwater, Phys. Rev. D **73**, 014008 (2006) [hep-ph/0509267]
209. G. Eilam, J.L. Hewett, A. Soni, Phys. Rev. D **44**, 1473 (1991)
210. S. Willenbrock, The Standard Model and the Top Quark, Lectures presented at the 12th Advanced Study Institute on Techniques and Concepts of High Energy Physics, St. Croix, U.S. Virgin Islands, 2002 [hep-ph/0211067]
211. A. Juste, Y. Kiyo, F. Petriello, T. Teubner, K. Agashe, P. Batra, U. Baur, C.F. Berger, J.A.R. Cembranos, A. Gehrmann-De Ridder, T. Gehrmann, E.W.N. Glover, S. Godfrey, A. Hoang, M. Perelstein, Z. Sullivan, T. Tait, S. Zhu, Report of the 2005 Snowmass Top/QCD Working Group, Proceedings of the 2005 International Linear Collider Physics and Detector Workshop and 2nd ILC Accelerator Workshop, Snowmass, Colorado, 2005 (ECONF C0508141:PLEN0043,2005) [hep-ph/0601112]
212. S. Heinemeyer, G. Weiglein, J. High Energ. Phys. **0210**, 072 (2002) [hep-ph/0209305]
213. R. Barbieri, M. Beccaria, P. Ciafaloni, G. Curci, A. Vicere, Nucl. Phys. B **409**, 105 (1993)
214. J. Fleischer, F. Jegerlehner, O.V. Tarasov, Phys. Lett. B **319**, 249 (1993)
215. S. Frixione, B. Webber, The MC@NLO Event Generator (2004) [hep-ph/0402116]
216. S. Frixione, B. Webber, J. High Energ. Phys. **06**, 029 (2002)
217. S. Frixione, P. Nason, B. Webber, J. High Energ. Phys. **0308**, 007 (2003) [hep-ph/0305252]
218. S. Frixione, B. Webber, The MC@NLO 3.1 event generator p. 22 (2005) [hep-ph/0506182]
219. S. Frixione, E. Laenen, P. Motylinski, B. Webber, J. High Energ. Phys. **0603**, 092 (2006) [hep-ph/0512250]
220. S. Frixione, B. Webber, The MC@NLO 3.2 event generator p. 23 (2006) [hep-ph/0601192]
221. J. Campbell, R.K. Ellis, Phys. Rev. D **65**, 113007 (2002) [hep-ph/0202176]
222. J.M. Campbell, J. Huston, Phys. Rev. D **70**, 094021 (2004)
223. T. Sjöstrand et al., Comput. Phys. Commun. **135**, 238 (2001) [hep-ph/0010017]
224. M.L. Mangano et al., J. High Energ. Phys. **0307**, 001 (2003) [hep-ph/0206293]
225. R.K. Ellis, R. Field et al., Report on the QCD Tools Working Group, Proc. Workshop on 'QCD and Weak Boson Physics in Run II', Fermilab, March–November 1999., p. 47 (2000) [hep-ph/0011122]
226. G. Marchesini et al., Comput. Phys. Commun. **67**, 465 (1992)
227. G. Corcella et al., J. High Energ. Phys. **0101**, 010 (2001) [hep-ph/0011363]
228. J.M. Butterworth, J.R. Foreshaw, M.H. Seymour, Z. Phys. C **72**, 637 (1996) [hep-ph/9601371]
229. J.M. Butterworth, , M.H. Seymour, Multiparton Interactions in HERWIG: JIMMY, Proc. CERN-DESY workshop on HERA and the LHC, 2005
230. R. Brun et al., Simulation program for particle physics experiments, GEANT: user guide and reference manual, CERN DD 78-2, 1978 (unpublished)
231. Geant4, S. Agostinelli et al., Nucl. Instrum. Methods A **506**, 250 (2003)
232. S.R. Slabopitsky, Event generators for top quark production and decays, Proc. Int. Workshop Top Quark Physics, Coimbra, Portugal, 2006 (PoS TOP2006:019, 2006) [hep-ph/0603124]
233. F. Paige, S.D. Protopopescu, ISAJET, Brookhaven Report BNL-38034, 1986 (unpublished)
234. T. Gleisberg, S. Hoeche, F. Krauss, A. Schaelicke, S. Schumann, J. Winter, Nucl. Instrum. Methods A **559** (2006) [hep-ph/0508315]
235. The CompHep Collaboration, E. Boos et al., Nucl. Instrum. Methods A **534**, 250 (2004) [hep-ph/0403113]
236. F. Maltoni, Stelzer, J. High Energ. Phys. **02**, 027 (2003) [hep-ph/0208156]
237. T. Stelzer, W.F. Long, Comput. Phys. Commun. **81**, 337 (1994) [hep-ph/9401258]
238. B.P. Kersevan, E. Richter-Was, The Monte Carlo event generator AcerMC version 2.0 with interfaces to PYTHIA 6.2 and HERWIG 6.5 (2004) [hep-ph/0405247]
239. B.P. Kersevan, E. Richter-Was, Comput. Phys. Commun. **149**, 142 (2003) [hep-ph/0201302]
240. E. Boos, L. Dudko, V. Savrin, CMS Note 2000/065, 2000 (unpublished)
241. S.R. Slabopitsky, L. Sonnenschein, Comput. Phys. Commun. **148**, 87 (2002) [hep-ph/0201292]
242. J. Campbell, R.K. Ellis, MCFM Monte Carlo for FeMtobarn Processes, http://mcfm.fnal.gov (2002) [hep-ph/0105226]
243. D.J. Lange et al., Nucl. Instrum. Methods A **462**, 152 (2001)
244. P. Avery, K. Read, G. Trahern, QQ Monte Carlo Program, CLEO Report No. CSN-212, 1985 (unpublished)
245. S. Jadach, Z. Was, R. Decker, J.H. Kühn, Comput. Phys. Commun. **76**, 361 (1993)

246. Z. Was, P. Golonka, TAUOLA as tau Monte Carlo for future applications, CERN-PH-TH/2004-23, HNINP-V-04-05 (2004) [hep-ph/0411377]
247. J.M. Campbell, R.K. Ellis, Phys. Rev. D **60**, 113006 (1999) [hep-ph/9905386]
248. S. Mrenna, P. Richardson, J. High Energ. Phys. **0405**, 040 (2004) [hep-ph/0312274]
249. T. Gleisberg et al., Monte Carlo Models at the LHC, Proc. DIS2004, Strbske Pleso, Slovakia, 2004 [hep-ph/040 7365]
250. A. Krauss, F. Schälicke, S. Schumann, G. Soff, Phys. Rev. D **70**, 114009 (2004) [hep-ph/0409106]
251. S. Catani, F. Krauss, R. Kuhn, B.R. Webber, J. High Energ. Phys. **0111**, 063 (2001) [hep-ph/0109231]
252. F. Krauss, J. High Energ. Phys. **0208**, 015 (2002) [hep-ph/0205283]
253. T. Gleisberg, S. Hoeche, F. Krauss, A. Schälicke, S. Schumann, J. Winter, J. High Energ. Phys. **0402**, 056 (2004) [hep-ph/0311263]
254. A. Schälicke, T. Gleisberg, S. Hoeche, S. Schumann, J. Winter, F. Krauss, G. Soff, Prog. Part. Nucl. Phys. **53**, 329 (2004) [hep-ph/0311270]
255. T. Gleisberg, S. Hoeche, F. Krauss, A. Schälicke, S. Schumann, G. Soff, J. Winter, Predictions for multi-particle final states with SHERPA, Talk at Physics at LHC, Vienna, Austria, 2004 [hep-ph/0409122]
256. M.L. Mangano, A Review of MLM's Prescription for Removal of Double Counting, http://cepa.fnal.gov/patriot/mc4run2/MCTuning/061104/mlm.pdf (2004)
257. H.T. Edwards, Ann. Rev. Nucl. Part. Sci. **35**, 605 (1985)
258. The Fermilab Beams Division, TeV-I Group, Design Report Tevatron 1 Project, FERMILAB-DESIGN-1984-01 (unpublished) (1984)
259. D. Möhl, G. Petrucci, L. Thorndahl, S. van der Meer, Phys. Rep. **58**, 73 (1980)
260. Fermilab Beams Division, http://www-bdnew.fnal.gov/pbar/AEMPlots
261. P.C. Bhat, W.J. Spalding, Fermilab Collider Run II: Accelerator Status and Upgrades, Proc. 15th Topical Conference on Hadron Collider Physics, HCP2004, Michigan State University, East Lansing, MI, June 14–18 (2004) [hep-ex/0410046]
262. V. Shiltsev, Status of Tevatron Collider Run II and Novel Technologies for Luminosity Upgrades, Proc. 2004 European Accelerator Conference, Lucern, Switzerland **1**, 239 (2004)
263. G. Jackson, Fermilab Recycler Ring Technical Design Report, Fermilab-TM-1991, Fermilab (unpublished) (1996)
264. Fermilab Beams Division Run II, Run II Handbook, http://www-bd.fnal.gov/runII/index.html, http://www-bd.fnal.gov/lug (2001)
265. A.M. Budker, Sov. Atom. Energ. **22**, 246 (1967)
266. The Particle Data Group, K. Hagiwara et al., Phys. Rev. D **66**, 010001 (2002)
267. The CDF Collaboration, F. Abe et al., Nucl. Instrum. Methods A **271**, 387 (1988)
268. The CDF Collaboration, D. Amidei et al., Nucl. Instrum. Methods A **350**, 73 (1994)
269. The CDF Collaboration, J. Antos et al., Nucl. Instrum. Methods A **360**, 118 (1995)
270. The DØ Collaboration, S. Abachi et al., Nucl. Instrum. Methods A **338**, 185 (1994)

271. The CDF Collaboration, C. Newman-Holmes et al., The CDF Upgrade, Fermilab-Conf-96-218-E, 1996 (unpublished)
272. The CDF Collaboration, R. Blair et al., The CDF-II Detector: Technical Design Report, Fermilab-Pub-96-390-E, 1996 (unpublished)
273. The DØ Collaboration, S. Abachi et al., The DØ Upgrade: The Detector and its Physics, Fermilab-PUB-96-357-E, 1996 (unpublished)
274. T. LeCompte, T. Diehl, Ann. Rev. Nucl. Part. Sci. **50**, 71 (2000)
275. The CDF Collaboration, D. Acosta et al., Nucl. Instrum. Methods A **461**, 540 (2001)
276. The CDF Collaboration, A. Sill et al., Nucl. Instrum. Methods A **447**, 1 (2000)
277. The CDF Collaboration, T. Affolder et al., Nucl. Instrum. Methods A **526**, 249 (2004)
278. The CDF Collaboration, D. Acosta et al., Nucl. Instrum. Methods A **518**, 605 (2004)
279. The CDF Collaboration, L. Balka et al., Nucl. Instrum. Methods A **267**, 272 (1988)
280. The CDF Collaboration, S. Bertolucci et al., Nucl. Instrum. Methods A **267**, 301 (1988)
281. The CDF Collaboration, M. Albrow et al., Nucl. Instrum. Methods A **480**, 524 (2002)
282. The DØ Collaboration, V.M. Abazov et al., Nucl. Instrum. Methods A **565**, 463 (2006) [physics/0507191]
283. The DØ Collaboration, T. Edwards et al., Determination of the effective inelastic $p\bar{p}$ cross-section for the DØ Run II luminosity measurement., Fermilab-TM-2278-E, Fermilab (unpublished) (2004)
284. The DØ Collaboration, V.M. Abazov et al., DØ Run IIB Upgrade Technical Design Report, Fermilab-PUB-02-327-E, Fermilab (unpublished) (2002)
285. B. Quinn, Int. J. Mod. Phys. A **20**, 3793 (2005) [hep-ex/0501055]
286. ATLAS Detector and Physics Performance TDR, Volume I+II, CERN/LHCC 99-14/15, 1999 (unpublished)
287. CMS Detector and Physics Performance TDR, Volume I+II (unpublished, in preparation), CERN/LHCC, 2006 (unpublished)
288. M. Beneke, I. Efthymiopoulos, J. Mangano, M.L. Womersley et al., Top Quark Physics, in: Proc. 1999 CERN Workshop on Standard Model Physics (and more) at the LHC, 2000, ed. by G. Altarelli, M.L. Mangano [hep-ph/0003033]
289. The CDF Collaboration, F. Abe et al., Phys. Rev. D **45**, 1448 (1992)
290. G.C. Blazey et al., Run II Jet Physics, Proc. Workshop: QCD and Weak Boson Physics in Run II, ed. by U. Bauer, R.K. Ellis, D. Zeppenfeld, FERMILAB-PUB-00-297, 2000 (unpublished)
291. J.R. Vlimant, Mesure de la section efficace de production de pairs de quarks top/anti-top dans des collisions protons/anti-proton à $\sqrt{s}$ égale 1.96 TeV auprès de l'expérience DØ, Ph.D. thesis (l'Université Paris VI – Pierre et Marie Curie, 2005)
292. The DØ Collaboration, V.M. Abazov et al., Phys. Rev. D **71**, 072004 (2005) [hep-ex/0412020]
293. The CDF Collaboration, T. Affolder et al., Phys. Rev. D **64**, 032002 (2001) [hep-ex/0101036]

294. S. Greder, Étiquetage des quarks beaux et mesure de la section efficace de production de paires de quarks top à $\sqrt{s} = 1.96$ TeV dans l'expérience DØ, Ph.D. thesis (Université Louis Pasteur, Strasbourg IRES 05-006 No. d'ordre ULP 4652, 2005)

295. H.P. Nilles, Phys. Rep. **110**, 1 (1984)

296. H.E. Haber, G.L. Kane, Phys. Rep. **117**, 75 (1985)

297. J.F. Gunion et al., The Higgs Hunters Guide, (Addison-Wesley, Redwood City, California, 1990)

298. H.J. He, N. Polonsky, S. Su, Phys. Rev. D **64**, 053004 (2001) [hep-ph/0102144]

299. V.A. Novikov, L.B. Okun, A.N. Rozanov, M.I. Vysotsky, Phys. Lett. B **529**, 111 (2002) [hep-ph/0111028]

300. D. Choudhury, T.M.P. Tait, C.E.M. Wagner, Phys. Rev. D **65**, 053002 (2002) [hep-ph/0109097]

301. H.-C. Cheng, I. Low, J. High Energ. Phys. **0309**, 051 (2003) [hep-ph/0308199]

302. The CDF Collaboration, D. Acosta et al., Phys. Rev. Lett. **93**, 142001 (2004) [hep-ex/0404036]

303. The CDF Collaboration, D. Acosta et al., A global analysis of the high-p_T dilepton sample using 200 pb^{-1} of Run 2 data, CDF note 7192 (2004)

304. The CDF Collaboration, D. Acosta et al., Phys. Rev. D **72**, 052003 (2005) [hep-ex/0504053]

305. C. Peterson, T. Rögnvaldsson, L. Lönnblad, Comput. Phys. Commun. **81**, 185 (1994)

306. The CDF Collaboration, D. Acosta et al., Measurement of the Cross Section for $t\bar{t}$ Production in $p\bar{p}$ Collisions using the Kinematics of Lepton+Jets Events, CDF note 7753 (2005)

307. The CDF Collaboration, D. Acosta et al., Phys. Rev. D **71**, 052003 (2005) [hep-ex/0410041]

308. The CDF Collaboration, D. Acosta et al., Measurement of the $t\bar{t}$ Production Cross Section in $p\bar{p}$ collisions at $\sqrt{s} = 1.96$ TeV using Lepton+Jets events with secondary vertex b-Tagging, CDF note 7801 (2005)

309. The CDF Collaboration, D. Acosta et al., Phys. Rev. D **71**, 072005 (2005) [hep-ex/0409029]

310. The CDF Collaboration, D. Acosta et al., Measurement of the $t\bar{t}$ production cross section in the missing E_T+jets channel, CDF note 7792, 2005 (unpublished)

311. The CDF Collaboration, D. Acosta et al., Phys. Rev. D **72**, 032002 (2005) [hep-ex/0506001]

312. The CDF Collaboration, D. Acosta et al., Measurement of the $t\bar{t}$ Production Cross Section in the Jet Probability Tagged Sample in $p\bar{p}$ Collisions at $\sqrt{s} = 1.96$ TeV, CDF note 7236, 2004 (unpublished)

313. The CDF Collaboration, D. Acosta et al., Measurement of the $t\bar{t}$ production cross section in the all-hadronic channel, CDF note 7075, 2004 (unpublished)

314. The CDF Collaboration, D. Acosta et al., Measurement of the $t\bar{t}$ production cross section in the all-hadronic channel, CDF note 7793, 2005 (unpublished)

315. L. Lyons, D. Gibaut, P. Clifford, Nucl. Instrum. Methods A **270**, 110 (1988)

316. L. Lyons, A.J. Martin, D.H. Saxon, Phys. Rev. D **41**, 982 (1990)

317. The CDF Collaboration, A. Abulencia et al., Combination of CDF top pair production cross section measurements, CDF note 7794, 2005 (unpublished)

318. The DØ Collaboration, V.M. Abazov et al., Phys. Lett. B **626**, 55 (2005) [hep-ex/0505082]

319. The DØ Collaboration, V.M. Abazov et al., Phys. Rev. D **71**, 072004 (2005) [hep-ex/0412020]

320. The CDF Collaboration, A. Abulencia et al., Phys. Rev. Lett. **95**, 022001 (2005) [hep-ex/0412042]

321. The DØ Collaboration, V.M. Abazov et al., Measurement of the $t\bar{t}$ Production Cross Section at $\sqrt{s} = 1.96$ TeV in Dilepton Final States Using 370 pb^{-1} of DØ Data, DØ-Note 4850-CONF, 2005 (unpublished)

322. The DØ Collaboration, V.M. Abazov et al., Phys. Lett. B **626**, 45 (2005) [hep-ex/0504043]

323. T.F. Golling, Measurements of the Top Quark Pair Production cross Section in Lepton+Jets Final States using a Topological Multivariate Technique as well as Lifetime b-Tagging in Proton–Antiproton Collisions at $\sqrt{s} = 1.96$ TeV with the DØ Detector at the Tevatron, Ph.D. thesis (Rheinische Friedrich-Wilhelms-Universität Bonn, FERMILAB-THESIS-2005-01, 2005)

324. M. Klute, A Measurement of the $t\bar{t}$ Production Cross-Section in Proton Anti–Proton Collisions at $\sqrt{s} = 1.96$ TeV with the DØ Detector at the Tevatron Using Final States with a Muon and Jets, PhD thesis (Rheinische Friedrich-Wilhelms-Universität Bonn, FERMILAB-THESIS-2004-20, 2004)

325. S. Park, Measuring the $t\bar{t}$ production Cross-Section in the Electron+Jets Channel in $p\bar{p}$ Collisions at $\sqrt{s} = 1.96$ TeV with the DØ Detector at the Tevatron: A Monte Carlo Study, Diploma thesis (Rheinische Friedrich-Wilhelms-Universität Bonn, FERMILAB-MASTERS-2004-03, 2004)

326. J.M. Meyer, Monte Carlo Study of the Measurement of the $t\bar{t}$ Production Cross-Section in the Muon+Jets Channel with the DØ-Detector at $\sqrt{s} = 1.96$ TeV, Diploma thesis (Rheinische Friedrich-Wilhelms-Universität Bonn, FERMILAB-MASTERS-2004-05, 2004)

327. The DØ Collaboration, B. Abbott, Phys. Rev. D **61**, 072001 (2000) [hep-ex/9906025]

328. V. Barger, J. Ohnemus, R.J.N. Philips, Phys. Rev. D **48**, 3953 (1993) [hep-ph/9308216]

329. V.M. Abazov et al., Combined $t\bar{t}$ Production Cross Section at $\sqrt{s} = 1.96$ TeV in the Lepton+Jets and Dilepton Final States using Event Topology, DØ-note 4906-CONF (2005)

330. The DØ Collaboration, V.M. Abazov et al., Phys. Lett. B **626**, 35 (2005) [hep-ex/0504058]

331. M.C. Smith, S. Willenbrock, Phys. Rev. D **54**, 6696 (1996)

332. P. Sinervo, Definition and Treatment of Systematic Uncertainties in High Energy Physics and Astrophysics, Proc. Conference on Statistical Problems in Particle Physics, Astrophysics, and Cosmology, ed. by L. Lyons, R. Mount, R. Reitemeyer (SLAC, Stanford, 2003), SLAC-R-703

333. The DØ Collaboration, V.M. Abazov et al., Measurement of the $t\bar{t}$ Production Cross Section in $p\bar{p}$ Collisions at $\sqrt{s} = 1.96$ TeV Using Lepton+Jets Events, DØ Note 4888-Conf, 2005 (unpublished)

334. The DØ Collaboration, V.M. Abazov et al., Simultaneous measurement of $\mathcal{B}(t \to Wb)/\mathcal{B}(t \to Wq)$ and $\sigma(p\bar{p} \to t\bar{t})$ at DØ, DØ Note 4833-CONF, 2005 (unpublished)

335. The DØ Collaboration, V.M. Abazov et al., Measurement of the $t\bar{t}$ Production Cross-section at $\sqrt{s} = 1.96$ TeV in the $e\mu$ Channel Using Secondary Vertex b-tagging, DØ Note 4528-CONF, 2004 (unpublished)

336. The DØ Collaboration, V.M. Abazov et al., Measurement of cross section times branching ratio for $Z \to \mu^+\mu^-$ in $p\bar{p}$ collisions at 1.96 TeV, DØ note 4284, 2004 (unpublished)

337. The DØ Collaboration, V.M. Abazov et al., Measurement of the $t\bar{t}$ production cross section in the all-jets channel., DØ Note 4428-Conf, 2004 (unpublished)

338. The DØ Collaboration, B. Abbott et al., Phys. Rev. D **60**, 012001 (1999) [hep-ex/9808034]

339. The DØ Collaboration, V.M. Abazov et al., Measurement of the $t\bar{t}$ Production Cross Section in $p\bar{p}$ Collisions at $\sqrt{s} = 1.96$ TeV in the all hadronic final state, DØ Note 4879-CONF, 2005 (unpublished)

340. The DØ Collaboration, B. Abbott et al., Phys. Rev. Lett. **83**, 1908 (1999) [hep-ex/9901023]

341. The DØ Collaboration, V. Abazov et al., Phys. Rev. D **67**, 012004 (2003) [hep-ex/0205019]

342. S. Mrenna, C.-P. Yuan, Phys. Lett. B **416**, 200 (1998) [hep-ph/9703224]

343. T.M.P. Tait, C.-P. Yuan, Phys. Rev. D **63**, 014018 (2001) [hep-ph/0007298]

344. E.H. Simmons, Phys. Rev. D **55**, 5494 (1997) [hep-ph/9612402]

345. A.P. Heinson, A.S. Belyaev, E.E. Boos, Phys. Rev. D **56**, 3114 (1997) [hep-ph/9612424]

346. R.S. Chivukula, E.H. Simmons, J. Terning, Phys. Lett. B **331**, 383 (1994) [hep-ph/9404209]

347. D.J. Muller, S. Nandi, Phys. Lett. B **383**, 345 (1996) [hep-ph/9602390]

348. E. Malkawi, T. Tait, C.P. Yuan, Phys. Lett. B **385**, 304 (1996) [hep-ph/9603349]

349. D.J. Muller, S. Nandi, Nucl. Phys. Proc. Suppl. **52A**, 192 (1997) [hep-ph/9607328]

350. H.-J. He, T. Tait, C.P. Yuan, Phys. Rev. D **62**, 011702 (2000) [hep-ph/9911266]

351. The CDF and DØ Higgs Working Group, L. Babukhadia et al., Results of the Tevatron Higgs sensitivity study, Fermilab-Pub-03-320-E, 2003 (unpublished)

352. The CDF Collaboration, D. Acosta et al., Phys. Rev. D **65**, 091102 (2002) [hep-ex/0110067]

353. The CDF Collaboration, D. Acosta et al., Phys. Rev. D **69**, 052003 (2004)

354. The DØ Collaboration, V.M. Abazov et al., Phys. Lett. B **517**, 282 (2001) [hep-ex/0106059]

355. The DØ Collaboration, V.M. Abazov et al., Phys. Rev. D **63**, 031101 (2000) [hep-ex/0008024]

356. The CDF Collaboration, D. Acosta et al., Phys. Rev. D **71**, 012005 (2005) [hep-ex/0410058]

357. The DØ Collaboration, V.M. Abazov et al., Phys. Lett. B **622**, 265 (2005) [hep-ex/0505063]

358. The DØ Collaboration, V.M. Abazov et al., Improved Search for Single Top Quark Production at DØ in Run II, DØ note 4722, (2005) (unpublished)

359. E. Boos, L. Dudko, Nucl. Instrum. Methods A **502**, 486 (2003) [hep-ph/0302088]

360. J. Schwindling, MLPFit: A Tool for Designing and Using Multi-Layer Perceptrons, http://schwind.home.cern.ch/schwind/MLPfit.html (2000)

361. L. Breiman, J.H. Friedman, R.A. Olshen, C.J. Stone, Classification and Regression Trees (Wadsworth, Belmont, California, 1984)

362. The DØ Collaboration, V.M. Abazov et al., Search for Single Top Quark Production Using Likelihood Discriminants at DØ in Run II, DØ note 4871 (2005) (unpublished)

363. B. Abbott et al., Phys. Rev. Lett. **85**, 256 (2000) [hep-ex/0002058]

364. M. Jeżabek, Nucl. Phys. B (Proc. Suppl.) **37**, 197 (1994) [hep-ph/9406411]

365. I. Bigi, Phys. Lett. B **175**, 233 (1986)

366. M. Jeżabek, J. Kühn, Phys. Lett. B **329**, 317 (1994) [hep-ph/9403366]

367. R.H. Dalitz, G.R. Goldstein, Phys. Rev. D **45**, 1531 (1992) [hep-ph/9205246]

368. W. Bernreuther, M. Fücker, Y. Umeda, Phys. Lett. B **582**, 32 (2004) [hep-ph/0308296]

369. V. Barger, J. Ohnemus, R.J.N. Phillips, Int. J. Mod. Phys. A **4**, 617 (1989)

370. The DØ Collaboration, B. Abbott et al., Phys. Rev. Lett. **80**, 2063 (1998) [hep-ex/9706014]

371. The CDF Collaboration, D. Acosta et al., Phys. Rev. Lett. **95**, 102002 (2005) [hep-ex/0505091]

372. G. Feldman, R. Cousins, Phys. Rev. D **57**, 3873 (1998) [physics/9711021]

373. The DØ Collaboration, V.M. Abazov et al., Measurement of $\mathcal{B}(t \to Wb)/\mathcal{B}(t \to Wq)$ at DØ, DØ Note 4586-CONF, 2005 (unpublished)

374. The CDF Collaboration, T. Affolder et al., Phys. Rev. Lett. **86**, 3233 (2001) [hep-ex/0012029]

375. The CDF Collaboration, F. Abe et al., Phys. Rev. Lett. **79**, 3585 (1997) [hep-ex/9704007]

376. C. Yue, H. Zong, L. Liu, Mod. Phys. Lett. A **18**, 2187 (2003) [hep-ph/0309255]

377. T. Han, M.B. Magro, Phys. Lett. B **476**, 79 (2000) [hep-ph/9911442]

378. J. Guasch, J. Sola, Phys. Lett. B **416**, 353 (1998) [hep-ph/9707535]

379. A. Rasin, Phys. Rev. D **57**, 3977 (1998) [hep-ph/9705210]

380. The CDF Collaboration, A. Acosta, A Measurement of $BR(t \to \tau\nu)$, CDF note 7179 (2004)

381. V. Barger, R.J.N. Philips, Phys. Rev. D **41**, 884 (1990)

382. The CDF Collaboration, T. Affolder et al., Phys. Rev. D **62**, 012004 (2000) [hep-ex/9912013]

383. M. Fischer, S. Groote, J.G. Körner, M.C. Mauser, Phys. Rev. D **65**, 054036 (2002) [hep-ph/0101322]

384. G.L. Kane, G.A. Ladinsky, C.P. Yuan, Phys. Rev. D **45**, 124 (1992)

385. C.A. Nelson, B.T. Kress, M. Lopes, T.P. McCauleyothers, Phys. Rev. D **56**, 5928 (1997) [hep-ph/9707211]

386. W. Bernreuther, O. Nachtmann, P. Overmann, T. Schröder, Nucl. Phys. B **388**, 53 (1992)

387. M. Fischer, S. Groote, J.G. Körner, M.C. Mauser, Phys. Rev. D **63**, 031501 (2001) [hep-ph/0011075]

388. M.A.B. Beg, R.V. Budny, R. Mohapatra, A. Sirlin, Phys. Rev. Lett. **38**, 1252 (1977)

389. S.H. Nam, Phys. Rev. D **66**, 055008 (2002) [hep-ph/0206037]

390. The CDF Collaboration, T. Affolder et al., Phys. Rev. Lett. **84**, 216 (2000) [hep-ex/9909042]

391. The DØ Collaboration, V.M. Abazov et al., Phys. Lett. B **617**, 1 (2005) [hep-ex/0404040]

392. K. Fujikawa, A. Yamada, Phys. Rev. D **49**, 5890 (1994)

393. P. Cho, M. Misiak, Phys. Rev. D **49**, 5894 (1994) [hep-ph/9310332]

394. C. Jessop, A world average for $B \to X_s \gamma$, SLAC-PUB-9610 (1994)
395. The CDF Collaboration, A. Abulencia et al., Measurement of the Helicity of W Bosons in Top-Quark Decays, CDF note 7804 (2005)
396. The DØ Collaboration, V.M. Abazov et al., Nature **429**, 638 (2004) [hep-ex/0406031]
397. The DØ Collaboration, V.M. Abazov et al., New Measurement of the Top Quark Mass in Lepton+Jets $t\bar{t}$ Events at DØ, FERMILAB Pub-04/102-E, 2004 (unpublished) [hep-ex/0407 005]
398. The DØ Collaboration, B. Abbott et al., Phys. Rev. D **60**, 052001 (1999) [hep-ex/9808029]
399. R.H. Dalitz, G.R. Goldstein, Proc. R. Soc. London A **445**, 2803 (1999) [hep-ph/9802249]
400. K. Kondo, T. Chikamatsu, S.H. Kim, J. Phys. Soc. Japan **62**, 1177 (1993)
401. The DØ Collaboration, V.M. Abazov et al., Phys. Rev. D **72**, 011104 (2005) [hep-ex/0505031]
402. The DØ Collaboration, V.M. Abazov et al., Search for Right-handed W Bosons in Dilepton Top Quark Pair Candidates, DØ note 4829 (2005)
403. The CDF Collaboration, D. Acosta et al., Phys. Rev. D **71**, 031101 (2005) [hep-ex/0411070]
404. H. Fritsch, Phys. Lett. B **224**, 423 (1989)
405. T. Han, R.D. Peccei, X. Zhang, Nucl. Phys. B **454**, 527 (1995) [hep-ph/9506461]
406. S. Parke, Summary of top quark physics, Proc. 1994 Meeting of the American Physical Society, Division of Particles and Fields (1994) [hep-ph/9409312]
407. B. Grzadkowski, J.F. Gunion, P. Krawczyk, Phys. Lett. B **268**, 106 (1991)
408. G. Eilam, J.L. Hewett, A. Soni, Phys. Rev. D **59**, 039901 (1999) [Erratum]
409. D. Atwood, L. Reina, A. Soni, Phys. Rev. D **53**, 1199 (1996) [hep-ph/9606243]
410. G.M. de Divitiis, R. Petronzio, L. Silvestrini, Nucl. Phys. B **504**, 45 (1997) [hep-ph/9704244]
411. R.D. Peccei, X. Zhang, Nucl. Phys. B **337**, 269 (1990)
412. B.A. Arbuzov, M.Y. Osipov, Phys. Atom. Nucl. **62**, 485 (1999) [hep-ph/9802392]
413. H. Fritsch, D. Holtmannspötter, Phys. Lett. B **457**, 186 (1999) [hep-ph/9901411]
414. T. Han, J.L. Hewett, Phys. Rev. D **60**, 074015 (1999) [hep-ph/9811237]
415. F. del Aguilar, J.A. Aguilar-Saavedra, R. Miquel, Phys. Rev. Lett. **82**, 1628 (1999) [hep-ph/9808400]
416. The CDF Collaboration, F. Abe et al., Phys. Rev. Lett. **80**, 2525 (1998)
417. V. Barger, T. Han, J. Ohnemus, D. Zeppenfeld, Phys. Rev. D **41**, 2782 (1990)
418. J.F. Obraztsov, S.R. Slabopitsky, O.P. Yushchenko, Phys. Lett. B **426**, 393 (1998) [hep-ph/9712394]
419. S. Moretti, K. Odagiri, Phys. Rev. D **57**, 3040 (1998) [hep-ph/9709435]
420. The H1 Collaboration, C. Adloff et al., Eur. Phys. J. C **5**, 575 (1998) [hep-ex/9806009]
421. The H1 Collaboration, V. Andreev et al., Phys. Lett. B **561**, 241 (2003) [hep-ex/0301030]
422. K.P. Diener, C. Schwanenberger, M. Spira, Eur. Phys. J. C **25**, 405 (2002) [hep-ph/0203269]
423. K.P. Diener, C. Schwanenberger, M. Spira, Photoproduction of W-Bosons at HERA: Reweighting Method for implementing QCD Corrections in Monte Carlo Programs (2003) [hep-ex/0302040]
424. The H1 Collaboration, A. Aktas et al., Eur. Phys. J. C **33**, 9 (2004) [hep-ex/0310032]
425. The ZEUS Collaboration, S. Chekanov et al., Phys. Lett. B **559**, 153 (2003) [hep-ex/0302010]
426. A. Pukhov et al., CompHep, Preprint LSUHE-145-1993, 1993 (unpublished)
427. A. Pukhov et al., CompHEP – a package for evaluation of Feynman diagrams and integration over multi-particle phase space. User's manual for version 33 (1999) (unpublished) [hep-ex/9908288]
428. O. Panella, G. Pancheri, Y.N. Srivastava, Phys. Lett. B **318**, 241 (1993)
429. K. Hagiwara, M. Tanaka, T. Stelzer, Phys. Lett. B **325**, 521 (1994) [hep-ph/9401295]
430. E. Boos et al., Phys. Lett. B **326**, 190 (1996)
431. C.-S. Huang, X.-H. Wu, S.-H. Zhu, Phys. Lett. B **452**, 143 (1999) [hep-ph/9901369]
432. The ALEPH Collaboration, A. Heister et al., Phys. Lett. B **543**, 173 (2002) [hep-ex/0206070]
433. The DELPHI Collaboration, J. Abdallah et al., Phys. Lett. B **590**, 21 (2004) [hep-ex/0404014]
434. The L3 Collaboration, P. Achard et al., Phys. Lett. B **549**, 290 (2002) [hep-ex/0210041]
435. The OPAL Collaboration, G. Abbiendi et al., Phys. Lett. B **521**, 181 (2001) [hep-ex/0110009]
436. D. Dannheim, Exclusion summary plot of the flavour-changing neutral current couplings, 2005 (private communication)
437. The CDF Collaboration, F. Abe et al., Phys. Rev. Lett. **73**, 225 (1994) [hep-ex/9405005]
438. E. Simmon, Thinking About Top: Looking Outside The Standard Model, Proc. Thinkshop on Top Quark Physics at Run-II, Batavia, Il (1999) [hep-ph/9908511]
439. K. Kondo, J. Phys. Soc. Japan **57**, 4126 (1988)
440. K. Kondo, J. Phys. Soc. Japan **60**, 836 (1991)
441. K. Kondo, K. Kusakabe, submitted to J. Phys. Soc. Japan (2005)
442. K. Kondo, Dynamical Likelihood Method for Reconstruction of Quantum Process (2005) [hep-ex/0508035]
443. R.H. Dalitz, G.R. Goldstein, Phys. Lett. B **287**, 225 (1992)
444. The DELPHI Collaboration, P. Abreu et al., Eur. Phys. J. C **2**, 581 (1998)
445. M. Mulders, Int. J. Mod. Phys. A **16S1A**, 284 (2001)
446. The CDF Collaboration, F. Abe et al., Phys. Rev. Lett. **82**, 271(E) (1999) [hep-ex/9810029]
447. The CDF Collaboration, A. Abulencia et al., Measurement of the Top Quark Mass using the ϕ Weighting Method In Dilepton Events at CDF, CDF note 7759, 2005 (unpublished)
448. The CDF Collaboration, D. Acosta et al., Measurement of the top mass using full kinematic template method in dilepton channel at CDF, CDF note 7797, 2005 (unpublished)
449. D. Acosta et al., Measurement of the Top Quark Mass using the Template Method in the Lepton plus Jets channel at CDF, CDF note 7153, 2004 (unpublished)
450. F. James, Comput. Phys. Commun. **10**, 343 (1975)
451. The CDF Collaboration, D. Acosta et al., Updated Top Quark Mass Measurement in the Lepton+Jets Channel

Using the Jet Probability Algorithm, CDF note 7637, 2005 (unpublished)

452. The CDF Collaboration, D. Acosta et al., Top Quark Mass Measurement using the Dynamical Likelihood Method in the lepton plus jets channel at CDF RunII, CDF note 7754, 2005 (unpublished)

453. The CDF Collaboration, D. Acosta et al., Top Quark Mass Measurement using the Dynamical Likelihood Method in the lepton plus jets channel at CDF RunII, CDF note 7056, 2004 (unpublished)

454. The CDF Collaboration, A. Abulencia et al., Top Mass Measurement in the Lepton+jets Channel using the Matrix Element Method, CDF note in preparation, 2005 (unpublished)

455. The CDF Collaboration, D. Acosta et al., Top Mass Measurement in the Lepton+jets Channel using a Multivariate Template Method, CDF note 7102, 2004 (unpublished)

456. D. Scott, Multivariate Density Estimation: Theory, Practice, and Visualization (Wiley-Interscience, New York, 1992)

457. The CDF Collaboration, A. Abulencia et al., First Measurement of the Top Quark Mass in the Lepton+Jets channel using the Decay Length Technique, CDF note 7781, 2005 (unpublished)

458. C.S. Hill, J.R. Incandela, J.M. Lamb, Phys. Rev. D **71**, 054 029 (2005) [hep-ex/0501043]

459. The CDF Collaboration, A. Abulencia et al., Measurement of the Top Quark Mass in the Dilepton Channel using the Leading-Order Differential Cross-Section at CDF II, CDF note 7718, 2005 (unpublished)

460. The CDF Collaboration, D. Acosta et al., Measurement of the Top Quark Mass using the Neutrino Weighting Algorithm on Dilepton Events at CDF, CDF note 7303, 2004 (unpublished)

461. The CDF Collaboration, A. Abulencia et al., Measurement of the Top Quark Mass using the Neutrino Weighting Algorithm on Dilepton Events at CDF, CDF note 7303, 2005 (unpublished)

462. The CDF Collaboration, D. Acosta et al., Measurement of the Top Quark Mass using the Minuit Fitter In Dilepton Events at CDF, CDF note 7239, 2004 (unpublished)

463. The CDF Collaboration, D. Acosta et al., Measurement of the top mass using full kinematic template method in dilepton channel at CDF, CDF note 7194, 2004 (unpublished)

464. The CDF Collaboration, F. Abe et al., Phys. Rev. Lett. **79**, 1992 (1997)

465. The DØ Collaboration, B. Abbott et al., Phys. Rev. D **58**, 052 001 (1998) [hep-ex/9801025]

466. The DØ Collaboration, V.M. Abazov et al., Measurement of the Top Quark Mass in the Lepton+Jets Channel using DØ Run II Data, DØ note 4574, 2004 (unpublished)

467. The DØ Collaboration, V.M. Abazov et al., Measurement of the Top Quark Mass in the Lepton+Jets Channel using DØ Run II Data: The Low Bias Template Method, DØ note 4728, 2005 (unpublished)

468. F.A. Berends, C.G. Papadopoulos, R. Pittau, Phys. Lett. B **417**, 385 (1998) [hep-ph/9709257]

469. The DØ Collaboration, B. Abbott et al., Phys. Rev. D **60**, 072 002 (1999) [hep-ex/9905005]

470. The CTEQ Collaboration, H.L. Lai, J. Botts, J. Huston, J.G. Morfin, J.F. Owens, J.W. Qiu, W.K. Tung, H. Weerts, Phys. Rev. D **51**, 4763 (1995) [hep-ph/9410404]

471. CERN Program Library, http://wwwasdoc.web.cern.ch/wwwasdoc/Welcome.html (1981)

472. F.A. Berends, H. Kuijf, B. Tausk, W.T. Giele, Nucl. Phys. B **357**, 32 (1991)

473. The Tevatron Electroweak Working Group, Demortier, R. Hall, R. Hughes, B. Klima, Combining the top quark mass results for Run I from CDF and DØ, Fermilab-TM-2084, 1999 (unpublished)

474. K.A. Kröninger, A Measurement of the Top Quark Mass with the DØ Detector at $\sqrt{s} = 1.96$ TeV Using the Matrix Element Method, Diploma thesis (Rheinische Friedrich-Wilhelms-Universität Bonn, FERMILAB-MASTERS-2004-04, BONN-IB-2004-06, 2004)

475. P. Schieferdecker, Measurement of the Top Quark Mass at DØ Run II with the Matrix Element Method in the Lepton+Jets Final State, PhD thesis (Ludwig-Maximilians-Universität München, 2005)

476. G.P. Lepage, J. Comput. Phys. **27**, 192 (1978)

477. G.P. Lepage, VEGAS: An adaptive Multi-dimensional Integration Program, Cornell preprint CLNS:80-447, 1980 (unpublished)

478. M. Galassi, J. Davies, J. Theiler, B. Gough, G. Jungman, M. Booth, F. Rossi, GNU Scientific Library Reference Manual, 2nd edn., (2005) available at: http://www.gnu.org/software/gsl/

479. The DØ Collaboration, V.M. Abazov et al., Measurement of the Top Quark Mass in the Dilepton Channel, DØ note 4725 (2005)

480. The DØ Collaboration, V.M. Abazov et al., Phys. Lett. B **606**, 25 (2005) [hep-ex/0410086]

481. A. Valassi, Nucl. Instrum. Methods A **500**, 391 (2003)

482. The CDF Collaboration, T. Affolder et al., Phys. Rev. D **63**, 032 003 (2001) [hep-ex/0006028]

483. The Tevatron Electroweak Working Group, J.F. Arguin et al., Combination of CDF and DØ Results on the Top-Quark Mass, FERMILAB-TM-2321-E, 2005 (unpublished) [hep-ex/0507006]

484. The DØ Collaboration, S. Abachi et al., Phys. Rev. Lett. **79**, 1197 (1997) [hep-ex/9703008]

485. The CDF Collaboration, F. Abe et al., Phys. Rev. Lett. **80**, 2767 (1998) [hep-ex/9801014]

486. D. Chang, W.F. Chang, E. Ma, Phys. Rev. D **59**, 091 503 (1999) [hep-ph/9810531]

487. D. Chang, W.F. Chang, E. Ma, Phys. Rev. D **61**, 037 301 (2000) [hep-ph/9909537]

488. The DØ Collaboration, V.M. Abazov et al., Measurement of the charge of the top quark with the DØ experiment, DØ note 4876-CONF, 2005 (unpublished)

489. The CDF Collaboration, D. Acosta et al., $t\bar{t}$ Production Cross-Section Ratio and Limits on Non-Standard Model Decays of Top, http://www-cdf.fnal.gov/physics/top/RunIIWjets/webpages/xs_ratio/xs_ratio.html, 2004 (unpublished)

490. The CDF Collaboration, T. Affolder et al., Phys. Rev. Lett. **80**, 2779 (1998) [hep-ex/9802017]

491. The DØ Collaboration, S. Abachi et al., Phys. Rev. Lett. **79**, 1203 (1997) [hep-ex/9704015]

492. R.M. Barnett, L.J. Hall, Phys. Rev. Lett. **77**, 3506 (1996) [hep-ph/9607342]

493. T.G. Rizzo, Top Quark Production at the Tevatron: Probing Anomalous Chromomagnetic Moments and Theories of Low Scale Gravity (1999) [hep-ph/9902273]

494. K.D. Kane, Phys. Rev. D **52**, 1546 (1995) [hep-ph/9501260]

495. The CDF Collaboration, T. Affolder et al., Phys. Rev. Lett **87**, 102001 (2001)

496. The CDF Collaboration, T. Affolder et al., Phys. Rev. D (Erratum-ibid) **67**, 119901(E) (2003) [hep-ex/0101036]

497. R.M. Harris, C.T. Hill, S. Parke, Cross Section for Top-color Z' decaying to $t\bar{t}$, Fermilab-FN-687 (1999) (unpublished) [hep-ph/9911288]

498. S. Weinberg, Phys. Rev. D **13**, 974 (1976)

499. L. Susskind, Phys. Rev. D **20**, 2619 (1979)

500. S. Dimopoulos, L. Susskind, Nucl. Phys. B **155**, 237 (1979)

501. E. Eichten, K. Lane, Phys. Lett. B **90**, 125 (1980)

502. The CDF Collaboration, F. Abe et al., Phys. Rev. Lett. **82**, 2038 (1999) [hep-ex/9809022]

503. The CDF Collaboration, T. Affolder et al., Phys. Rev. Lett. **85**, 2062 (2000) [hep-ex/0003005]

504. The DØ Collaboration, V.M. Abazov et al., Phys. Rev. Lett. **92**, 221801 (2004) [hep-ex/0307079]

505. The DØ Collaboration, V.M. Abazov et al., Search for a $t\bar{t}$ Resonance in $p\bar{p}$ Collisions at $\sqrt{s} = 1.96$ TeV in the Lepton+Jets Final State, DØ note 4880-CONF (2005)

506. S. Martin, A Supersymmetry Primer (1997) [hep-ph/9709356]

507. M. Carena, H.E. Haber, Prog. Part. Nucl. Phys. **50**, 63 (2003) [hep-ph/0208209]

508. The CDF Collaboration, F. Abe et al., Phys. Rev. Lett. **79**, 357 (1997) [hep-ex/9704003]

509. The DØ Collaboration, B. Abbott et al., Phys. Rev. Lett. **82**, 4975 (1999) [hep-ex/9902028]

510. M. Carena, D. Garcia, U. Nierste, C.E.M. Wagner, Nucl. Phys. B **577**, 88 (2000) [hep-ph/9912516]

511. The CDF Collaboration, A. Abulencia et al., Search for charged Higgs boson in $t\bar{t}$ decay products, CDF note 7712 (2005)

512. J.S. Lee et al., Comput. Phys. Commun. **156**, 283 (2001) [hep-ph/0307377]

513. The LEP Working Group for Higgs Boson Searches, Search for the Neutral MSSM Higgs Bosons at LEP, ALEPH 2004-008, DELPHI 2004-042, L3 Note 2820, OPAL TN-744, LHWG-Note/2004-01, contrib. paper to summer conferences 2004 (2004)

514. G.L.B. Ananthanarayan, G. Lazarides, Q. Shafi, Phys. Rev. D **44**, 1613 (1991)

515. S. Dimopoulos, L.J. Hall, S. Raby, Phys. Rev. D **45**, 4192 (1992)

516. G.W. Anderson, S. Raby, S. Dimopoulos, L.J. Hall, Phys. Rev. D **47**, 3702 (1993) [hep-ph/9209250]

517. The CDF Collaboration, D. Acosta et al., Search for $t' \rightarrow Wq$ Using Lepton Plus Jets Events, CDF note 7113, 2004 (unpublished)

518. T. Han, H.E. Logan, B. McElrath, L.T. Wang, Phys. Lett. B **563**, 191 (2003) [hep-ph/0302188]

519. The Large Hadron Collider LHC at CERN, http://lhc-new-homepage.web.cern.ch (2000)

520. L.R. Evans, Eur. Phys. J. C **34**, S11 (2004)

521. L.R. Evans, Eur. Phys. J. C **34**, 57 (2004)

522. L.R. Evans, IEEE Trans. Appl. Supercond. **14**, 147 (2004)

523. ATLAS Technical Proposal for a General-Purpose pp Experiment at the Large Hadron Collider at CERN, CERN/LHCC/94-43, LHCC/P2, 1994 (unpublished)

524. CMS – Technical Proposal, CERN/LHCC/94-38, 1994 (unpublished)

525. F. Gianotti, M.L. Mangano, LHC physics: the first one–two year(s), Proc. 2nd Italian Workshop on the physics of AT-LAS and CMS, 2005 (unpublished) [hep-ph/0504221]

526. The CMS Collaboration, D. Green, A Study of Single Top at CMS, CMS Note 1999-048 (1999) (unpublished)

527. C. Weiser, Top Physics at the LHC, Proc. XXXX-th Rencontres de Moriond, QCD and High Energy Hadronic Interactions, La Thuile, Italy (2005) [hep-ex/0506024]

528. S. Bentvelsen, Studies of Top Quark Properties at the LHC, Proc. 39th Rencontres de Moriond on QCD and High-Energy Hadronic Interactions, La Thuile, Italy (2004) [hep-ph/0408111]

529. The ATLAS Collaboration, I. Borjanovic et al., Eur. Jour. Phys. C **39S2**, 47 (2004) [hep-ex/0403021]

530. The CMS Collaboration, L. Sonnenschein, Top Quark Physics at the LHC, Proc. 14th Int. Spin Physics Symposium (SPIN2000), Osaka, Japan (2000)

531. M. Barisonzi, Top Physics at ATLAS, Proc. IFAE – Incontri di Fisica delle Alte Energie – Catania (2005) [hep-ex/0508008]

532. The CMS Collaboration, L. Sonnenschein, Top mass determination in the $t\bar{t}$ semileptonic decay channel, CMS Note 2001/001 (2001) (unpublished)

533. The CMS Collaboration, A. Kharchivala, Top Mass Determination in leptonic final states with J/ψ, CMS Note 1999-065 (1999) (unpublished)

534. A. Kharchivala, Phys. Lett. B **476**, 73 (2000) [hep-ph/9912320]

535. M.L. Nekrasov, Eur. Phys. J. C **44**, 233 (2005) [hep-ph/0412219]

536. M. Ciljak, M. Jurcovicova, S. Tokar, U. Baur, Top charge measurement at ATLAS detector, ATL-PHYS-2003-035 (2003) (unpublished)

537. L. Sonnenschein, The $t\bar{t}$ Production in pp Collisions at $\sqrt{s} = 14$ TeV, Ph.D. thesis (Rheinisch-Westfälische Technische Hochschule Aachen, PITHA 01/04, 2001)

538. The ATLAS Collaboration, F. Hubaut, E. Monnier, P. Pralavorio, K. Smolek, V. Simak, Eur. Phys. J. C **44S2**, 13 (2005) [hep-ex/0508061]

539. J. Cao, G. Liu, J.M. Yang, Phys. Rev. D **70**, 114035 (2004) [hep-ph/0409334]

540. K. Smolek, V. Simak, Measurement of Spin Correlations of top-antitop pairs, ATL-PHYS-2003-012, 2003 (unpublished)

541. F. Hubaut et al., ATLAS Sensitivity to $t\bar{t}$ Spin Correlations, ATL-PHYS-PUB-2005-001, 2005 (unpublished)

542. U. Baur, A. Juste, L.H. Orr, D. Rainwater, Phys. Rev. D **71**, 054013 (2005) [hep-ph/0412021]

543. U. Baur, Probing Electroweak Top Quark Couplings at Hadron and Lepton Colliders, Proc. 2005 Int. Linear Collider Workshop, Stanford, USA, 2005 [hep-ph/0508151]

544. F. Beaudette, Top physics prospects at LHC, Proc. XXXX-th Rencontres de Moriond, Electroweak Interactions and Unified Theories, La Thuile, Italy, 2005 [hep-ex/0506056]

545. J. Lévêque, Recherche d'un boson de Higgs léger produit en association avec une paire de quarks top dans l'expérience ATLAS, Ph.D. thesis, Centre de Physique des Particules de Marseille et de l'Université de la Mediterranée – Aix-Marseille II, 2003

546. J. Lévêque, V. Kostyukhin, A. Rozanov, J.B. De Vivie de Régie, Search for the Standard Model Higgs boson in the $t\bar{t}H^0$, $H^0 \rightarrow WW^{(*)}$ channel, ATL-PHYS-2002-019, 2002 (unpublished)

547. J. Cammin, Study of a Light Standard Model Higgs Boson in the $t\bar{t}H^0$ Channel with ATLAS at LHC and Decay-Mode Independent Searches for Neutral Higgs Bosons with OPAL at LEP, Ph.D. thesis (Rheinische Friedrich-Wilhelms-Universität Bonn, 2003)

548. J. Cammin, M. Schumacher, The ATLAS discovery potential for the channel $t\bar{t}H$, $H \rightarrow b\bar{b}$, ATL-PHYS-2003-024, 2003 (unpublished)

549. S. Abdullin et al., Summary of the CMS Potential for the Higgs Boson Discovery, CMS Note 2003/033, 2003 (unpublished)

550. K. Desch, M. Schumacher, Eur. Phys. J. C **46**, 527 (2006) [hep-ph/0407159]

551. M. Dührssen, Prospects for the measurement of Higgs boson coupling parameters in the mass range from 110–190 GeV/c^2, ATL-PHYS-2003-030, 2003 (unpublished)

552. M. Dührssen, S. Heinemeyer, H. Logan, D. Rainwater, G. Weiglein, D. Zeppenfeld, Phys. Rev. D **70**, 113 009 (2004) [hep-ph/0406323]

553. M. Cobal, Expectations from LHC and LC on Top Physics, Proc. 15th Topical Conf. in Hadron Collider Physics (HCP2004), Michigan State University, East Lansing, USA, 2004 [hep-ex/0412053]

554. TeV-2000 Study Group, Future Electroweak Physics at the Fermilab Tevatron: Report of the TEV-2000 Study Group, SLAC-REPRINT-1996-085, FERMILAB-PUB-96-082, DØ-NOTE-2589, CDF-NOTE-3177, Batavia, ed. by D. Amidei, R. Brock p. 227 (1996)

555. T. Han, K. Whisnant, B.L. Young, X. Zhang, Phys. Lett. B **385**, 311 (1996) [hep-ph/9606231]

556. J. Carvalho, N. Castro, A. Onofre, F. Veloso, Study of ATLAS sensitivity to FCNC top decays, ATL-PHYS-PUB-2005-009, 2005 (unpublished)

Printing: Krips bv, Meppel
Binding: Stürtz, Würzburg